Mathematics Study Resources

Volume 12

Series Editors

Kolja Knauer, Departament de Matemàtiques Informàtic, Universitat de Barcelona, Barcelona, Barcelona, Spain

Elijah Liflyand, Department of Mathematics, Bar-Ilan University, Ramat-Gan, Israel

This series comprises direct translations of successful foreign language titles, especially from the German language.

Powered by advances in automated translation, these books draw on global teaching excellence to provide students and lecturers with diverse materials for teaching and study.

Jürgen Müller

Concepts of Function Theory

Real and Complex Analysis of one Variable

 Springer

Jürgen Müller
Fachbereich IV, Mathematik
Universität Trier
Trier, Germany

ISSN 2731-3824   ISSN 2731-3832   (electronic)
Mathematics Study Resources
ISBN 978-3-662-69114-4   ISBN 978-3-662-69115-1   (eBook)
https://doi.org/10.1007/978-3-662-69115-1

Translation from the German language edition: "Konzepte der Funktionentheorie" by Jürgen Müller, ©
Springer-Verlag GmbH Deutschland, ein Teil von Springer Nature 2018. Published by Springer Berlin
Heidelberg. All Rights Reserved.

This book is a translation of the original German edition "Konzepte der Funktionentheorie" by Jürgen
Müller, published by Springer-Verlag GmbH, DE in 2018. The translation was done with the help of an
artificial intelligence machine translation tool. A subsequent human revision was done primarily in terms
of content, so that the book will read stylistically differently from a conventional translation. Springer
Nature works continuously to further the development of tools for the production of books and on the
related technologies to support the authors.

This Springer imprint is published by the registered company Springer-Verlag GmbH, DE, part of
Springer Nature.
The registered company address is: Heidelberger Platz 3, 14197 Berlin, Germany

If disposing of this product, please recycle the paper.

The theory of differentiable functions of a complex variable is one of the most fascinating areas of mathematics. The transition from real to complex analysis proves to be a step into a mathematical world where one can recognize a multitude of relationships that remain hidden in the real one.

Students typically learn complex analysis following the real analysis of several variables. This textbook presents a direct path from real analysis of one variable to complex analysis. In doing so, classic topics of real analysis such as the differential and integral calculus of one variable are largely presented from a complex perspective. Short sections attached to each chapter about the concepts discussed there provide insights into higher-dimensional analysis. In this way, an impression of their universal significance for all of mathematics is conveyed.

The elaboration is self-contained, that is, basically no prior knowledge is assumed. However, one should bring some experience in dealing with basic mathematical concepts. The main goal is the compact presentation of classic results of one-dimensional real and complex analysis, but also of more advanced results of complex analysis such as the Riemann mapping theorem, the theorems of Montel and Picard, and Runge's theorems. Building on the great theorem of Montel and the extremely efficient Zalcman lemma, an introduction to the exciting theory of the dynamic behavior of entire functions is also given.

Naturally, in view of further applications of analysis, one cannot do without multidimensional theory. Knowledge about partial differential equations or integral theorems is almost mandatory for many applications. However, I am convinced that the concepts conveyed within the framework of complex analysis also have a certain bridging function to higher-dimensional analysis and thus represent a first step in this direction. Possibly, in view of the fact that some curricula—such as in the case of teacher training courses—hardly provide for further analysis, an interesting alternative opens up. In addition, the integration of individual parts of complex analysis at a comparatively early stage of study can certainly promote understanding of multidimensional analysis.

The first three chapters of the book essentially correspond thematically to what introductory analysis lectures contain, with the presentation being adapted to the objective from the outset. The integrated approach is reflected, for example, in the mean value inequality, the comparatively early introduction of analytic functions, and the discussion about the existence of primitive functions—all topics that also promote a deeper understanding of real analysis.

Chapter 4 marks the entry into complex analysis. Initially, the focus is on functions on circles and on circular disks. The Cauchy integral formula for circles leads directly to the first classical results such as the analyticity of holomorphic functions, Liouville's theorem, and the maximum principle. Subsequently, a rudimentary Fourier series theory is used as a foundation for the elaboration of the theory of isolated singularities.

Chapter 5 begins with further concepts of topology in order to be able to formulate central statements about the global behavior of holomorphic functions with the Cauchy theorem. The fusion with the local theory of isolated singularities in the residue theorem is undoubtedly one of the highlights of complex analysis and enables a multitude of interesting applications. In the consequences formulated further, such as the Rouché theorem, the inversion theorem, and the Hurwitz theorem, local aspects are again in the foreground.

In the sixth chapter, advanced topics of complex analysis such as the Riemann mapping theorem, Picard's theorems, and the dynamics of entire functions are discussed. The essential theoretical foundation is provided by Montel's normality theorems. An explicit goal is to develop this foundation in a self-contained manner. In a first step, Montel's (small) normality theorem for families of holomorphic functions is derived, which follows directly from the Arzelà-Ascoli theorem using the Cauchy inequality. Therefore, the chapter begins with equicontinuous families and related compactness statements. The path to Montel's big theorem is then chosen via the Zalcman lemma. In preparation for this, a suitable bound theorem for spherical derivatives is derived.

Largely independent of the previous chapter, the Cauchy theorem is taken up in the last part within the framework of Runge's theory to prove various results about uniform and locally uniform approximation of holomorphic functions by rational functions or polynomials. Runge's theorem for polynomial approximation is used, among other things, in the first part of the last section. There, an impression is given of how complicated the situation typically becomes when considering the boundary behavior of holomorphic functions.

The Chaps. 1 to 5 are largely built on each other, so that the order here is more or less predetermined. The material presented up to this point essentially corresponds to what an introductory complex analysis typically provides. The remaining parts allow for some flexibility in working through. The presentation tends to be tighter than in the first ones. As already indicated, the Chaps. 6 and 7 are largely independent of each other (only basic features of topology and the metric space $H(\Omega)$, as introduced in the first section of Chap. 6, are relevant for both).

I approached this book project with a certain naivety—had this not been the case, I probably would not have started at all. The fact that the project actually became a book, I owe to a number of participants.[1] First, I would like to thank

---

[1]All graphics were created with Mathematica 11, Version 11.0.1.0, by Wolfram Research (Wolfram Research, Inc., 100 Trade Center Drive, Champaign, IL 61820-7237, USA).

Mrs. Ruhmann from Springer-Verlag for her flexibility in allowing me a completely different project than she originally had in mind, and at the same time for the very pleasant cooperation, with this thanks also going to Mrs. Herrmann. Great thanks go to Professor Dr. Dr. h. c. Wolfgang Luh and Dr. Thierry Meyrath for their extremely valuable help in implementing the concept and shaping the manuscript. Elke Gawronski also made a significant contribution to the second point, to whom my special thanks also go. Finally, I would like to thank the Analysis working group, our doctoral students and the mathematics students at the University of Trier for the pleasant atmosphere and the always very stimulating cooperation over many years—and my dear wife Ulla for her support in all of this.

Trier   Jürgen Müller
in October 2017

Contents

Mathematics is simple—or more precisely: dual. Essentially, one deals with only two types of objects, namely sets and mappings. In the first section, the corresponding terms are introduced in a rather informal way and some basic properties are proven. Subsequently, general structures of mathematics such as groups, rings, and fields are discussed, with arithmetic aspects initially being in the foreground. This essentially sets the rules of the game. Classical formulas such as the geometric sum formula and the binomial formula are formulated in a comparatively general framework. The extension by an order structure represents the transition to analysis, where primarily estimates and approximations are at the center of the investigations.

If one is looking for a minimal ordered field, one naturally ends up with the rational numbers. However, to be able to efficiently perform analysis, one primarily needs another structural property: completeness. Only the completeness of a space ensures that reasonable limit processes do not fizzle out, but result in (limit) values. The search for complete ordered fields not only naturally leads to the real numbers, but even inevitably. The last section outlines a possible constructive approach to the real numbers.

Compared to the extension of the rational numbers to the real ones, the extension of the real numbers to the complex ones is much less problematic. One only has to "invent" a meaningful multiplication of two pairs of real numbers. The price one pays for this is the loss of the order structure. On the other hand, the advantages are manifold, as the following chapters will show.

1.1   Sets and Mappings

We start with some introductory definitions and results from the theory of sets and mappings, which are the basis of all mathematics. Our presentation is based on the (so-called naive) set concept coined by G. Cantor:

J. Müller, *Concepts of Function Theory*, Mathematics Study Resources 12,
https://doi.org/10.1007/978-3-662-69115-1_1

> A **set** M is a collection of certain, well-distinguished objects of our perception or our thinking into a whole.

Such an object x is called an **element** of the set M (notation: $x \in M$; if x is not an element of M, we write $x \notin M$). The set without elements is called the **empty set** (notation: $\emptyset$ or $\{\,\}$).

There are various ways of representing sets, such as the enumerative notation or the descriptive one, i.e., a characterization of the elements. The descriptive variant generally has the form

$$M := \{x : x \text{ has the property } E\},$$

where E is a given "property". Alternatively, instead of x : one also writes $x|$.

In the following, we will use the symbol $:=$ as a defining equality sign, that is, the left side is defined by the right one. We will also assume that integers and rational numbers along with their arithmetic properties are known, but we will briefly discuss an axiomatic introduction in Sect. 1.6. One defines

$$\mathbb{N} := \{x : x \text{ positive integer}\},$$
$$\mathbb{N}_0 := \{x : x \text{ nonnegative integer}\},$$
$$\mathbb{Z} := \{x : x \text{ integer}\},$$
$$\mathbb{Q} := \{x : x \text{ rational number}\}.$$

Definition 1.1.1 Let A, B be sets.

1. A is called a **subset** of B (notation: $A \subset B$ or also $A \subseteq B$), if from $x \in A$ also follows $x \in B$. Then B is also called a **superset** of A and is written as $B \supset A$.
2. A and B are called **equal** (notation $A = B$), if $A \subset B$ and $B \subset A$ hold. Specifically, if $A := \{x\}$ and $B := \{y\}$ are single-point, we call x and y **equal** (notation: $x = y$; if x and y are unequal, we write $x \neq y$).
3. The set

$$B \setminus A := \{x : x \in B \text{ and } x \notin A\}$$

is called the **difference set** of B and A. If $A \subset B$, then

$$A^c := C_B(A) := B \setminus A$$

is called the **complement** of A (with respect to B).

Similar to the above introduction of sets, we want to rely on a rather informal definition of the second fundamental concept of mathematics:

Let X and Y be non-empty sets. A **function** or **mapping** f from X to (or into) Y is a "rule" that assigns to each $x \in X$ *exactly one* element $f_x = f(x) \in Y$. Here, X is called the **domain** and Y the **codomain** of f. Also, we refer to x as the (independent) variable. We write $f : X \to Y$ or alternatively $X \ni x \mapsto f(x) \in Y$ (or shorter $x \mapsto f(x)$) or $(f_x)_{x \in X}$. In the case of the notation $(f_x)_{x \in X}$, we also speak of a **family** in Y and then call X the **index set**. Further, we set

$$Y^X := \mathrm{Map}(X, Y) := \{f : f \text{ mapping from } X \text{ to } Y\}.$$

If $f, g \in Y^X$, then f and g are called **equal**, if $f(x) = g(x)$ for all $x \in X$. If $M \subset X$, then the function $f|_M : M \to Y$, defined by $f|_M(x) := f(x)$ for all $x \in M$, is called the **restriction** of f to M.

Definition 1.1.2 If $n \in \mathbb{N}$ and X is a non-empty set and $x : \{1, \ldots, n\} \to X$ is a function, one usually writes $(x_1, \ldots, x_n)$ or $(x_j)_{j=1,\ldots,n}$ and then speaks of an n-**tuple** in X. In the case $n = 2$, one also speaks of (ordered) pairs and in the case $n = 3$ of triples. In these cases, one often uses an index-free notation such as (u, v) instead of (x_1, x_2), or (u, v, w) instead of (x_1, x_2, x_3).

Furthermore, one sets

$$X^n := X^{\{1,\ldots,n\}}$$

and for any sets $A_1, \ldots, A_n \subset X$

$$A_1 \times \cdots \times A_n := \{(x_1, \ldots, x_n) \in X^n : x_j \in A_j \text{ for } j \in \{1, \ldots, n\}\}\,.$$

If $f : A_1 \times \cdots \times A_n \to Y$ is a function, one saves brackets by writing briefly $f(x_1, \ldots, x_n)$ instead of $f(x_1, \ldots, x_n)$. A subset R of $X \times X$ is called a **relation** on (or in) X. One then also writes uRv, if $(u, v) \in R$ applies.

Definition 1.1.3 Let M be a non-empty set. A function $f : M \times M \to M$ is called a **binary operation** on M. One often chooses a non-alphabetical symbol like $\cdot, \circ, *$, $\times, +$ for f and again writes xfy instead of $f(x, y)$ for $x, y \in M$, so for example $x \cdot y$, $x \circ y, x * y, x \times y, x + y$. In the case of the multiplication sign $\cdot$, one usually writes briefly xy instead of $x \cdot y$.

As we will further discuss, as already indicated, we will assume knowledge of the "usual" operations $+$ and $\cdot$ on $\mathbb{N}, \mathbb{N}_0, \mathbb{Z}$ and $\mathbb{Q}$ as well as the relations $<$ or $\leq$ on $\mathbb{Q}$ along with corresponding calculation rules. More details can be found in Sect. 1.6.

Example 1.1.4 Let $X := Y := \mathbb{N}$, and let $f : \mathbb{N} \to \mathbb{N}$ be defined by

$$f(x) := \begin{cases} x, & \text{if } x \text{ is even} \\ 2x, & \text{if } x \text{ is odd} \end{cases}.$$

If $M := \{x \in \mathbb{N} : x \text{ odd}\}$, then[1]

$$f|_M(x) = 2x \qquad (x \in M).$$

Definition 1.1.5 If X, Y are sets and $f : X \to Y$, then for $B \subset Y$

$$f^{-1}(B) := \{x \in X : f(x) \in B\}$$

---

[1] The notation $(x \in X)$ is to be read as a shorthand for "for all $x \in X$" in the following.

is called the **preimage** of B under f and for $A \subset X$

$$f(A) := \{f(x) : x \in A\} = \{y \in Y : y = f(x) \text{ for some } x \in A\}$$

is called the **image** of A under f. Specifically,

$$W(f) := f(X)$$

is called the **range** of f. If $W(f)$ is single-pointed, then f is called **constant**.

Example 1.1.6 In the situation of Example 1.1.4, approximately

$$f^{-1}(\{2,4,6\}) = f^{-1}(\{1,2,3,4,5,6\}) = \{1,2,3,4,6\}$$

and

$$f(\{1,2,3\}) = \{2,6\}.$$

Also, $W(f) = \{y \in \mathbb{N} : y \text{ even}\}$.

Definition 1.1.7 Let X, Y be sets and $f : X \to Y$.

1. f is called **surjective** (or mapping from X **onto** Y), if $W(f) = Y$.[2]
2. f is called **injective**, if from $x_1, x_2 \in X$ and $f(x_1) = f(x_2)$ already $x_1 = x_2$ follows.
3. f is called **bijective**, if f is injective and surjective.

Example 1.1.8 Let f be as in Example 1.1.4. Then f is neither surjective nor injective (for example, $1 \notin W(f)$ and $f(2) = f(1)$), on the other hand, $f|_M$ is injective.

Definition 1.1.9 Let X, Y, Z be sets and $f : X \to Y$ and $g : Y \to Z$ be mappings. Then $g \circ f : X \to Z$, defined by

$$(g \circ f)(x) := g(f(x)) \qquad (x \in X),$$

is called the **composition** of f with g (or **concatenation** of f and g).

Example 1.1.10 If $X = Y = Z = \mathbb{N}$ and $f, g : \mathbb{N} \to \mathbb{N}$ are defined by

$$f(x) := x^2, \quad g(x) := x + 1 \quad (x \in \mathbb{N}),$$

then $g \circ f : \mathbb{N} \to \mathbb{N}$ is given by

$$(g \circ f)(x) = x^2 + 1 \quad (x \in \mathbb{N}).$$

---

[2]At this point a small note on the question of whether *mapping* and *function* are different terms: According to the above (informal) definition, this is not the case. However, the term mapping is often used when the codomain also plays a significant role, such as in the case of surjectivity. One rarely speaks of a surjective function and almost never of a function from X *onto* Y. We will occasionally—as usual and very practical—take the liberty to identify two functions $f : X \to Y$ and $g : X \to Z$ when $f(x) = g(x)$ for all $x \in X$ and the codomain does not matter.

Remark: Here, $f \circ g : \mathbb{N} \to \mathbb{N}$ is also defined and it holds

$$(f \circ g)(x) = (x + 1)^2 \quad (x \in \mathbb{N}).$$

Here, $g \circ f \neq f \circ g$ (since for example $(g \circ f)(1) = 2 \neq 4 = (f \circ g)(1)$).

Theorem 1.1.11 *Let X, Y, Z, U be sets and $f : X \to Y$, $g : Y \to Z$ and $h : Z \to U$ be mappings. Then the following holds*

$$h \circ (g \circ f) = (h \circ g) \circ f \ .$$

Proof We have $h \circ (g \circ f) : X \to U$ and $(h \circ g) \circ f : X \to U$ and for all $x \in X$ we have

$$(h \circ (g \circ f))(x) = h((g \circ f)(x)) = h(g(f(x))) = (h \circ g)(f(x)) = ((h \circ g) \circ f)(x) \ .$$

Therefore, the two functions are equal.   □

Remark and Definition 1.1.12 Let X, Y be sets and let $f : X \to Y$ be bijective. Then for every $y \in Y$ there exists *exactly* one $x \in X$ such that $f(x) = y$. We define

$$f^{-1}(y) := x \qquad (y \in Y) \ ,$$

where $y = f(x)$. The mapping $f^{-1} : Y \to X$ is called the **inverse mapping** of f. It holds that $f^{-1} \circ f : X \to X$ and

$$(f^{-1} \circ f)(x) = x \qquad (x \in X) \ ,$$

that is $f^{-1} \circ f = \mathrm{id}_X$, where $\mathrm{id}_X : X \to X$, defined by $\mathrm{id}_X(x) := x$ for $x \in X$, denotes the **identity mapping** on X. Similarly, $f \circ f^{-1} = \mathrm{id}_Y$, and also $f^{-1} : Y \to X$ is bijective.

Remark and Definition 1.1.13 Let X be a set. Then

$$\mathscr{P}(X) := \{A : A \subset X\}$$

is called the **power set** of X. In the following, we always understand a family of sets in X to be a family in $\mathscr{P}(X)$.

Definition 1.1.14 Let $I \neq \emptyset$ be a set, X a set and $(A_\alpha) := (A_\alpha)_{\alpha \in I}$ a family of sets in X. Then

$$\bigcup_{\alpha \in I} A_\alpha := \{x : x \in A_\alpha \text{ for some } \alpha \in I\}$$

is called union of (A_α) and

$$\bigcap_{\alpha \in I} A_\alpha := \{x : x \in A_\alpha \text{ for all } \alpha \in I\}$$

intersection of (A_α).

In particular, for a set of sets (a so-called set system) $\mathscr{F}$ also

$$\bigcup_{M \in \mathscr{F}} M \qquad \text{and} \qquad \bigcap_{M \in \mathscr{F}} M$$

is defined (here specifically $I = \mathscr{F}$ and $A_M = M$). One then also writes briefly $\bigcup \mathscr{F}$ or $\bigcap \mathscr{F}$.

If I is given in enumerative form, then one places $\bigcup$ or $\bigcap$ usually between the individual sets, for example in the case of $I = \{1, 2, 3\}$

$$A_1 \cup A_2 \cup A_3 := \bigcup_{\alpha \in I} A_\alpha, \qquad A_1 \cap A_2 \cap A_3 := \bigcap_{\alpha \in I} A_\alpha.$$

Also in the case of a set system, one alternatively writes $\bigcup$ and $\bigcap$ between the individual sets, if $\mathscr{F}$ is given in enumerative form, for example in the case of $\mathscr{F} = \{A, B, C\}$

$$A \cup B \cup C \qquad \text{and} \qquad A \cap B \cap C.$$

Example 1.1.15 If $A := \{2k : k \in \mathbb{Z}\}$ and $B := \{3k : k \in \mathbb{Z}\}$, then

$$A \cap B = \{6k : k \in \mathbb{Z}\}.$$

By definition, two sets are equal if the first is a subset of the second and the second is a subset of the first. Therefore, one usually proves equality by separately demonstrating the two inclusions. We will indicate this in the following by the notation $\subset:$ and $\supset:$ in the corresponding proofs.

Theorem 1.1.16 *Let X be a set, $(A_\alpha)_{\alpha \in I}$ a family of sets in X and $B \subset X$. Then*

1. $B \cap \left(\bigcup_{\alpha \in I} A_\alpha \right) = \bigcup_{\alpha \in I} (B \cap A_\alpha) \quad and \quad B \cup \left(\bigcap_{\alpha \in I} A_\alpha \right) = \bigcap_{\alpha \in I} (B \cup A_\alpha).$

2. **(De Morgan's Laws):**

$$B \setminus \left(\bigcup_{\alpha \in I} A_\alpha \right) = \bigcap_{\alpha \in I} (B \setminus A_\alpha) \quad and \quad B \setminus \left(\bigcap_{\alpha \in I} A_\alpha \right) = \bigcup_{\alpha \in I} (B \setminus A_\alpha).$$

Proof We will exemplify the proofs of the first statement in 1. and 2. The second ones result in a similar way. We write briefly $\bigcup$ instead of $\bigcup_{\alpha \in I}$.

1. $\subset:$ Let $x \in B \cap (\bigcup A_\alpha)$. Then $x \in B$ and $x \in \bigcup A_\alpha$, so $x \in B$ and $x \in A_\beta$ for a $\beta \in I$. Thus, $x \in B \cap A_\beta$, so also $x \in \bigcup (B \cap A_\alpha)$.
    $\supset:$ Let $x \in \bigcup (B \cap A_\alpha)$. Then there exists a $\beta \in I$ with $x \in B \cap A_\beta$. Thus, $x \in B$ and $x \in A_\beta$, so also $x \in B$ and $x \in \bigcup A_\alpha$, that is $x \in B \cap (\bigcup A_\alpha)$.
2. $\subset:$ Let $x \in B \setminus (\bigcup A_\alpha)$ be given. Then $x \in B$ and $x \notin \bigcup A_\alpha$, so $x \in B$ and $x \notin A_\alpha$ for all $\alpha \in I$. Thus, $x \in B \setminus A_\alpha$ for all $\alpha \in I$, so $x \in \bigcap (B \setminus A_\alpha)$.
    $\supset:$ Let $x \in \bigcap (B \setminus A_\alpha)$ be given. Then $x \in B \setminus A_\alpha$ for all $\alpha \in I$, so $x \in B$ and $x \notin A_\alpha$ for all $\alpha \in I$. Thus, $x \in B$ and $x \notin \bigcup A_\alpha$, that is $x \in B \setminus (\bigcup A_\alpha)$. $\qquad \square$

Theorem 1.1.17 *Let X, Y be sets and $f: X \to Y$.*

1. *If $(B_\alpha)_{\alpha \in I}$ is a family of sets in Y, then*

$$f^{-1}\left(\bigcup_{\alpha \in I} B_\alpha\right) = \bigcup_{\alpha \in I} f^{-1}(B_\alpha) \quad and \quad f^{-1}\left(\bigcap_{\alpha \in I} B_\alpha\right) = \bigcap_{\alpha \in I} f^{-1}(B_\alpha).$$

2. *If $(A_\alpha)_{\alpha \in I}$ is a family of sets in X, then*

$$f\left(\bigcup_{\alpha \in I} A_\alpha\right) = \bigcup_{\alpha \in I} f(A_\alpha) \quad and \quad f\left(\bigcap_{\alpha \in I} A_\alpha\right) \subset \bigcap_{\alpha \in I} f(A_\alpha).$$

Proof We again limit ourselves to the respective first statements.

1. $\subset$: Let it be $x \in f^{-1}\left(\bigcup B_\alpha\right)$. Then $f(x) \in \bigcup B_\alpha$ is true, that is, there exists a $\beta \in I$ with $f(x) \in B_\beta$. Thus, $x \in f^{-1}(B_\beta)$ and therefore also $x \in \bigcup f^{-1}(B_\alpha)$.
   $\supset$: If $\beta \in I$, then $B_\beta \subset \bigcup B_\alpha$ is true, and therefore also $f^{-1}(B_\beta) \subset f^{-1}\left(\bigcup B_\alpha\right)$. Since $\beta \in I$ was arbitrary, $\bigcup f^{-1}(B_\alpha) \subset f^{-1}\left(\bigcup B_\alpha\right)$ applies.
2. $\subset$: Let it be $y \in f\left(\bigcup A_\alpha\right)$. Then there exists a $x \in \bigcup A_\alpha$ with $f(x) = y$. If $\beta \in I$ with $x \in A_\beta$, then $y = f(x) \in f(A_\beta)$. Thus, $y \in \bigcup f(A_\alpha)$ is true.
   $\supset$: If $\beta \in I$, then $A_\beta \subset \bigcup A_\alpha$ is true, and therefore also $f(A_\beta) \subset f\left(\bigcup A_\alpha\right)$. Since $\beta \in I$ was arbitrary, $\supset$ applies.

Remark 1.1.18 Please note that there is no equality in the last statement of the second part of Theorem 1.1.17. In fact, equality for all families (A_α) occurs exactly when f is injective (see Exercise 1.1.19.4).

Exercises 1.1.19

1. Examine the following functions for injectivity, surjectivity, and bijectivity:
   a) $f : \mathbb{Z} \to \mathbb{N}_0, f(x) := x^2 \quad (x \in \mathbb{Z})$,
   b) $g := f|_{\mathbb{N}_0}$,
   c) $h : \mathbb{Z} \times \mathbb{Z} \to \mathbb{Z}, h(x, y) := x + y \quad (x, y \in \mathbb{Z})$.
2. Find bijective functions $f, g : \{1, 2, 3\} \to \{1, 2, 3\}$ with $g \circ f \neq f \circ g$.
3. Let $\mathbb{Q}_+ := \{x \in \mathbb{Q} : x > 0\} = \{p/q : p, q \in \mathbb{N}, p, q \text{ coprime}\}$.
   a) Show: The equation $x^2 + x = 1$ has no solution in $\mathbb{Q}_+$.
      Hint: Use that for two numbers $m, n \in \mathbb{Z}$ the product nm is odd exactly when both numbers are odd.
   b) Examine the function $f : \mathbb{Q}_+ \to \mathbb{Q}_+$ with

   $$f(x) := x^2 + x \qquad (x \in \mathbb{Q}_+)$$

   for injectivity and surjectivity.
4. Let X, Y be sets and $f : X \to Y$. Prove:

$$f\left(\bigcap_{\alpha \in I} A_\alpha\right) = \bigcap_{\alpha \in I} f(A_\alpha)$$

holds for all families $(A_\alpha)_{\alpha \in I}$ in $\mathscr{P}(X)$ if and only if f is injective.

5. Let X, Y be sets and $f : X \to Y$. Prove:
   a) For all $B \subset Y$,

   $$f\left(f^{-1}(B)\right) \subset B$$

   and for all $A \subset X$,

   $$A \subset f^{-1}\left(f(A)\right).$$

   b) In a), equality holds for all B if and only if f is surjective.
   c) In a), equality holds for all A if and only if f is injective.

1.2   Monoids, Groups, and Rings

The aim of this section is to formalize algebraic structures. To do this, we consider sets that are provided with certain operations.

Definition 1.2.1 Let M be a non-empty set and $\cdot$ a binary operation on M.

1. The operation is called **associative**, if $x(yz) = (xy)z$ holds for $x, y, z \in M$, and **commutative**, if $xy = yx$ for $x, y \in M$. If $\cdot$ is associative, then $(M, \cdot)$ is called a **semigroup**. If $\cdot$ is also commutative, then the semigroup is called **abelian** (or also **commutative**).
2. An $e \in M$ is called **neutral** (with respect to $\cdot$), if $ex = xe = x$ for all $x \in M$. If a neutral element e exists in a semigroup $(M, \cdot)$, then $(M, \cdot, e)$ is called a **monoid**.

For associative operations, parentheses are usually omitted, so for example $xyz := (xy)z = x(yz)$. The symbol plus + is typically only used for commutative operations. Finally, one also writes M instead of $(M, \cdot)$ or $(M, \cdot, e)$.

Example 1.2.2

1. The pair $(\mathbb{N}, +)$ is an abelian semigroup, $(\mathbb{N}_0, +, 0)$, $(\mathbb{N}, \cdot, 1)$ and $(\mathbb{Z}, \cdot, 1)$ are abelian monoids.
2. If $(X, \cdot)$ is a semigroup, then

   $$A \cdot B := \{xy : x \in A, \, y \in B\} \qquad (A, B \subset X)$$

   defines an associative operation $\cdot$ on $\mathscr{P}(X)$, so $(\mathscr{P}(X), \cdot)$ is a semigroup. If $(X, \cdot, e)$ is a monoid, then $(\mathscr{P}(X), \cdot, \{e\})$ is also a monoid. In the case of a single-point set $A = \{x\}$, we usually write xB instead of $\{x\} \cdot B$ and in the case of the symbol plus as an operation on X, we naturally also write $A + B$ instead of $A \cdot B$ and $x + B$ instead of xB. The set $A + B$ is then called the **Minkowski sum** of A and B.

Definition 1.2.3 Let $(M, \cdot, e)$ be a monoid. If $x \in M$, then a $y \in M$ is called **left inverse** to x, if $yx = e$, and **right inverse** to x, if $xy = e$. Furthermore, y is briefly called **inverse** to x, if $yx = xy = e$. Accordingly, x is then called **left invertible**

or **right invertible** or **invertible**. If every $x \in M$ is invertible, then M is called a **group**.

Remark 1.2.4 Let $(M, \cdot, e)$ be a monoid.

1. Neutral elements are unique, because if e and e' are neutral, then

$$e' = ee' = e.$$

Also, inverse elements are unique in the case of their existence. More precisely: If $x, y_1, y_2 \in M$ with y_1 being left-inverse and y_2 right-inverse to x, then

$$y_1 = y_1 e = y_1(xy_2) = (y_1 x)y_2 = ey_2 = y_2.$$

The inverse element to x is denoted by x^{-1}. When using the operation symbol $+$, one usually writes $-x$ (and then also briefly $x - y$ instead of $x + (-y)$).
2. Let $x, y \in M$ be invertible. From $x^{-1}x = xx^{-1} = e$ and

$$xyy^{-1}x^{-1} = xx^{-1} = e = y^{-1}y = y^{-1}x^{-1}xy$$

it follows that also x^{-1} and xy are invertible with

$$\left(x^{-1}\right)^{-1} = x \quad \text{and} \quad (xy)^{-1} = y^{-1}x^{-1}.$$

If U is the set of invertible elements in M, then $(U, \cdot, e)$ is a group (with $\cdot$ restricted to $U \times U$ and range U).
3. If every $x \in M$ is left-invertible, then M is a group (Exercise 1.2.16.1). The same applies if every x is right-invertible.
4. If $a, b \in M$ and a is invertible, then the equations $ax = b$ and $ya = b$ are uniquely solvable, namely by $x = a^{-1}b$ or $y = ba^{-1}$ (Exercise 1.2.16.1). If M is a group, then the equations are uniquely solvable for all a, b.

Example 1.2.5

1. The triples $(\mathbb{Z}, +, 0)$, $(\mathbb{Q}, +, 0)$ and $(\mathbb{Q} \setminus \{0\}, \cdot, 1)$ are abelian groups. In the monoid $(\mathbb{Z}, \cdot, 1)$ only ± 1 are invertible.
2. Let $X \neq \emptyset$ be a set. Then $\mathrm{Map}(X) := \mathrm{Map}(X, X)$ with the composition $\circ$ of functions as a link is a monoid with neutral element id_X. Here, $f : X \to X$ is exactly invertible when f is bijective (Exercise 1.2.16.2). According to Remark 1.2.4.2,

$$S(X) := \{f \in \mathrm{Map}(X) : f \text{ bijective}\}$$

is therefore a group. The inverse element to $f \in S(X)$ is the inverse function, which is luckily already denoted by f^{-1}. $S(X)$ is called the **symmetric group** of X, and an element $f \in S(X)$ is called a **permutation** of X.
For $n \in \mathbb{N}$, specifically, $S_n := S(\{1, \ldots, n\})$ is the n-th symmetric group. For $n \geq 3$, S_n is not abelian (Exercise 1.1.19.2).

We now want to define products and sums of more than two factors or summands, respectively.

Remark and Definition 1.2.6 Let $(M, \cdot, e)$ be a monoid and $N \in \mathbb{N}$. If $x_1, \ldots, x_N \in M$, then we set $\prod_{k=1}^{0} x_k := e$ and

$$\prod_{k=1}^{n} x_k := \left(\prod_{k=1}^{n-1} x_k \right) \cdot x_n$$

for $n = 1, \ldots, N$. More generally, if $x_{m+1}, \ldots, x_n \in M$ for $m, n \in \mathbb{Z}$ and $n \geq m$, then we also set

$$\prod_{j=m+1}^{n} x_j := \prod_{k=1}^{n-m} x_{k+m}.$$

In addition, in the case $x_1 = \ldots = x_n = x$ we write briefly $x^n := \prod_{k=1}^{n} x$. In particular, this means that $x^0 = e$. If x is invertible, we also set $x^{-n} := (x^{-1})^n$ for $n \in \mathbb{N}$.

In the case of the symbol plus as binary operation, we write $\sum$ instead of $\prod$. We also write nx instead of x^n. Note that the mapping $(n, x) \mapsto nx$ is in general not an operation on M.

Closely connected with the principle of recursive or inductive definition just used is an important proof method: the **induction**.

For all $n \in \mathbb{N}$ a statement $A(n)$ is given. To prove the assertion

$$A(n) \text{ holds for all } n \in \mathbb{N}$$

one often proceeds as follows:

1. One shows that $A(1)$ is correct (induction start).
2. One assumes that $A(n)$ or also $A(1), \ldots, A(n)$ for an arbitrary $n \in \mathbb{N}$ is correct (induction assumption) and shows that from the induction assumption the correctness of the statement $A(n + 1)$ follows (induction step).

From 1. and 2. it follows that $A(n)$ is correct for all $n \in \mathbb{N}$.

Sometimes one wants to show the assertion $A(n)$ for all $n \in \mathbb{N}_0$, $n \geq n_0$ instead of for $n \geq 1$. Then one proves the induction start not for $n = 1$, but for $n = n_0$, and carries out the induction step from n to $n + 1$ for any $n \geq n_0$.

Remark 1.2.7 Let $q \in \mathbb{N}$ be. We prove by induction the following statement, which implies the uniqueness of the q-adic representation of natural numbers (cf. Sect. 1.6).

For all $n \in \mathbb{N}$ holds: If $a_j \in \{0, \ldots, q - 1\}$ for $j = 0, \ldots, n - 1$, then

$$\sum_{j=0}^{n-1} a_j q^j < q^n .$$

1. Induction start: For $n = 1$ and $a_0 \in \{0, \ldots, q - 1\}$ holds $a_0 q^0 = a_0 < q$.
2. Induction assumption: For a $n \in \mathbb{N}$ holds: If $a_0, \ldots, a_{n-1} \in \{0, \ldots, q - 1\}$, then

$$\sum_{j=0}^{n-1} a_j q^j < q^n .$$

Induction step: If $a_0, \ldots, a_n \in \{0, \ldots, q - 1\}$, then by induction assumption

$$\sum_{j=0}^{n} a_j q^j = \left(\sum_{j=0}^{n-1} a_j q^j \right) + a_n q^n < q^n + a_n q^n \leq q^n + (q - 1)q^n = q^{n+1}.$$

Therefore, the assertion holds for $n + 1$ and thus for all $n \in \mathbb{N}$.

Remark and Definition 1.2.8 Let $(M, \cdot, e)$ be an abelian monoid. Then one can (inductively) show that for $x_1, x_2, x \in M$ and $m, m_1, m_2 \in \mathbb{N}_0$ the following power laws apply:

$$x^{m_1} x^{m_2} = x^{m_1 + m_2} ,$$
$$x_1^m x_2^m = (x_1 x_2)^m ,$$
$$(x^{m_1})^{m_2} = x^{m_1 m_2} .$$

If M is an abelian group, these power laws also apply for $m, m_1, m_2 \in \mathbb{Z}$.

Furthermore, one can (inductively and not easily) show that for $\varphi \in S_n$ and $x_1, \ldots, x_n \in M$

$$\prod_{k=1}^{n} x_{\varphi(k)} = \prod_{k=1}^{n} x_k$$

applies. This makes the following notation meaningful: If $n \in \mathbb{N}$ and if I is any n-element set, we set for families $(x_j)_{j \in I}$ in M

$$\prod_{j \in I} x_j := \prod_{k=1}^{n} x_{\psi(k)} ,$$

where $\psi : \{1, \ldots, n\} \to I$ is any bijective mapping. If $(y_j)_{j \in I}$ is another family in M, then

$$\prod_{j \in I} (x_j y_j) = \prod_{j \in I} x_j \prod_{j \in I} y_j$$

applies.

We now turn to algebraic structures with two operations.

Remark and Definition 1.2.9 Let R be a set and let $+$ and $\cdot$ be operations on R with:

(R1) $(R, +, 0)$ is an abelian group.

(R2) $(R, \cdot, 1)$ is a monoid.

(R3)   The operation $\cdot$ is **distributive** with respect to $+$, that is, for $x, y, z \in R$ the following holds

$$x(y + z) \; = \; (xy) + (xz) \qquad \text{and} \qquad (x + y)z \; = \; (xz) + (yz).$$

Then $(R, +, \cdot)$ is called a **ring**, the neutral element 0 for $+$ is called **zero element** or simply **zero** and the neutral element 1 for $\cdot$ is called **one element** or **one**. If $(R, \cdot, 1)$ is abelian, the ring is called **commutative**. Sometimes one writes more clearly 0_R and 1_R for the neutral elements of a ring. On the other hand, one often writes simply R instead of $(R, +, \cdot)$. Standard examples of commutative rings are $(\mathbb{Z}, +, \cdot)$ and $(\mathbb{Q}, +, \cdot)$.

Remark 1.2.10 As in $(\mathbb{Q}, +, \cdot)$, in general rings R we use point-before-line notations, so for example $x + yz := x + (yz)$. Inductively, for $x \in R$ and finite families $(x_j)_{j \in I}$ in R the general distributive laws result

$$x \sum_{j \in I} x_j = \sum_{j \in I} x x_j \quad \text{and} \quad \left(\sum_{j \in I} x_j \right) x = \sum_{j \in I} x_j x.$$

Remark and Definition 1.2.11 Let R be a ring. Then (Exercise 1.2.16.4)

1. 0 is **absorbing** for R, that is $0 \cdot x = x \cdot 0 = 0$ for $x \in R$.
2. $(-x)y = x(-y) = -xy$ and $(-x)(-y) = xy$ for $x, y \in R$.
3. $x(y - z) = xy - xz$ and $(x - y)z = xz - yz$ for $x, y, z \in R$.

Remark 1.2.12 Let R be a ring and X a non-empty set. We define for $f, g \in R^X$ the functions $f \pm g \in R^X$ and $f \cdot g \in R^X$ argument-wise by

$$(f \pm g)(x) := f(x) \pm g(x) \quad \text{and} \quad (f \cdot g)(x) := f(x) \cdot g(x) \quad (x \in X).$$

Thus, $R^X = (R^X, +, \cdot)$ is a ring with zero 0_{R^X} and one 1_{R^X}, defined by $0_{R^X}(x) := 0_R$ and $1_{R^X}(x) := 1_R$ for $x \in X$. If R is commutative, then R^X is also commutative.

Definition 1.2.13 A ring $(R, +, \cdot)$ with zero 0_R and one 1_R is called a **field**, if $(R^*, \cdot, 1_R)$ with

$$R^* := R \setminus \{0_R\}$$

is an abelian group. We also write $1/x$ instead of x^{-1} for the inverse element of $x \neq 0_R$ with respect to multiplication and x/y instead of $xy^{-1}(= y^{-1}x)$.

Remark 1.2.14 An important property of fields is the **absence of zero divisors**: If $x, y \in R$ with $xy = 0$, then $x = 0$ or $y = 0$ (since $\cdot$ is an operation on R^* !).

Example 1.2.15

1. Let $\mathbb{F}_2 := \{\heartsuit, \clubsuit\}$, where the addition and multiplication are defined by the following commutative operation tables:

It can be easily calculated that $(\mathbb{F}_2, +, \cdot)$ is a field, called the **binary field**. It holds that $\heartsuit = 0 = 0_{\mathbb{F}_2}$ and $\clubsuit = 1 = 1_{\mathbb{F}_2}$, so in binary arithmetic $1 + 1 = 0$.

2. The ring $(\mathbb{Q}, +, \cdot)$ is a field, the ring $(\mathbb{Z}, +, \cdot)$ is not.

Exercises 1.2.16

1. Let $(M, \cdot, e)$ be a monoid. Show:
   a) If every $x \in M$ is left-invertible, then $(M, \cdot, e)$ is a group.
   b) If $a \in M$ is invertible, then the equation $ax = b$ has exactly one solution, namely $x = a^{-1}b$.
2. Let X, Y be sets and $f : X \to Y$, $g : Y \to X$ such that

$$g \circ f = \mathrm{id}_X.$$

   Consider that f is injective and g is surjective. Are f and g always bijective?
3. Prove: For all $n \in \mathbb{N}$ the following holds

$$\sum_{v=1}^{n} v^3 = \frac{1}{4} n^2 (n+1)^2.$$

4. Let $(R, +, \cdot)$ be a ring. Show:
   a) $0 \cdot x = x \cdot 0 = 0 \ (x \in R)$.
   b) $(-x)y = x(-y) = -(xy) \ (x, y \in R)$.
5. Show that the ring $(\mathbb{Z}^2, +, \cdot) \ \left(= (\mathbb{Z}^{\{1,2\}}, +, \cdot) \right)$ is not free of zero divisors.

1.3   Geometric Sum Formula and Binomial Formula

We now come to various basic formulas that apply in commutative rings.

Theorem 1.3.1 *Let $(R, +, \cdot)$ be a commutative ring. Then for all $a, b \in R$ and all $n \in \mathbb{N}$*

$$a^n - b^n = (a - b) \sum_{v=0}^{n-1} a^v b^{n-1-v}. \qquad (1.3.1)$$

Proof One has

$$(a - b) \sum_{v=0}^{n-1} a^v b^{n-1-v} = \sum_{v=0}^{n-1} a a^v b^{n-1-v} - \sum_{v=0}^{n-1} b a^v b^{n-1-v}$$

$$= \sum_{v=0}^{n-1} a^{v+1} b^{n-(v+1)} - \sum_{v=0}^{n-1} a^v b^{n-v}$$

$$= \sum_{\mu=1}^{n} a^\mu b^{n-\mu} - \sum_{v=0}^{n-1} a^v b^{n-v} = a^n - b^n.$$

Remark 1.3.2 (**geometric sum formula**) If $(R, +, \cdot)$ is a field, then for $x \neq 1$ according to Theorem 1.3.1

$$\sum_{\nu=0}^{n-1} x^{\nu} = \frac{x^n - 1}{x - 1} . \tag{1.3.2}$$

In addition to the geometric sum formula, there is another formula in commutative rings, the binomial formula. It is a sum formula for the expressions $(a + b)^n$, where $a, b \in R$ and $n \in \mathbb{N}$ is. To be able to state the general formula, we need

Definition 1.3.3 For $n \in \mathbb{N}_0$ one defines n-**factorial** by

$$n! := \prod_{\nu=1}^{n} \nu$$

and for $n, \nu \in \mathbb{N}_0$ the **binomial coefficient** n over ν by

$$\binom{n}{\nu} := \frac{1}{\nu!} \prod_{k=1}^{\nu} (n - k + 1) .$$

Note that: According to Remark 1.2.6 are

$$0! = 1 \qquad \text{and} \qquad \binom{n}{0} = 1.$$

We compile some properties of the binomial coefficients.

Theorem 1.3.4 *For $n, \nu \in \mathbb{N}_0$*

$$\binom{n}{\nu} = \begin{cases} \dfrac{n!}{\nu!(n - \nu)!} = \dbinom{n}{n - \nu}, & \textit{if } \nu \leq n \\ = 0, & \textit{if } \nu > n \end{cases} .$$

Proof It holds for $\nu \leq n$

$$\binom{n}{\nu} = \frac{\prod\limits_{k=1}^{\nu} (n - k + 1)}{\nu!} = \frac{\prod\limits_{k=1}^{\nu} (n - k + 1)}{\nu!} \cdot \frac{(n - \nu)!}{(n - \nu)!} = \frac{n!}{\nu!(n - \nu)!} .$$

Thus,

$$\binom{n}{\nu} = \frac{n!}{\nu!(n - \nu)!} = \frac{n!}{(n - (n - \nu))!(n - \nu)!} = \binom{n}{n - \nu} .$$

For $\nu > n$, $n - \nu + 1 \leq 0$ and thus $\prod\limits_{k=1}^{\nu} (n - k + 1) = 0$, so also $\binom{n}{\nu} = 0$.

The following recursion formula is particularly important:

Theorem 1.3.5 *For $n, v \in \mathbb{N}$ one has*

$$\binom{n+1}{v} = \binom{n}{v-1} + \binom{n}{v}.$$

Proof According to Theorem 1.3.4, for $v \in \{1, \ldots, n\}$

$$\binom{n}{v-1} + \binom{n}{v} = \frac{n!}{(v-1)!(n-v+1)!} + \frac{n!}{v!(n-v)!}$$

$$= \frac{n!}{v!(n+1-v)!} \left(v + (n+1-v) \right) = \frac{(n+1)!}{v!(n+1-v)!} = \binom{n+1}{v}.$$

For $v = n + 1$, according to Theorem 1.3.4

$$\binom{n}{v-1} + \binom{n}{v} = \binom{n}{n} + 0 = 1 = \binom{n+1}{v},$$

and for $v > n + 1$, both sides have the value 0.

If you arrange the binomial coefficients $\binom{n}{v}$ in a triangular scheme, where in the n-th row (starting with row 0) the coefficients $\binom{n}{0}, \ldots, \binom{n}{n}$ appear, the **Pascal's Triangle**:

$$\binom{0}{0}$$
$$\binom{1}{0} \quad \binom{1}{1}$$
$$\binom{2}{0} \quad \binom{2}{1} \quad \binom{2}{2}$$
$$\vdots \quad \vdots \quad \vdots$$
$$\vdots \quad \vdots \quad \vdots$$
$$\binom{n}{0} \quad \binom{n}{1} \cdots \binom{n}{v-1} \quad \binom{n}{v} \cdots \binom{n}{n}$$
$$\binom{n+1}{0} \quad \binom{n+1}{1} \cdots \binom{n+1}{v} \cdots \binom{n+1}{n+1}$$

The first rows are calculated by using Theorem 1.3.5 to

$$1$$
$$1 \quad 1$$
$$1 \quad 2 \quad 1$$
$$1 \quad 3 \quad 3 \quad 1$$
$$1 \quad 4 \quad 6 \quad 4 \quad 1$$
$$1 \quad 5 \quad 10 \quad 10 \quad 5 \quad 1$$
$$1 \quad 6 \quad 15 \quad 20 \quad 15 \quad 6 \quad 1$$

Theorem 1.3.6 (Binomial theorem) *Let $(R, +, \cdot)$ be a commutative ring. Then for all $a, b \in R$ and all $n \in \mathbb{N}_0$*

$$(a + b)^n = \sum_{\nu=0}^{n} \binom{n}{\nu} a^\nu b^{n-\nu} \, .$$

Proof

1. For $n = 0, (a + b)^0 = 1 = \sum_{\nu=0}^{0} \binom{0}{\nu} a^\nu b^{0-\nu}$ holds.

2. For a $n \in \mathbb{N}_0$, assume $(a + b)^n = \sum_{\nu=0}^{n} \binom{n}{\nu} a^\nu b^{n-\nu}$. Then, by Theorem 1.3.5,

$$
\begin{aligned}
(a + b)^{n+1} &= (a + b)(a + b)^n = (a + b) \sum_{\nu=0}^{n} \binom{n}{\nu} a^\nu b^{n-\nu} \\
&= \sum_{\nu=0}^{n} \binom{n}{\nu} a^{\nu+1} b^{n-\nu} + \sum_{\nu=0}^{n} \binom{n}{\nu} a^\nu b^{n-\nu+1} \\
&= \sum_{\mu=1}^{n+1} \binom{n}{\mu - 1} a^\mu b^{n+1-\mu} + \sum_{\nu=0}^{n} \binom{n}{\nu} a^\nu b^{n+1-\nu} \\
&= a^{n+1} + \sum_{\nu=1}^{n} \binom{n+1}{\nu} a^\nu b^{n+1-\nu} + b^{n+1} \\
&= \sum_{\nu=0}^{n+1} \binom{n+1}{\nu} a^\nu b^{n+1-\nu}
\end{aligned}
$$

follows. Therefore, the claim also holds for $n + 1$ and thus by induction for all $n \in \mathbb{N}_0$.

Example 1.3.7 For $n = 6$,

$$
\begin{aligned}
(a + b)^6 &= \sum_{\nu=0}^{6} \binom{6}{\nu} a^\nu b^{6-\nu} \\
&= 1 \cdot b^6 + 6 \cdot ab^5 + 15a^2b^4 + 20a^3b^3 + 15a^4b^2 + 6a^5b + 1 \cdot a^6
\end{aligned}
$$

holds.

Remark 1.3.8 As special cases from Theorem 1.3.6, interesting relationships for the Pascal's triangle arise: For $R = \mathbb{Z}$ and $a = 1, b = 1$, we get

$$2^n = (1 + 1)^n = \sum_{\nu=0}^{n} \binom{n}{\nu} 1^\nu 1^{n-\nu} = \sum_{\nu=0}^{n} \binom{n}{\nu} ,$$

that is, the sum of the binomial coefficients in the n-th row of Pascal's triangle always results in 2^n. For $a = -1, b = 1$, we get for $n \in \mathbb{N}$

$$0 = 0^n = ((-1) + 1)^n = \sum_{\nu=0}^{n} \binom{n}{\nu} (-1)^\nu \,,$$

that is, if we alternately assign the sign $+$ and $-$ to the binomial coefficients in the n-th row, the sum will be 0. For $n = 6$, for example,

$$1 + 6 + 15 + 20 + 15 + 6 + 1 = 64 = 2^6$$

and

$$1 - 6 + 15 - 20 + 15 - 6 + 1 = 0 \,.$$

In conclusion, we briefly deal with the meaning of factorials and binomial coefficients in combinatorics.

Definition 1.3.9 Let A, B be arbitrary sets.

1. A and B are called **equinumerous**, if a bijective mapping $\varphi : A \to B$ exists.
2. A has **cardinality** $n \in \mathbb{N}$, if A is equinumerous to $\{1, \ldots, n\}$, that is, if a tuple $(a_1, \ldots, a_n)$ exists with $a_j \neq a_k$ for $j \neq k$ and $A = \{a_1, \ldots, a_n\}$ (it can be shown that n is unique). The empty set is assigned the cardinality 0. Thus, A is called **finite**, if A has a cardinality $n \in \mathbb{N}_0$. We then write $\#A := n$. If A is not finite, then A is called **infinite** and we briefly write $\#A = \infty$.

Remark and Definition 1.3.10 A family $(A_\alpha)_{\alpha \in I}$ of sets is called **disjoint**, if $A_\alpha \cap A_\beta = \emptyset$ for $\alpha, \beta \in I$, $\alpha \neq \beta$. If A is a set and $(A_\alpha)_{\alpha \in I}$ is a disjoint family of non-empty sets with $A = \bigcup_{\alpha \in I} A_\alpha$, then $(A_\alpha)_{\alpha \in I}$ is called a **decomposition** of A. If A is finite and $(A_j)_{j \in J}$ is a decomposition of A, then

$$\#A = \sum_{j \in J} \#A_j.$$

Theorem 1.3.11 *For $n \in \mathbb{N}$ the following applies: If X is a n-element set, then for the symmetric group $S(X)$*

$$\#(S(X)) = n! \,.$$

Proof We prove by induction on n.

1. Induction start: For $n = 1$, the assertion is clear.
2. Induction step: Let X be a set with $(n + 1)$ elements. Without loss of generality, we can assume $X = \{1, \ldots, n + 1\}$ (thus $S(X) = S_{n+1}$). We define

$$T_j := \{\sigma \in S_{n+1} : \sigma(j) = n + 1\} \qquad (j = 1, \ldots, n + 1) \,.$$

Then

$$\bigcup_{j=1}^{n+1} T_j = S_{n+1} \quad \text{and} \quad T_j \cap T_k = \emptyset \quad (j \neq k).$$

Therefore, $\#(S_{n+1}) = \sum_{j=1}^{n+1} \#(T_j)$.

If we define for $\sigma \in T_j$ the function $\tau_\sigma \in S_n$ by

$$\tau_\sigma(k) := \begin{cases} \sigma(k), & (k = 1, \ldots, j-1) \\ \sigma(k+1), & (k = j, \ldots, n) \end{cases},$$

then $T_j \ni \sigma \mapsto \tau_\sigma \in S_n$ is a bijective mapping. By induction hypothesis, $\#(S_n) = n!$ and thus also $\#(T_j) = n!$. Therefore,

$$\#(S_{n+1}) = \sum_{j=1}^{n+1} n! = (n+1)n! = (n+1)! \, .$$

Remark 1.3.12 If X is an n-element set and $\mathscr{A}_k(X) \subset \mathscr{P}(X)$ for $k \in \{0, \ldots, n\}$ is the set of k-element subsets of X, then (see Exercise 1.3.13.4)

$$\#(\mathscr{A}_k(X)) = \binom{n}{k}.$$

According to Remark 1.3.8,

$$\#(\mathscr{P}(X)) = \sum_{\nu=0}^{n} \#(\mathscr{A}_\nu(X)) = 2^n$$

is also valid.

Exercises 1.3.13

1. Prove: For all $n \in \mathbb{N}$ 13 is a divisor of $17^n - 4^n$.
2. a) Show: For all $n, m \in \mathbb{N}$ is

$$\sum_{\nu=1}^{n} \binom{m+\nu-1}{m} = \binom{m+n}{m+1}.$$

   b) What results in a) as a special case for $m = 1$ and $m = 2$?
   c) Find a closed representation for $\sum_{\nu=1}^{n} \nu^2$.
3. (**Abel's partial summation**) Let $(R, +, \cdot)$ be a ring and $a_k, b_k \in R$ for $k = 1, \ldots, m$. Prove: With $B_k := \sum_{\nu=1}^{k} b_\nu$

$$\sum_{k=1}^{m} a_k b_k + \sum_{k=1}^{m-1} (a_{k+1} - a_k) B_k = a_m B_m.$$

4. Let $n \in \mathbb{N}_0$ and X be a n-element set. Show: If $k \in \{0, \ldots, n\}$ and

$$\mathscr{A}_k(X) := \{A \subset X : \#A = k\},$$

then

$$\#(\mathscr{A}_k(X)) = \binom{n}{k}.$$

5. a) How many ways are there to choose "6 out of 49" if the order of the drawn numbers does not matter? How many possibilities are there if the order in which the numbers are drawn is taken into account?

   b) In a society of n people, everyone shakes hands with everyone once. How many handshakes are there?

1.4 Ordered Fields and Real Numbers

Definition 1.4.1 Let $X \neq \emptyset$ be a set. A relation $<$ on X is called an **order** (on X), if the following holds

(O1) For all $x, y \in X$, either $x = y$ or $x < y$ or $y < x$ holds (Trichotomy).
(O2) For $x, y, z \in X$, if $x < y$ and $y < z$, then $x < z$ follows (Transitivity).

The pair $(X, <)$ is then called an **ordered set**. Moreover, $x \leq y$ means that either $x < y$ or $x = y$ holds. Finally, we also write $y > x$ instead of $x < y$ and $y \geq x$ instead of $x \leq y$.

Definition 1.4.2 Let $(X, <)$ be ordered and $M \subset X$.

1. M is called **bounded above**, if there exists an $s \in X$ with

   $$x \leq s \quad \text{for all } x \in M .$$

   Such an s is then called an **upper bound** of M. If $s \in M$, then s is called the **maximum** of M (notation: $\max M := s$).
2. M is called **bounded below**, if there exists an $s \in X$ with

   $$x \geq s \quad \text{for all } x \in M .$$

   Such an s is then called a **lower bound** of M. If $s \in M$, then s is called the **minimum** of M (notation: $\min M := s$).
3. M is called **bounded**, if M is bounded above and below.

Example 1.4.3 Let $(X, <) = (\mathbb{Q}, <)$. Then the set $M = \{1/n : n \in \mathbb{N}\}$ is bounded with $\max M = 1$, but M has no minimum!

Remark 1.4.4 By induction, one can show: If $(X, <)$ is ordered, then every non-empty, finite set $M \subset X$ has a maximum and a minimum. Furthermore, it can be shown that every non-empty, upper-bounded set $M \subset \mathbb{Z}$ has a maximum and every non-empty, lower-bounded set $M \subset \mathbb{Z}$ has a minimum.

Definition 1.4.5 Let $K = (K, +, \cdot)$ be a field. If $<$ is an order on K, then $K = (K, +, \cdot, <)$ is called an **ordered field**, if for $x, y \in K$ the following properties are fulfilled with respect to addition and multiplication:

(O3)   From $x < y$ it follows that $x + z < y + z$ for all $z \in K$ (1st law of monotonicity).
(O4)   From $x < y$ and $z > 0$ it follows that $xz < yz$ (2nd law of monotonicity).

We call $x \in K$ **positive**, if $x > 0$ hold and **negative**, if $x < 0$ hold. In addition, we set $K_+ := \{x \in K : x > 0\}$ and $K_- := \{x \in K : x < 0\}$.

Theorem 1.4.6 *Let $K = (K, +, \cdot, <)$ be an ordered field and $x, y \in K$. Then the following holds*

1. $x > 0$ *if and only if* $-x < 0$ *is true.*
2. *From* $x, y < 0$ *or* $x, y > 0$ *it follows that* $xy > 0$.
3. *For* $x \neq 0$ *it is* $x^2 > 0$, *in particular* $1 = 1^2 > 0$.
4. *From* $0 < x < y$ *it follows* $0 < 1/y < 1/x$.

Proof

1. From $0 < x$ it follows that $-x = 0 + (-x) < x + (-x) = 0$, with (O3), that is $-x < 0$. Correspondingly, from $-x < 0$ also follows $0 = x + (-x) < x + 0 = x$.
2. If $x, y > 0$, then with (O4) immediately follows $0 = 0y < xy$. If on the other hand $x, y < 0$, then $-y > 0$ according to 1. Because $x < 0$, with (O4)

$$-(xy) = x(-y) < 0(-y) = 0 ,$$

   it follows that $xy > 0$ with 1.
3. This follows directly from 2. and (O1).
4. Firstly, $1/x > 0$; because assuming, it holds $1/x \leq 0$ and thus $1/x < < 0$. Then follows $1 = x/x < x \cdot 0 = 0$ with (O4), in contradiction to 3. Similarly, $1/y > 0$. From $x < y$ it follows that $x/y < y/y = 1$ with (O4) and again with (O4)

$$\frac{1}{y} = \frac{1}{y} \cdot \frac{x}{x} = \frac{x}{y} \cdot \frac{1}{x} < 1 \cdot \frac{1}{x} = \frac{1}{x} .$$

Example 1.4.7

1. $(\mathbb{Q}, +, \cdot, <)$ is an ordered field.
2. In the binary field $(\mathbb{F}_2, +, \cdot)$ there is no order relation with the properties from Definition 1.4.5.

   Because, assuming the contrary. Then $1 > 0$ according to Theorem 1.4.6.3, so also $0 = 1 + 1 > 1 + 0 = 1$, contradicting (O1).

Remark 1.4.8 Let K be an ordered field. By induction, it is easy to see:

1. If $n \in \mathbb{N}$ and $x < y$, then $nx < ny$ and in the case $x > 0$ also $0 < x^n < y^n$.
2. If $x > 0$ and $n, m \in \mathbb{N}$ with $n > m$, then $nx > mx > 0$.

In particular, it follows from 2. that K contains infinitely many elements. More precisely, if $x, y \in K$ with $x < y$, then the set $\{z \in K : x < z < y\}$ is infinite.

Because: For all $n, m \in \mathbb{N}$ with $n > m$, $n1 > m1 > 0$, so $1/(m1) > 1/(n1) > 0$ and consequently

$$x < x + (y - x)/(n1) < x + (y - x)/(m1) \leq y.$$

In general, in ordered fields, equations of the form

$$x^n = c,$$

where $c \in K, n \in \mathbb{N}, n > 1$ are not solvable. If $c < 0$ and n is even, this is excluded anyway according to Theorem 1.4.6.3. But even in the case of $c > 0$, there is generally no solution.

Theorem 1.4.9 *For all $x \in \mathbb{Q}, x^2 \neq 2$.*

Proof Assume there exists a $x = p/q \in \mathbb{Q}$ with $(p/q)^2 = 2$. We can assume without loss of generality that $p \in \mathbb{Z}, q \in \mathbb{N}$ are coprime and therefore not both even. From $p^2 = 2q^2$ it follows that p^2 is even. Thus, p is also even, that is $p = 2p_0$ for a $p_0 \in \mathbb{Z}$. Then

$$2q^2 = p^2 = 4p_0^2 \, ,$$

that is $q^2 = 2p_0^2$, so q^2 and thus also q is even. This contradicts the fact that p and q are not both even. Therefore, the assumption is incorrect.

Our further goals are:

- Expanding $(\mathbb{Q}, +, \cdot, <)$ to an ordered field $(\mathbb{R}, +, \cdot, <)$ so that $x^n = c$ for all $n \in \mathbb{N}$ and $c \geq 0$ is solvable.
- Expanding $(\mathbb{R}, +, \cdot)$ to a field $(\mathbb{C}, +, \cdot)$ so that $x^n = c$ for all $n \in \mathbb{N}$ and $c \in \mathbb{C}$ is solvable.

Remark and Definition 1.4.10 Let $(X, <)$ be ordered and let $M \subset X$. With an upper bound s of M, every $t \in X$ with $t > s$ is also an upper bound for M. The question naturally arises for the smallest upper (and largest lower) bounds.

An upper bound $s^* \in X$ of M is called the **smallest upper bound** or **supremum** of M, if $s \geq s^*$ applies for every upper bound s of M. A lower bound $s_* \in X$ of M is called the **largest lower bound** or **infimum** of M, if $s \leq s_*$ applies for every lower bound s of M.

From the definition, it immediately follows that for each M at most one supremum and one infimum exist. In the case of existence, we write

$$\sup M := s^*$$

or

$$\inf M := s_* \, .$$

If $\max M$ exists, then $\sup M = \max M$ applies. In the case of the existence of $\min M$, $\inf M = \min M$ holds.

Example 1.4.11 Let $(X, <) = (\mathbb{Q}, <)$.

1. If $M = \{1/n : n \in \mathbb{N}\}$, then $1 = \max M = \sup M$ holds. Although M has no minimum, $\inf M$ exists and $\inf M = 0$ holds.

   Because: Firstly, 0 is a lower bound of M. If $s > 0$, thus $s = p/q$ with $p, q \in \mathbb{N}$, then $1/(q + 1) < s$ and $1/(q + 1) \in M$. Thus, s is not a lower bound of M. Therefore, every lower bound $s \leq 0$.

2. If $M := \{x \in \mathbb{Q} : x \geq 0, \ x^2 \leq 2\}$, then $\inf M = \min M = 0$ and M is also bounded above.

   Because: If $x > 3/2$, then it follows $x^2 > (3/2)^2 = 9/4 > 2$, thus $x \notin M$. Therefore, 3/2 is an upper bound of M. However, there is *no supremum* of M, as will follow from Theorem 1.4.13.

Remark 1.4.12 Let K be an ordered field. Then for all $x > -1$ the **Bernoulli inequality**

$$(1 + x)^n \geq 1 + nx$$

(Exercise 1.4.27.1) holds. Moreover, for $0 \leq b \leq a$ according to Theorem 1.3.1

$$a^n - b^n = (a - b) \sum_{\nu=0}^{n-1} a^\nu b^{n-\nu-1} \leq n(a - b)a^{n-1} .$$

Theorem 1.4.13 *Let K be an ordered field and $n \in \mathbb{N}$ as well as $0 \leq c \in K$. Then*

$$M := \{x \in K : x \geq 0, x^n \leq c\}$$

has the following properties

1. *M is non-empty and bounded above.*
2. *If $s := \sup M$ exists, then $s^n = c$.*

Proof

1. It holds that $0 \in M$, so $M \neq \emptyset$. Also, $1 + c$ is an upper bound of M, because if $x \in K$ with $x > 1 + c$, then according to the Bernoulli inequality

$$x^n > (1 + c)^n \geq 1 + nc > nc \geq c$$

   and thus $x \notin M$.
2. We show that neither $s^n > c$ nor $s^n < c$ can hold (thus $s^n = c$).

Assume that $s^n > c$. Then for $\delta := \frac{s^n - c}{ns^{n-1}}$ according to Remark 1.4.12

$$s^n - (s - \delta)^n \le n\delta s^{n-1} \le s^n - c,$$

thus $(s - \delta)^n \ge c$. If $x \in M$, then $x^n \le c \le (s - \delta)^n$ and thus also $x \le s - \delta$. Therefore, $s - \delta$ is an upper bound of M contradicting that s is the smallest upper bound.

Assume that $s^n < c$. Then there exists a $\delta > 0$ with $(s + \delta)^n \le c$ (according to Remark 1.4.12 with $a = s + \delta$ and $b = s$)

$$\delta := \min\left\{1, \frac{c - s^n}{n(s + 1)^{n-1}}\right\}$$

is suitable). But then $s + \delta \in M$ and thus s is not an upper bound of M. Contradiction.

Definition 1.4.14 An ordered set $(X, <)$ is called **order complete** or briefly **complete**, if every non-empty, upper-bounded subset M of X has a supremum. An ordered field $(K, +, \cdot, <)$ is called **completely ordered**, if $(K, <)$ is order complete.

Remark and Definition 1.4.15 If K is an ordered field, then for each $c \in K, c \ge 0$ and each $n \in \mathbb{N}$ the equation

$$x^n = c$$

has at most one solution $s \ge 0$, because from $s_1, s_2 \in K$ with $0 \le s_1 < s_2$ follows $s_1^n < s_2^n$. If K is *complete,* then according to Theorem 1.4.13 exactly one solution $s \in K$ with $s \ge 0$ exists. We set

$$\sqrt[n]{c} := s.$$

For $c, d \in K$ with $c, d \ge 0$ as well as $m, n \in \mathbb{N}$ from the corresponding power laws it easily follows that

$$\sqrt[n]{cd} = \sqrt[n]{c}\sqrt[n]{d} \qquad \text{and} \qquad \sqrt[m]{\sqrt[n]{c}} = \sqrt[nm]{c}$$

and for $0 \le c < d$ also $\sqrt[n]{c} < \sqrt[n]{d}$.

Of fundamental importance to analysis is the following result:

There exists a completely ordered field $(\mathbb{R}, +, \cdot, <)$ such that $\mathbb{Q}$ is embedded in $\mathbb{R}$.

It can be shown that in a certain sense only one completely ordered field exists. The elements of $\mathbb{R}$ are called **real numbers**. A possible construction of the real numbers and the proof of the above statement (including a clarification of what we mean by "embedded") will be discussed in Sect. 1.6.

Remark 1.4.16 As an important consequence of completeness, it follows (Exercises 1.4.27.6 and 1.4.27.7)

1. $\mathbb{N}$ is unbounded above in $\mathbb{R}$, that is, for all $x \in \mathbb{R}$ there exists a $n \in \mathbb{N}$ with $n > x$ (Archimedean property of $\mathbb{R}$).
2. If $x, y \in \mathbb{R}$ with $x < y$, then there exists a $r \in \mathbb{Q}$ with $x < r < y$ (Density of $\mathbb{Q}$ in $\mathbb{R}$).

Remark and Definition 1.4.17 Sometimes it is practical and sensible to extend the ordered set $(\mathbb{R}, <)$ by two points $+\infty$ (or shortly ∞) and $-\infty$ such that by definition $-\infty < x < \infty$ holds for all $x \in \mathbb{R}$. For $M \subset \mathbb{R}$ this implies $\sup M = \infty$, if M is unbounded above, and $\inf M = -\infty$, if M is unbounded below.

A non-empty set $I \subset \mathbb{R}$ is called an **interval**, if $x \in I$ holds for all $x \in \mathbb{R}$ with $\inf I < x < \sup I$. For $a, b \in \mathbb{R} \cup \{\pm\infty\}$ we set

$$[a, b] := \{x \in \mathbb{R} : a \leq x \leq b\}, \text{ if } -\infty < a \leq b < \infty,$$
$$(a, b) := \{x \in \mathbb{R} : a < x < b\}, \text{ if } -\infty \leq a < b \leq \infty,$$
$$[a, b) := \{x \in \mathbb{R} : a \leq x < b\}, \text{ if } -\infty < a < b \leq \infty,$$
$$(a, b] := \{x \in \mathbb{R} : a < x \leq b\}, \text{ if } -\infty \leq a < b < \infty.$$

Every interval has exactly one such form, where always $a = \inf I$ and $b = \sup I$.

We will show in the following that in a certain sense the majority of real numbers consist of irrational numbers.

Definition 1.4.18 If $N \subset \mathbb{N}_0$ is infinite (and thus unbounded above in $\mathbb{R}$) and X is a non-empty set, we call a function $(a_n)_{n \in N} : N \to X$ a **sequence** (in X). In the case of $N = \{n \in \mathbb{N}_0 : n \geq n_0\}$ we also write $(a_n)_{n=n_0}^{\infty}$ or $(a_n)_{n \geq n_0}$ or simply (a_n), if n_0 is clear or irrelevant. If $J \subset N$ is infinite, we call $(a_n)_{n \in J}$ a **subsequence** of $(a_n)_{n \in N}$.

Definition 1.4.19 Let A be a set.

1. A is called **countably infinite**, if A is equipotent to $\mathbb{N}$, that is, if a sequence $(a_n)_{n \in \mathbb{N}}$ exists with $a_j \neq a_k$ for $j \neq k$ and $A = \{a_n : n \in \mathbb{N}\}$.
2. A is called **countable**, if A is finite or countably infinite. Otherwise, A is called **uncountable**.

Theorem 1.4.20 *A non-empty set A is countable if and only if a sequence $(a_n)_{n \in \mathbb{N}}$ exists with $A = \{a_n : n \in \mathbb{N}\}$.*

Proof $\Rightarrow$: If A is countably infinite, such a sequence exists by definition. If A is finite, say $A = \{a_1, ..., a_n\}$, then set $a_j := a_n$ for $j > n$.

$\Leftarrow$: If A is finite, then A is countable. So let $A = \{a_n : n \in \mathbb{N}\}$ be infinite. We set $n_1 := 1$. If $n_1, ..., n_k$ are already defined, then

$$n_{k+1} := \min\{n \in \mathbb{N} : a_n \notin \{a_{n_1}, ..., a_{n_k}\}\}$$

exists. If we set $b_k := a_{n_k}$, then by construction $b_j \neq b_k$ for $j \neq k$ and $\{b_k : k \in \mathbb{N}\} = A$. Therefore, A is countably infinite.

Remark 1.4.21 From Theorem 1.4.20 it immediately follows that images of countable sets under any mappings are again countable and thus also that subsets of countable sets are countable.

Theorem 1.4.22 *Let* $J \neq \emptyset$ *be countable. If* $(A_j)_{j \in J}$ *is a family of countable sets, then* $\bigcup_{j \in J} A_j$ *is also countable.*

Proof Without restriction, we can assume $A_j \neq \emptyset$ for all $j \in J$. According to Theorem 1.4.20, sequences $(a_n^{(j)})_{n \in \mathbb{N}}$ exist with

$$A_j = \{a_n^{(j)} : n \in \mathbb{N}\} \qquad (j \in \mathbb{N}).$$

Furthermore, there exists a sequence (j_k) with $J = \{j_k : k \in \mathbb{N}\}$. We set

$$B_m := \{a_n^{(j_k)} : k, n = 1, \dots, m\} \qquad (m \in \mathbb{N}).$$

Since $\#(B_m) \leq m^2$ and $B_m \subset B_{m+1}$ for all $m \in \mathbb{N}$, the existence of a sequence $(b_\ell)_{\ell \in \mathbb{N}}$ with $B_m = \{b_1, \dots, b_{m^2}\}$ for all $m \in \mathbb{N}$ is inductively derived. By construction,

$$\bigcup_{j \in J} A_j = \bigcup_{m \in \mathbb{N}} B_m = \{b_\ell : \ell \in \mathbb{N}\}.$$

With Theorem 1.4.20, the assertion follows.

Remark 1.4.23 For $n \in \mathbb{Z}$ the mapping $\mathbb{Z} \ni j \mapsto (j, n) \in \mathbb{Z} \times \{n\}$ is bijective. Since $\mathbb{Z} = \mathbb{N}_0 \cup (-\mathbb{N}_0)$ is countable, $\mathbb{Z} \times \{n\}$ is also countable. Therefore, according to Theorem 1.4.22,

$$\mathbb{Z} \times \mathbb{Z}^* = \bigcup_{n \in \mathbb{Z}^*} (\mathbb{Z} \times \{n\})$$

is also countable. According to Remark 1.4.21, $\mathbb{Q}$ is finally also countable, because $\mathbb{Q} = \varphi(\mathbb{Z} \times \mathbb{Z}^*)$ holds for $\varphi(a, b) := a/b$.

We now want to show that $\mathbb{R}$ is uncountable. To do this, we first prove:

Theorem 1.4.24 (Interval Nesting Principle) *Let* (I_n) *be a sequence of intervals of the form* $I_n = [a_n, b_n]$ *with* $I_{n+1} \subset I_n$ $(n \in \mathbb{N})$. *Then*

$$c := \sup\{a_n : n \in \mathbb{N}\} \leq \inf\{b_n : n \in \mathbb{N}\} =: d$$

and

$$\bigcap_{n \in \mathbb{N}} I_n = [c, d].$$

Proof $\subset$: If $x \in \bigcap_{n \in \mathbb{N}} I_n$, then from $a_n \leq x \leq b_n$ for all n also $c \leq x \leq d$ follows by definition of the supremum and the infimum.

$\supset$: If $m \in \mathbb{N}$, then by assumption $a_n \leq b_n \leq b_m$ for $n \geq m$ and $a_k \leq a_m$ for $k \leq m$, so $a_n \leq b_m$ for all $n \in \mathbb{N}$. By definition of the supremum, $c \leq b_m$. Since $m \in \mathbb{N}$ was arbitrary, $c \leq d$ is also true by definition of the infimum, so overall

$$a_n \le c \le d \le b_n \qquad (n \in \mathbb{N})$$

and thus $[c, d] \subset I_n$ for all n.

Theorem 1.4.25 *Every non-single-point interval $I \subset \mathbb{R}$ is uncountably infinite.*

Proof It is easy to see that it is sufficient to prove the assertion for $I = [0, 1]$. Suppose $[0, 1]$ is countable, that is

$$[0, 1] = \{x_n : n \in \mathbb{N}\}.$$

We divide $I_0 := [0, 1]$ into the three intervals $[0, 1/3]$, $[1/3, 2/3]$ and $[2/3, 1]$. Then x_1 is not contained in one or two of these intervals. We choose such an interval and call the starting point a_1 and the endpoint b_1. Now we divide $I_1 := [a_1, b_1]$ into three equal length intervals (thus of length $1/9 = 1/3^2$). Then x_2 is not contained in one of these intervals (I_2 named). Inductively, we obtain a sequence $I_n = [a_n, b_n]$ of intervals in $[0, 1]$ with $I_{n+1} \subset I_n$ and $x_n \notin I_n$. According to the interval nesting principle, there exists an $x \in \bigcap_{n \in \mathbb{N}} I_n$. If $k \in \mathbb{N}$, then by construction $x_k \notin \bigcap_{n \in \mathbb{N}} I_n$ holds. Thus, $x_k \ne x \in [0, 1]$ for all $k \in \mathbb{N}$. Contradiction!

Remark 1.4.26 If X is uncountable and A is countable, then $X \setminus A$ is also uncountable, because otherwise, according to Remark 1.4.21 and Theorem 1.4.22,

$$X = (X \cap A) \cup (X \setminus A)$$

would also be countable. In particular, according to Remark 1.4.23 and Theorem 1.4.25 for every non-single-point interval I, the set $I \setminus \mathbb{Q}$ is uncountable; the irrational numbers form the overwhelming majority within the real numbers.

Exercises 1.4.27

1. Let K be an ordered field and $x \in K$. Show:
   a) (Bernoulli Inequality) For all $n \in \mathbb{N}$ and $x \ge -1$,
   $$(1 + x)^n \ge 1 + nx.$$
   b) For all $n \in \mathbb{N}$,
   $$(1 + x)^n \left\{ \begin{array}{c} \ge \\ \le \end{array} \right\} 1 + nx + \frac{n(n-1)}{2} x^2 \left\{ \begin{array}{l} \text{if } x \ge 0 \\ \text{if } -1 \le x < 0 \end{array} \right..$$

2. a) Show: For all $n \in \mathbb{N}$, $(n + 1)^n \ge 2n^n$.
   b) For which $n \in \mathbb{N}$ is $n! \le (n/2)^n$?
3. Let $(K, +, \cdot, <)$ be an ordered field. Prove: For all $x, y \in K$,
   $$4xy \le (x + y)^2.$$
4. Let $(X, <) = (\mathbb{Q}, <)$ and $M := \{x \in \mathbb{Q} : x \ge 0, x < 1\}$. Investigate whether $\inf M$, $\min M$, $\sup M$ or $\max M$ exist and determine them if possible.

5. Let K be an ordered field and A, $B \subset K$ be upper bounded. Prove: If $\sup A$ and $\sup B$ exist, then $\sup(A + B)$ also exists and

$$\sup(A + B) = \sup(A) + \sup(B).$$

holds.

6. (Archimedean property of $\mathbb{R}$) Prove: For all $x \in \mathbb{R}$, there exists a $n \in \mathbb{N}$ with $n > x$.

7. Let $x, y \in \mathbb{R}$ with $x < y$ and $q \in \mathbb{N}$, $q \geq 2$. Prove:
   a) There exists a $r \in \bigcup_{m \in \mathbb{Z}} q^{-m}\mathbb{Z} \subset \mathbb{Q}$ with $x < r < y$.

   b) There exists a $s \in \mathbb{R} \setminus \mathbb{Q}$ with $x < s < y$.

8. (Geometric mean versus arithmetic mean) Show: For $x, y \in \mathbb{R}$, $x, y \geq 0$,

$$\sqrt{xy} \leq \frac{1}{2}(x + y).$$

9. Let

$$M = \bigcup_{k \in \mathbb{N}} \left[\frac{1}{2k}, \frac{1}{2k-1} \right) \subset \mathbb{R}.$$

Examine whether $\sup M$, $\max M$, $\inf M$ or $\min M$ exist and determine them if possible.

10. Show: If A and B are countable, then $A \times B$ is also countable.

11. Prove:
    a) For all $d \in \mathbb{N}$, $\mathbb{N}^d$ is countable.
    b) $\{0, 1\}^{\mathbb{N}}$ is uncountable.

1.5 Complex Numbers

As we have seen in the last section, in $\mathbb{R}$ every equation $x^n = c$ for $n \in \mathbb{N}$ and $c \geq 0$ has a solution. Unfortunately, this is no longer the case when $c < 0$ and n is even (since $x^n \geq 0$ for even n and any $x \in \mathbb{R}$ according to Theorem 1.4.6.3). Our goal now is to extend the field of real numbers so that $x^2 = c$ is also solvable for $c < 0$ (for example, $x^2 = -1$). We will later see that then also $x^n = c$ is solvable for any c.

Remark and Definition 1.5.1 We consider the abelian group

$$(\mathbb{R}^2, +, (0, 0)) = (\mathbb{R}^{\{1,2\}}, +, 0)$$

from Remark 1.2.12. With the argument-wise multiplication defined there in general, $\mathbb{R}^2$ is indeed a commutative ring, but not free of zero divisors and therefore in particular not a field. We alternatively define for $x = (s, t)$ and $y = (u, v)$ in $\mathbb{R}^2$

$$x \cdot y = (s, t) \cdot (u, v) := (su - tv, sv + tu).$$

One verifies that with this, $(\mathbb{R}^2, +, \cdot)$ is a field with $1 = (1_{\mathbb{R}}, 0_{\mathbb{R}})$. If one bases this multiplication, one writes $\mathbb{C}$ instead of $\mathbb{R}^2$ and calls the elements of $\mathbb{C}$ **complex numbers**. Traditionally, one usually uses z or w as a designation for a complex number. For example, if $z = (3, -1)$ and $w = (1, 2)$, then Furthermore, we call Re

$$z \cdot w = (3, -1) \cdot (1, 2) = (3 - (-2), 6 - 1) = (5, 5) \, .$$

For $z = (s, t) \neq 0$,

$$\frac{1}{z} = \left(\frac{s}{s^2 + t^2}, \frac{-t}{s^2 + t^2} \right).$$

Remark and Definition 1.5.2 From the definition of addition and multiplication, we have $(s, 0) + (u, 0) = (s + u, 0)$ and $(s, 0)(u, 0) = (su, 0)$, that is, the addition and multiplication of the complex numbers $(s, 0)$ and $(u, 0)$ correspond to the addition and multiplication of s and u in $\mathbb{R}$. By identifying the complex number $(s, 0)$ with the real s, we can consider the field $\mathbb{C}$ as an extension of the field $\mathbb{R}$. We then also write briefly s instead of $(s, 0)$. Furthermore,

$$i := (0, 1) \in \mathbb{C}$$

is called the **imaginary unit** in $\mathbb{C}$. For i, we have

$$i^2 = (0, 1) \cdot (0, 1) = (-1, 0) = -1 \, .$$

With these designations, we can write every $z = (s, t) \in \mathbb{C}$ in the form

$$z = (s, t) = (s, 0) + (0, 1)(t, 0) = s + it$$

This representation is called **normal form** (or **Cartesian form**) of z. For example, we have

$$z = (3, -1) = 3 + i(-1) = 3 - i.$$

Furthermore, we call Re $z := s$ the **real part** of z and Im $z := t$ the **imaginary part** of z.

Remark 1.5.3 In $(\mathbb{C}, +, \cdot)$ it is not possible to define an order relation $<$ (with the properties from Definition 1.4.5).

Because: Assume, however. Then $1 > 0$ would be according to Theorem 1.4.6.3, so $- 1$ < 0 according to Theorem 1.4.6.1. For $z = i$ with Theorem 1.4.6.3 also $0 < i^2 = -1$, so contradiction to (O1). The proof shows that no field in which the equation $x^2 = -1$ has a solution can be made into an ordered field.

Remark and Definition 1.5.4 Let $z = s + it$ be a complex number in normal form.

1. The complex number $\bar{z} := s - it$ is called the **complex conjugate** z.
2. The number $|z| := \sqrt{s^2 + t^2} \in [0, \infty)$ is called **modulus** of z.

Fig. 1.1 z and $\bar{z}$

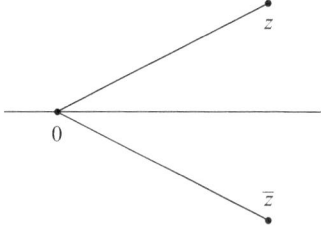

Geometrically, $\bar{z}$ is created by reflecting z on the real axis (see Fig. 1.1). The magnitude $|z|$ describes—according to the Pythagorean theorem—the length of the line from 0 to z in the Euclidean plane.

For $z, w \in \mathbb{C}$ it is easy to derive

$$\overline{z + w} = \bar{z} + \bar{w}, \qquad \overline{zw} = \bar{z} \cdot \bar{w}, \qquad \overline{(\bar{z})} = z$$

as well as

$$\mathrm{Re}(z) = \frac{1}{2}(z + \bar{z}) \quad \text{and} \quad \mathrm{Im}(z) = \frac{1}{2i}(z - \bar{z}).$$

Theorem 1.5.5 *Let $z, w \in \mathbb{C}$. Then the following holds*

1. $|z| \geq 0$ and $|z| = 0$ if and only if $z = 0$ is.
2. $|z| = |\bar{z}| = |-z|$, $|\mathrm{Re}\, z| \leq |z|$, $|\mathrm{Im}\, z| \leq |z|$.
3. $|z|^2 = z\bar{z}$ and $1/z = \bar{z}/|z|^2$, if $z \neq 0$.
4. $|zw| = |z|\, |w|$.
5. $|z \pm w|^2 = |z|^2 \pm 2\mathrm{Re}(z\bar{w}) + |w|^2$.
6. (**Triangle inequality**) $|z \pm w| \leq |z| + |w|$.

Proof 1. and 2. follow directly from the definition of the modulus and 3. is Exercise 1.5.9.1.
4. According to 3.

$$|zw|^2 = (zw)(\overline{zw}) = (z\bar{z})(w\bar{w}) = |z|^2|w|^2 = (|z||w|)^2 .$$

The assertion follows by taking the square root.

5. Again with 3.

$$|z \pm w|^2 = (z \pm w)(\bar{z} \pm \bar{w}) = z\bar{z} \pm z\bar{w} \pm w\bar{z} + w\bar{w} = |z|^2 \pm 2\,\mathrm{Re}\,(z\bar{w}) + |w|^2 .$$

6. After 5. as well as 2. and 4. is

$$|z \pm w|^2 \leq |z|^2 + 2|z\bar{w}| + |w|^2 = |z|^2 + 2|z||w| + |w|^2 = (|z| + |w|)^2 .$$

The assertion follows by taking the square root.

Fig. 1.2 Polar form $z = r\zeta$.

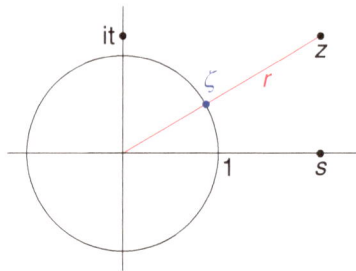

Example 1.5.6 For $z = 3 - i$ one has $|z| = \sqrt{9+1} = \sqrt{10}, \bar{z} = 3 - i(-1) = 3 + i$ and

$$z\bar{z} = (3 - i)(3 + i) = 9 + 1 = |z|^2.$$

Remark and Definition 1.5.7 We write

$$\mathbb{S} := \{z \in \mathbb{C} : |z| = 1\}$$

for the **unit circle** in $\mathbb{C}$. If $z \in \mathbb{C}^*$, then $z = r\zeta$ with $r = |z| > 0$ and $\zeta = z/|z| \in \mathbb{S}$. If $r' > 0$ and $\zeta' \in \mathbb{S}$ with $z = r'\zeta'$, then $r' = r$ and $\zeta' = \zeta$. Thus, every $z \in \mathbb{C}^*$ has exactly one multiplicative decomposition $z = r\zeta$ with $r > 0$ and $\zeta \in \mathbb{S}$. This representation of z is called the **polar form** of z (see Fig. 1.2).

Definition 1.5.8 In generalization of Definition 1.3.3 we set for $z \in \mathbb{C}$ and $\nu \in \mathbb{N}_0$

$$\binom{z}{\nu} := \frac{1}{\nu!} \prod_{k=1}^{\nu} (z - k + 1) = \begin{cases} \frac{z(z-1)\cdots(z-\nu+1)}{\nu!}, & \text{if } \nu \in \mathbb{N} \\ 1, & \text{if } \nu = 0 \end{cases}.$$

The complex number $\binom{z}{\nu}$ is called **binomial coefficient** z over ν.

Exercises 1.5.9

1. Let $z \in \mathbb{C}$ be in normal form $z = s + it$. Show:
   a) $|z|^2 = s^2 + t^2 = z\bar{z}$ and $z^2 = (s^2 - t^2) + i(2st)$,
   b) $\dfrac{1}{z} = \dfrac{1}{s^2 + t^2}(s - it) = \dfrac{\bar{z}}{|z|^2}$, if $z \neq 0$.
2. a) Show: For all $c \in \mathbb{C}$ the equation $z^2 = c$ has at most two solutions in $\mathbb{C}$.
   b) Calculate the solutions of the equation $z^2 = i$ in $\mathbb{C}$.

1.6   From Natural Numbers to Real Numbers

In this section, which is to be understood more as a kind of appendix, we will—as already indicated—discuss the axiomatic introduction of natural numbers and sketch a constructive approach to real numbers. With the exception of the concept

of an equivalence relation, we will not explicitly refer back to the concepts introduced here.

The **natural numbers** can be axiomatically described as a triple $(\mathbb{N}, 1, \nu)$ with
the three properties (**Peano axioms**):

(N1) $\mathbb{N}$ is a set with $1 \in \mathbb{N}$.
(N2) $\nu : \mathbb{N} \to \mathbb{N}$ is an injective function with $1 \notin \nu(\mathbb{N})$.
(N3) (Principle of complete induction) If $A \subset \mathbb{N}$ with $1 \in A$ and $\nu(A) \subset A$, then
      $A = \mathbb{N}$.

The number $\nu(n)$ is called the successor of n. With this, one can define the Arabic
numerals by $2 := \nu(1)$, $3 := \nu(2)$, $4 := \nu(3)$, $5 := \nu(4)$, $6 := \nu(5)$, $7 := \nu(6)$, $8 :=$
$\nu(7)$ and $9 := \nu(8)$. Further, one can (with a lot of effort; see for Example [2])
show:

On $\mathbb{N}$ is defined by $n + 1 := \nu (n)$ and $n + \nu(m) := \nu(n + m)$ for $n, m \in \mathbb{N}$ an
associative and commutative operation $+$ is recursively defined. Using the addi-
tion, $n < m$ (by definition) exactly when $m = n + k$ for a $k \in \mathbb{N}$ gives an order relation $<$ on $\mathbb{N}$. Furthermore, it can be shown: On $\mathbb{N}$ there exists an associative and
commutative operation $\cdot$, which is recursively defined by $n \cdot 1 := 1$ and $n(m + 1)$
$:= nm + n$ for $n, m \in \mathbb{N}$. If one extends $\mathbb{N}$ by an element 0 to $\mathbb{N}_0$ with $0 < n$ for all
$n \in \mathbb{N}$ and so that $n + 0 := 0 + n := n$ and $n \cdot 0 := 0 \cdot n := 0$ for all $n \in \mathbb{N}_0$, then
$(\mathbb{N}_0, +, 0)$ and $(\mathbb{N}_0, \cdot, 1)$ are Abelian monoids with the **cancellation laws**

$$n + m = n + k \;\Rightarrow\; m = k \quad \text{and} \quad n \cdot m = n \cdot k, n \neq 0 \;\Rightarrow\; m = k.$$

Finally, the important **well-ordering property** of $\mathbb{N}$ follows from the principle of
complete induction: *Every non-empty set $M \subset \mathbb{N}$ has a minimum.*

With this, one can easily show: For every pair $(n, p) \in \mathbb{N} \times \mathbb{N}$ there exists
exactly one pair $(a, r) \in \mathbb{N}_0 \times \mathbb{N}_0$ with $r < p$ and $n = ap + r$ (division with
remainder).

If $(M, +, 0)$ is an abelian monoid, we define for not necessarily finite index sets
I and non-empty sets $A \subset M$

$$A^{(I)} := \{x = (x_\alpha)_{\alpha \in I} \in A^I : I_x := \{\alpha \in I : x_\alpha \neq 0\} \text{ finite}\}$$

and

$$\sum_{\alpha \in I} x_\alpha := \sum_{\alpha \in I_x} x_\alpha \quad (x = (x_\alpha)_{\alpha \in I} \in M^{(I)}).$$

This leads to the following important statement about the *representation* of natural
numbers:

Theorem 1.6.1 *Let there be $q \in \mathbb{N}$ with $q \geq 2$ and $A := \{a \in \mathbb{N}_0 : a < q\}$. Then
for each $n \in \mathbb{N}_0$ there exists exactly one tuple $a(n) = (a_j(n))_{j \in \mathbb{N}_0} \in A^{(\mathbb{N}_0)}$ with*

$$n = \sum_{j \in \mathbb{N}_0} a_j(n) q^j.$$

Proof The uniqueness follows from Remark 1.2.7. We prove the existence by induction on n.

1. Induction start $n = 0$: Set $a_j(0) := 0$ for $j \in \mathbb{N}_0$.
2. Induction step: Let $k \in \mathbb{N}_0$ with $q^k \leq n < q^{k+1}$. Division with remainder yields

$$n = aq^k + n'$$

with $0 < a < q$ and $0 \leq n' < q^k$, so in particular $n' < n$.

According to the induction hypothesis (the assertion holds for every $n' < n$), there exists a sequence $(a_j(n'))$ with

$$n' = \sum_{j \in \mathbb{N}_0} a_j(n') q^j.$$

Here, $a_j(n') = 0$ for $j \geq k$, since $n' < q^k$. If we set

$$a_j(n) := \begin{cases} a_j(n') & \text{for } j \neq k \\ a & \text{for } j = k \end{cases},$$

then

$$n = aq^k + n' = \sum_{j \in \mathbb{N}_0} a_j(n) q^j.$$

With $d := d(n) := \max\{j \in \mathbb{N}_0 : a_j(n) \neq 0\}$ for $n \neq 0$, $(a_d(n)a_{d-1}(n)...a_0(n))_q$ is called the q-**adic representation** of n. The numbers $0, ..., q - 1$ are the digits with respect to q. In the case of $q = 2$, we speak of the **binary representation**, in the case of $q = 2 \cdot 5$, we speak of the **decimal representation** and in the case of $q = 2^4$, we speak of the **hexadecimal representation**.[3] Finally, in the decimal case, we also write briefly $a_d(n)...a_0(n)$ instead of $(a_d(n)...a_0(n))_{2 \cdot 5}$. So, for example, for $n = 8 + 8 + 7$

$$n = 1 \cdot 2^4 + 0 \cdot 2^3 + 1 \cdot 2^2 + 1 \cdot 2^1 + 1 \cdot 2^0 = (10111)_2$$

and

$$n = 2 \cdot (2 \cdot 5)^1 + 3 \cdot (2 \cdot 5)^0 = (23)_{2 \cdot 5} = 23.$$

Definition 1.6.2 A relation $\sim$ in X is called an **equivalence relation** (on X), if for all $x, y, z \in X$ the following holds

(A1) $x \sim x$ (reflexivity),
(A2) from $x \sim y$ follows $y \sim x$ (symmetry),

---

[3] The notation $q = 2 \cdot 5$ or $q = 2^4$ may seem cumbersome, but it cannot be replaced by $q = 10$ or $q = 16$, as this would already presuppose the decimal representation to be defined.

(A3) from $x \sim y$ and $y \sim z$ follows $x \sim z$ (transitivity). If $\sim$ is an equivalence relation, then $[x] := [x]_\sim := \{x' \in X : x \sim x'\}$ is called the **equivalence class** gen-
erated by x and every $x \in [x]$ is a **representative** of the equivalence class $[x]$.
Moreover, $X/_\sim := \{[x] : x \in X\}$ is called the **quotient set** of X (**modulo** $\sim$).

Remark and Definition 1.6.3 1. Let $X = \mathbb{N}_0 \times \mathbb{N}_0$. If we define

$$(a,b) \sim (c,d) \quad :\Leftrightarrow \quad a + d = b + c,$$

then $\sim$ is an equivalence relation on X.

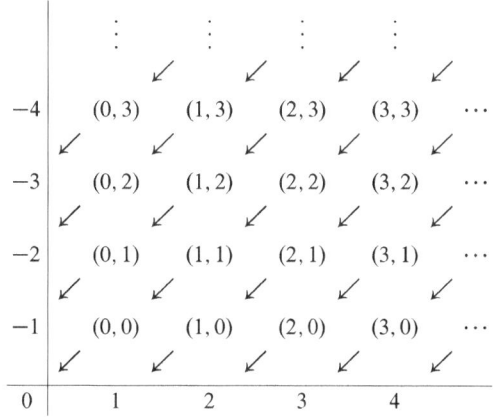

If $a, b \in \mathbb{N}_0$ with $a \geq b$, then there exists (exactly) one $n \in \mathbb{N}_0$ with $a = b + n$ and
therefore

$$[(a,b)] = \{(n + k, k) : k \in \mathbb{N}_0\} = [(n, 0)] \ .$$

If $a < b$, then $b = a + m$ for a $m \in \mathbb{N}$ and therefore

$$[(a,b)] = \{(k, k + m) : k \in \mathbb{N}_0\} = [(0, m)].$$

If one defines the set $\mathbb{Z}$ of **integers** as

$$\mathbb{Z} := X/_\sim = (\mathbb{N}_0 \times \mathbb{N}_0)/_\sim \ ,$$

then through

$$[(a,b)] + [(c,d)] := [(a + c, b + d)] \quad \text{and} \quad [(a,b)] \cdot [(c,d)] := [(ac + bd, ad + bc)]$$

operations $+$ and $\cdot$ on $\mathbb{Z}$ (regardless of the choice of the respective representatives)
are defined, with which $(\mathbb{Z}, +, \cdot)$ becomes a commutative ring. Then $[(0, n)] = -$
$[(n, 0)]$ for $n \in \mathbb{N}_0$. By identifying n with $[(n, 0)]$, $\mathbb{N}_0$ is embedded in $\mathbb{Z}$, and this
results in

$$[(a,b)] = a - b \quad (a, b \in \mathbb{N}_0).$$

Furthermore, through

$$a - b < c - d :\Leftrightarrow a + d < b + c$$

for $a, b, c, d \in \mathbb{N}_0$ an extension of the order $<$ from $\mathbb{N}_0$ to $\mathbb{Z}$ is defined.

Remark and Definition 1.6.4 Let it be $X := \mathbb{Z} \times \mathbb{Z}^*$. If one defines

$$(a, b) \sim (c, d) \quad :\Leftrightarrow \quad ad = bc,$$

for $(a, b), (c, d) \in X$, then $\sim$ is an equivalence relation on X. One can show: For $(a, b) \in X$ exist coprime $p \in \mathbb{Z}, q \in \mathbb{N}$ with

$$[(a, b)] = \{(mp, mq) : m \in \mathbb{Z}^*\} = [(p, q)].$$

If one defines the set $\mathbb{Q}$ of **rational numbers** as

$$\mathbb{Q} := X/_\sim = \left(\mathbb{Z} \times \mathbb{Z}^*\right)/_\sim$$

and operations $+$ and $\cdot$ by

$$[(a, b)] + [(c, d)] := [(ad + cb, bd)] \quad \text{and} \quad [(a, b)] \cdot [(c, d)] := [(ac, bd)],$$

then $(\mathbb{Q}, +, \cdot)$ becomes a field. Here, $1/[(a, b)] = [(b, a)]$ for $a \neq 0$. By identifying a and $[(a, 1)]$ again $\mathbb{Z}$ is embedded in $\mathbb{Q}$ and it holds

$$[(a, b)] = a/b \quad ((a, b) \in X).$$

Finally, the definition

$$\frac{a}{b} < \frac{c}{d} :\Leftrightarrow ad < bc$$

for $a, b \in \mathbb{Z}$ and $b, d \in \mathbb{N}$ (thus $b, d > 0$) extends the order $<$ of $\mathbb{Z}$ to $\mathbb{Q}$.

Remark and Definition 1.6.5 Let $q \in \mathbb{N}, q \geq 2$ and $A := \{0, 1, \ldots, q - 1\}$. We consider the set

$$A^{(\mathbb{Z})} := \{(a_j) = (a_j)_{j \in \mathbb{Z}} \in A^{\mathbb{Z}} : \text{ there is } d \in \mathbb{Z} \text{ with } a_j = 0 \text{ for } j > d\}.$$

A tuple $(a_j) \in A^{(\mathbb{Z})}$ we call a q-**adic expansion**. In the case of $q = 2$, we speak of **binary expansion** and in the case of $q = 10$, we speak of **decimal expansion**. The q-adic expansion (a_j) is called **terminating**, if $(a_j) \in A^{(\mathbb{Z})}$, i.e., if there exists an $m \in \mathbb{Z}$ with $a_j = 0$ for $j < m$. From Theorem 1.6.1 it follows that the mapping

$$A^{(\mathbb{Z})} \ni (a_j) \mapsto \sum_{j \in \mathbb{Z}} a_j q^j \in \bigcup_{m \in \mathbb{Z}} q^m \mathbb{N}_0 \subset \mathbb{Q}_+ \cup \{0\}$$

is bijective. We further identify the terminating expansion (a_j) with the corresponding rational number. Since both addition and multiplication are operations on $\bigcup_{m \in \mathbb{Z}} q^m \mathbb{N}_0$, we can thus add and multiply terminating expansions.

Next, we define an equivalence relation on $A^{(\mathbb{Z})}$ by $(a_j) \sim (b_j)$ if and only if $(a_j) = (b_j)$ or if there exists an $m \in \mathbb{Z}$ such that $a_j = b_j$ for $j > m$, $a_m = b_m + 1$

and $a_j = 0$ and $b_j = q - 1$ for $j < m$ (or correspondingly with swapped roles of a_j and b_j). Thus, all equivalence classes are either single or double elemented, with one of the two expansions being truncated in the double elemented case.

Remark 1.6.6 We now consider the case $q = 2$ and define

$$X := \{x = [(a_j)] : (a_j) \in \{0, 1\}^{\mathbb{Z}}\}.$$

If one chooses the truncated expansion as the representative in the case of two-element $[(a_j)]$ then each $x \in X$ corresponds exactly to one binary expansion (a_j). Unless otherwise stated, we commit to this representation and then also write $x = (a_j)$.

If $x = (a_j)$ (in this sense), we call for $m \in \mathbb{Z}$

$$\lfloor x \rfloor_m := \sum_{j \geq m} a_j 2^j \in 2^m \mathbb{N}_0$$

the m-th truncation of x. Using the relation $<$ on $\mathbb{Q}_+ \cup \{0\}$ we define for $x, y \in X$ with $x \neq y$

$$x < y :\Leftrightarrow \lfloor x \rfloor_m < \lfloor y \rfloor_m \text{ for some } m \in \mathbb{Z}.$$

From Remark 1.2.7 and $\lfloor x \rfloor_k \leq \lfloor y \rfloor_k$ for all $k \in \mathbb{Z}$ it also follows that $\lfloor x \rfloor_n < \lfloor y \rfloor_n$ for $n \leq m$. It can be shown that with this $(X, <)$ is ordered and *complete*.

We want to sketch the proof of completeness by providing a construction rule for $\sup M$. Let $M \subset X$ be non-empty and bounded above. We define a $s \in X$ recursively: Without loss of generality, we can assume that $M \neq \{0\}$. Firstly, the upper boundedness of M implies the upper boundedness of

$$\{m \in \mathbb{Z} : \lfloor x \rfloor_m \neq 0 \text{ for some } x \in M\}$$

in $\mathbb{Z}$. If d is the maximum of this set, we set $\xi_d := 1$ and $\xi_j := 0$ $(j > d)$. We further define

$$\xi_{d-1} := \begin{cases} 1, & \text{if } \lfloor x \rfloor_{d-1} > 2^d \text{ for some } x \in M \\ 0, & \text{else} \end{cases}$$

and correspondingly for $n \in \mathbb{N}$

$$\xi_{d-n-1} := \begin{cases} 1, & \text{if } \lfloor x \rfloor_{d-n-1} > 2^d + \xi_{d-1} 2^{d-1} + \ldots + \xi_{d-n} 2^{d-n} \text{ for some } x \in M \\ 0, & \text{else} \end{cases}.$$

This inductively defines a binary expansion $(\xi_j)_{j \in \mathbb{Z}}$. If $s := [(\xi_j)]$, then $s = \sup M$ results from the construction of s.

This allows the operations $+$ and $\cdot$ of $A^{(\mathbb{Z})}$ to be extended to X: Since $(X, <)$ is complete, for $x, y \in X$

$$x + y := \sup \left\{ \lfloor x \rfloor_m + \lfloor y \rfloor_m : m \in \mathbb{Z} \right\} \in X$$

and

$$x \cdot y := \sup \left\{ \lfloor x \rfloor_m \cdot \lfloor y \rfloor_m : m \in \mathbb{Z} \right\} \in X.$$

exist. The main problem now is to show that $(X, +, 0)$ is a monoid with the cancellation property and $(X \setminus \{0\}, \cdot, 1)$ is a group.[4] Once this is done, as with the extension from $\mathbb{N}_0$ to $\mathbb{Z}$, it can be seen that through

$$(a, b) \sim (c, d) \quad :\Leftrightarrow \quad a + d = b + c$$

for $(a, b), (c, d) \in X \times X$ is defined as an equivalence relation on $X \times X$. With this, we set $\mathbb{R} := (X \times X)/\sim$ and write again $a - b$ instead of $[(a, b)]$ and in the case $b = 0$ briefly a and in the case $a = 0$ briefly $- b$. The operations $+$ and $\cdot$ as well as the relation $<$ can be transferred to $\mathbb{R}$, in such a way that $(\mathbb{R}, +, \cdot, <)$ becomes a completely ordered field.[5] Here, $\mathbb{Q}$ is monotonically embedded in $\mathbb{R}$, that is, there exists an injective mapping $j : \mathbb{Q} \to \mathbb{R}$ such that $j(x + y) = j(x) + j(y)$ and $j(xy) = j(x)j(y)$ for all $x, y \in \mathbb{Q}$ and that $j(x) < j(y)$ holds exactly when $x < y$. By identifying $j(x)$ with x, one can consider $\mathbb{Q}$ as a subset of $\mathbb{R}$ interpret. It can be shown that in this sense the so-called periodic binary expansions correspond to the rational numbers.

References

1. Endl, K., Luh, W.: Analysis I, 8th ed. Aula-Verlag, Wiesbaden (1986)
2. Forster, O.: Algorithmische Zahlentheorie, 2nd ed. Springer Spektrum, Wiesbaden (2015)
3. Pöschel, J.: Etwas Analysis. Springer Spektrum, Wiesbaden (2015)
4. Rautenberg, W.: Messen und Zählen. Eine einfache Konstruktion der reellen Zahlen. Berliner Studienreihe zur Mathematik, Bd. 18. Heldermann Verlag, Lemgo (2007)
5. Rudin, W.: Analysis, 3rd ed. Oldenbourg Verlag, München (2005)

---

[4] A detailed presentation can be found in [4].

[5] In the literature, one often finds, in addition to axiomatic approaches (such as in [3]), the constructive approach via *Dedekind cuts* (as for example in [5]). The advantage of this approach lies in a simpler definition of the supremum and a less complex proof of the field axioms, but the objects considered in this context, which are ultimately referred to as real numbers, are less instructive than the binary developments introduced above. Moreover, the idea of a real number as a binary development usually ties in with prior knowledge from school and also hints at problems of floating point arithmetic and the structure of machine numbers. Another possible way to the real numbers as equivalence classes of Cauchy sequences of rational numbers (see for example [1]) leads directly to sequential completeness, which we will deal with in the next chapter.

Analysis can be seen as the mathematics of limits. As already indicated, completeness of the field of real numbers plays a crucial role. As a first fundamental consequence of completeness, the intermediate value theorem is proven, which essentially is a statement about the solvability of equations. To this end, limits and continuity of functions on suitable subsets of $\mathbb{R}$ or $\mathbb{C}$ are defined and discussed.

A central role for many further investigations is played by the monotone convergence theorem. Via the existence of the limit superior of bounded real sequences, the Bolzano-Weierstrass theorem is derived, and with it the Cauchy criterion. This enables us to prove the existence of elementary functions such as the exponential function(s), the trigonometric functions, and the corresponding inverse functions.

The Cauchy criterion guarantees in the language of metric spaces introduced later the (sequential) completeness of $\mathbb{R}$ and $\mathbb{C}$. In addition to completeness, topological concepts such as openness and closedness of subsets of metric spaces as well as continuity of mappings between metric spaces are examined. In this context, the much stricter concept of compactness, which goes far beyond completeness, is also introduced. It turns out that continuous functions on compact spaces have many favorable properties, in particular, the important one that continuous real-valued functions on compact sets become maximal and minimal. A first application is found in the proof of the Fundamental Theorem of Algebra.

In the last section, sequences and series of functions are examined. Here too, a variant of the Cauchy criterion plays a crucial role. As a result, statements about the completeness of suitable function spaces, which are of fundamental importance for the later chapters, are derived.

© The Author(s), under exclusive license to Springer-Verlag GmbH, DE, part of   37
Springer Nature 2025
J. Müller, *Concepts of Function Theory*, Mathematics Study Resources 12,
https://doi.org/10.1007/978-3-662-69115-1_2

2.1   Continuity and Limits

In the following, let always be $\mathbb{K} \in \{\mathbb{R}, \mathbb{C}\}$, thus $\mathbb{K}$ is the field of real or complex numbers. We initially mostly consider functions $f : X \to \mathbb{C}$, where $X \subset \mathbb{K}$ is.[1]

Definition 2.1.1 Let $X \subset \mathbb{K}$.

1. If $f : X \to \mathbb{C}$ and $a \in X$, then f is called **continuous at the point** a, if for every $\varepsilon > 0$ there exists a $\delta = \delta_\varepsilon > 0$ with

$$|f(x) - f(a)| < \varepsilon$$

for all $x \in X$ with $|x - a| < \delta$. Furthermore, f is called **continuous on the set** $M \subset X$, if f is continuous at every point $a \in M$. If $M = X$, then f is simply called **continuous**.
2. With $C(X)$ we denote the set of all continuous functions $f : X \to \mathbb{C}$.

Remark and Definition 2.1.2 From the definition, it follows immediately:

1. The identical mapping $f = \mathrm{id}_\mathbb{C}$ is continuous.
2. Constant functions $f : \mathbb{C} \to \mathbb{C}$ are continuous.
3. If f is continuous, then for every set $M \subset X$, $f|_M$ is also continuous.
4. If $a \in \mathbb{K}, \rho > 0$ and

$$U_\rho(a) := U_{\rho,X}(a) := \{x \in X : |x - a| < \rho\},$$

then $f : X \to \mathbb{C}$ is continuous at the point a if and only if $f|_{U_\rho(a)}$ is continuous at a. The set $U_\rho(a)$ is called ρ-**neighborhood** of a (with respect to X). In the case of $X = \mathbb{R}$, $U_\rho(a)$ is the interval $(a - \rho, a + \rho)$, and in the case of $X = \mathbb{C}$, $U_\rho(a)$ is the disk with center a and radius ρ (Exercise 2.1.28.1). A set $X \subset \mathbb{C}$ is called **bounded**, if there exists a $\rho > 0$ with $X \subset U_\rho(0)$.

We want to derive a characterization of continuity that is based on the central concept of the accumulation point.

Definition 2.1.3 Let $X \subset \mathbb{K}$. A point $a \in \mathbb{K}$ is called **accumulation point** of X, if for every $\varepsilon > 0$ there exists an $x \in X$ with $0 < |x - a| < \varepsilon$. We write X' for the set of all accumulation points of X. If $a \in X$ and is not an accumulation point, then a is called **isolated point** of X.

Example 2.1.4 For $X = \{1/k,\ k \in \mathbb{N}\}$ $0 \in X'$, but $0 \notin X$. In addition, every $a \in X$ is an isolated point of X.

---

[1] As already hinted earlier, we may occasionally take the liberty to also consider a function $f : X \to Y$ with $Y \subset \mathbb{C}$, such as $f : X \to \mathbb{R}$, as a function with range $\mathbb{C}$.

Remark and Definition 2.1.5

1. Let $X \neq \emptyset$ be a set and $g : X \to \mathbb{R}$. Then we write

$$\sup_X g := \sup_{x \in X} g(x) := \sup g(X) \; (\in \mathbb{R} \cup \{\infty\})$$

and

$$\inf_X g := \inf_{x \in X} g(x) := \inf g(X) \; (\in \mathbb{R} \cup \{-\infty\}).$$

If $f : X \to \mathbb{C}$ and $M \subset X$, then f is called **bounded on** M, if $\sup_M |f| < \infty$ holds, i.e., a $s \in \mathbb{R}$ exists with $|f(x)| \leq s$ for all $x \in M$. In the case $M = X$ we simply say, f is **bounded**.

2. If $X \subset \mathbb{K}$ and if $a \in X'$, then f is called **decaying** at a, if for every $\varepsilon > 0$ there exists a $\delta = \delta_\varepsilon > 0$ with $|f(x)| < \varepsilon$ for all $x \in X$ with $0 < |x - a| < \delta$. If there exists a constant $c \in \mathbb{K}$ such that $f - c$ decays at a, then f is called **convergent** at the point a and c is then the **limit** of f at the point a. In this case, we write briefly

$$f(x) \to c \qquad (x \to a).$$

Note that even in the case $a \in X$, i.e., even when $f(a)$ exists, the function value $f(a)$ plays no role here!

From $a \in X'$ it follows that at most one limit c of f at a exists (Exercise 2.1.28.4). In the case of existence, we also write

$$\lim_{x \to a} f(x) := c \, .$$

Remark 2.1.6 Let $X \subset \mathbb{K}$, $f, g : X \to \mathbb{C}$ and $a \in X'$.

1. If f and g decay at a, then $f \pm g$ also decay at a.

   Because: Let $\varepsilon > 0$ be given. Then there exist an $\eta > 0$ and a $0 < \delta \leq \eta$ with $|f(x)| < \varepsilon/2$ for $0 < |x - a| < \eta$ and $|g(x)| < \varepsilon/2$ for $0 < |x - a| < \delta$. Therefore, for $0 < |x - a| < \delta$ according to the triangle inequality

   $$|f(x) \pm g(x)| \leq |f(x)| + |g(x)| < \varepsilon/2 + \varepsilon/2 = \varepsilon.$$

2. If there exists a $\rho > 0$ such that f is bounded on $U_\rho(a)$, and g decays at a, then $f \cdot g$ also decays at a.

   Because: Let $s, \rho > 0$ be such that $|f(x)| \leq s$ for all $x \in U_\rho(a)$. If $\varepsilon > 0$ is given, there exists a $0 < \delta \leq \rho$ with $|g(x)| < \varepsilon/s$ for $0 < |x - a| < \delta$. Then

   $$|f(x)g(x)| \leq s|g(x)| < \varepsilon \quad (0 < |x - a| < \delta).$$

Definition 2.1.7 If X is a set and $M \subset X$, we define the **indicator function** $\mathbb{1}_M = \mathbb{1}_{M,X} : X \to \mathbb{R}$ of M (with respect to X) by

$$\mathbb{1}_M(x) := \begin{cases} 1, & \text{falls } x \in M \\ 0, & \text{falls } x \in X \setminus M \end{cases}.$$

Example 2.1.8 Let $X = \mathbb{R}$ and $M := \{1/k : k \in \mathbb{N}\}$.

1. For $f = \mathbb{1}_M$ it holds: For every $a \neq 0$ there exists a $\rho > 0$ with $f(x) = 0$ for all $0 < |x - a| < \rho$. In particular, this implies

$$f(x) \to 0 \quad (x \to a)$$

   for all $a \neq 0$. At the point $a = 0$, there is *no* limit (Exercise 2.1.28.6).
2. For $f = \mathbb{1}_M \cdot \mathrm{id}_{\mathbb{R}}$, thus

$$f(x) = \begin{cases} x, & \text{if } x \in M \\ 0, & \text{if } x \notin M \end{cases},$$

   holds

$$f(x) \to 0 \quad (x \to a)$$

   for all $a \in \mathbb{R}$ (for $a \neq 0$ one can argue as in 1., and for $a = 0$ the assertion follows from Remark 2.1.6, since $\mathbb{1}_M$ is bounded).

The following theorem shows that taking limits is compatible with the algebraic operations in $\mathbb{C}$. We define for this purpose, in addition to Remark 1.2.12, for any set X and a function without zeros $g : X \to \mathbb{C}$ (so $g(x) \neq 0$ for all $x \in X$) the function $1/g : X \to \mathbb{C}$ by

$$(1/g)(x) := 1/g(x) \qquad (x \in X)$$

and thus also $f/g := f \cdot (1/g)$ for $f : X \to \mathbb{C}$.

Theorem 2.1.9 *Let $X \subset \mathbb{K}$ and $f, g : X \to \mathbb{C}$. Further, let $a \in X'$ with*

$$f(x) \to b \quad \text{and} \quad g(x) \to c \quad (x \to a).$$

Then the following holds

1. $(f \pm g)(x) \to b \pm c (x \to a)$.
2. $(f \cdot g)(x) \to b \cdot c (x \to a)$.
3. *If g is without zeros and $c \neq 0$, it follows that $(f/g)(x) \to b/c$ ($x \to a$).*

Proof

1. The first statement follows immediately from Remark 2.1.6.1.
2. First, there exists a $\rho > 0$ with $|f(x) - b| < 1$ for $0 < |x - a| < \rho$ and thus also

$$|f(x)| = |f(x) - b + b| \leq 1 + |b|.$$

Thus, f is bounded on $U_\rho(a)$. Furthermore,

$$f(x)g(x) - bc = f(x)g(x) - cf(x) + cf(x) - bc = f(x)(g(x) - c) + c(f(x) - b).$$

According to Remark 2.1.6, the right side is decaying at a. Therefore, $(fg)(x) \to bc$ $(x \to a)$.

3. According to 2., it is sufficient to show:

$$(1/g)(x) \to 1/c \qquad (x \to a).$$

Since $c \neq 0$, there exists a $\rho > 0$ with $|g(x) - c| < |c|/2$ for $0 < |x - a| < \rho$. Therefore, with the reverse triangle inequality (Exercise 2.1.28.3)

$$|g(x)| = |c + g(x) - c| \geq |c| - |g(x) - c| > |c| - |c|/2 = |c|/2 > 0$$

and thus $|1/g(x)| \leq 2/|c|$ for $0 < |x - a| < \rho$. Consequently, $1/g$ is bounded on $U_\rho(a)$. According to Remark 2.1.6,

$$\frac{1}{g(x)} - \frac{1}{c} = \frac{1}{g(x)}(1 - g(x)/c)$$

is decaying at a, so $1/g(x) \to 1/c$ $(x \to a)$. $\qquad\qquad\square$

Remark 2.1.10 If $X \neq \emptyset$ is a set and $f, g : X \to \mathbb{R}$, we briefly write $f \leq g$, if $f(x) \leq g(x)$ for all $x \in X$. Accordingly, we write $f < g$, if $f(x) < g(x)$ for all $x \in X$. If under the conditions of the previous theorem f, g are real-valued and $f \leq g$, then $b \leq c$ applies (Exercise 2.1.28.4). In general, however, $f < g$ does *not* imply $b < c$ (but only $b \leq c$).

Remark 2.1.11 Let $X \subset \mathbb{K}$, $a \in X'$ and $f : X \to \mathbb{C}$. From the above definitions it follows directly: $f(x) \to c$ $(x \to a)$ holds exactly when the function $f_{a,c} : X \cup \{a\} \to \mathbb{C}$, defined by

$$f_{a,c}(x) := \begin{cases} c, & \text{if } x = a \\ f(x), & \text{if } x \neq a \end{cases},$$

is continuous at a. If $a \in X$, then it also holds:

f is continuous at a, if and only if $f(x) \to f(a)$ $(x \to a)$ holds.

From the definition of continuity, it also immediately follows: If a is an isolated point of X, then f is always continuous at a. With Theorem 2.1.9 it follows: If $X \subset \mathbb{K}$ and if f, g are continuous at $a \in X$, then $f \pm g, f \cdot g$ and for zero-free g also f/g are continuous at the point a.

Example 2.1.12 If $M = \{1/k : k \in \mathbb{N}\}$ and $f = \mathbb{1}_M$ as in Example 2.1.8, then for $k \in \mathbb{N}$

$$f(x) \to 0 \neq 1 = f(1/k) \quad (x \to 1/k)$$

and thus f is not continuous at $1/k$. Since no limit exists at 0, f is also not continuous at 0. The function $f = \mathbb{1}_M \mathrm{id}_\mathbb{R}$ from Example 2.1.8 is also discontinuous at all points $1/k$, but continuous at the point 0.

Definition 2.1.13 Let X be a set and $f : X \to \mathbb{C}$. If $f(x) = 0$, then x is called a **root** of f. With $Z(f)$ we denote the set of roots, thus

$$Z(f) := \{x \in X : f(x) = 0\}.$$

Example 2.1.14 A **polynomial function** (or simply **polynomial**) is a function $p : \mathbb{K} \to \mathbb{C}$ of the form

$$p(x) = \sum_{\nu=0}^{d} c_\nu x^\nu$$

with the **coefficients** $c_0, \ldots, c_d \in \mathbb{C}$. In the case $c_d \neq 0$, we call $\deg(p) := d$ the **degree** of p. If p, q are polynomials, then $p \cdot q$ is also a polynomial, specifically of degree $\deg(p) + \deg(q)$. Polynomials of degree $d \in \mathbb{N}$ have at most d zeros (Exercise 2.1.28.12). From Remark 2.1.2.1, it follows by repeated application of Remark 2.1.11 that every polynomial is continuous.

If $p, q \neq 0$ are polynomials, then $p/q : \mathbb{K} \setminus Z(q) \to \mathbb{C}$ is also continuous, again according to Remark 2.1.11. Functions of the form p/q are called **rational**.

Remark 2.1.15 Let $X \subset \mathbb{K}$ and $f : X \to \mathbb{C}$. Furthermore, let $U \subset \mathbb{K}$, $\varphi : U \to X$ and $\alpha \in U'$ with $\varphi(u) \to a$ $(u \to \alpha)$.

1. If $a \in X$ and f is continuous at a, then $(f \circ \varphi)(u) \to f(a)$ $(u \to \alpha)$. If additionally $\alpha \in U$ and φ is continuous at α, then $f \circ \varphi$ is also continuous at α.

   Because: If $\varepsilon > 0$ is given, there exists a $\eta > 0$ with

   $$|f(x) - f(a)| < \varepsilon \qquad (|x - a| < \eta).$$

   Furthermore, there exists a $\delta > 0$ such that

   $$|\varphi(u) - a| < \eta \qquad (0 < |u - \alpha| < \delta).$$

   Therefore, $|f(\varphi(u)) - f(a)| < \varepsilon$ for $0 < |u - \alpha| < \delta$.

2. If $a \in X' \backslash \varphi(U \backslash \{\alpha\})$ with $f(x) \to c$ $(x \to a)$, then $(f \circ \varphi)(u) \to c$ $(u \to \alpha)$.

   Because: The assertion results from 1. by applying to $f_{a,c}$ (see Remark 2.1.11).

Definition 2.1.16 Let $X \subset \mathbb{K}$ be unbounded. Then 0 is an limit point of $U := \{1/x : x \in X \backslash \{0\}\}$. If $f : X \to \mathbb{C}$ and $\varphi : U \to X$ is defined by $\varphi(u) := 1/u$ for $u \in U$, we write

$$f(x) \to c \quad (|x| \to \infty) \quad \text{or} \quad \lim_{|x| \to \infty} f(x) := c$$

if $f(1/u) = (f \circ \varphi)(u) \to c \ (u \to 0)$ holds.

Example 2.1.17

1. For $n \in \mathbb{N}$ holds $u^n \to 0 \ (u \to 0)$, thus

$$1/x^n \to 0 \quad (|x| \to \infty).$$

2. Let $p(x) = \displaystyle\sum_{\nu=0}^{d} a_\nu x^\nu$ and $q(x) = \displaystyle\sum_{\mu=0}^{d} b_\mu x^\mu$ be polynomials with $d = \deg(q) > 0$
    (and $\deg(p) \le d$). If $X := \mathbb{K} \setminus Z(q)$ and is $f := p/q : X \to \mathbb{C}$, then

$$f(x) \to a_d/b_d \quad (|x| \to \infty).$$

    Because: Since $b_d \ne 0$, for $0 \ne x \in \mathbb{K} \setminus Z(q)$

$$\frac{p(x)}{q(x)} = \frac{a_d + a_{d-1}x^{-1} + \cdots + a_0 x^{-d}}{b_d + b_{d-1}x^{-1} + \cdots + b_0 x^{-d}} \to \frac{a_d}{b_d} \quad (|x| \to \infty)$$

    according to 1. and Theorem 2.1.9.

Remark and Definition 2.1.18 Let be $X \subset \mathbb{K}$ and $f : X \to \mathbb{C}$. If $a \in \mathbb{K}$ and $M \subset X$
with $a \in M'$, we write

$$f(x) \to c \quad (x \to a, \, x \in M) \quad \text{or} \quad \lim_{M \ni x \to a} f(x) := c,$$

if $f|_M(x) \to c \ (x \to a)$. Specifically, if $X \subset \mathbb{R}$ and $M := X \cap (a, \infty)$, it is said that f
has the **right-sided limit** c at a, and then one writes $f(x) \to c \ (x \to a^+)$ and

$$f(a^+) := \lim_{x \to a^+} f(x) := c.$$

Correspondingly, in the case of $M = X \cap (-\infty, a)$, one speaks of the **left-sided
limit** c and then one writes $f(x) \to c \ (x \to a^-)$ and

$$f(a^-) := \lim_{x \to a^-} f(x) := c.$$

It is easy to see: If $a \in \mathbb{R}$ is a limit point of $X \cap (-\infty, a)$ and of $X \cap (a, \infty)$, then
$f(x) \to c \ (x \to a)$ applies exactly when $f(a^+)$ and $f(a^-)$ exist and

$$f(a^+) = f(a^-) = c$$

is fulfilled. If $f(a^+)$ and $f(a^-)$ exist with

$$f(a^+) \ne f(a^-),$$

then it is called a is a **jump discontinuity** of f. At jump discontinuities, f does not
have a (two-sided) limit.

Example 2.1.19

1. The **sign function** sgn $: \mathbb{R} \to \mathbb{R}$ is defined by

$$\text{sgn}(x) := \begin{cases} 1, & \text{if } x > 0 \\ 0, & \text{if } x = 0 \\ -1, & \text{if } x < 0 \end{cases}.$$

Here

$$\text{sgn}(0^+) = \lim_{x \to 0^+} \text{sgn}(x) = 1, \qquad \text{sgn}(0^-) = \lim_{x \to 0^-} \text{sgn}(x) = -1.$$

Thus, $a = 0$ is a jump discontinuity.

2. The indicator function $\mathbb{1}_{\mathbb{Q}}$ of $\mathbb{Q}$ (with respect to $\mathbb{R}$) is called **Dirichlet function**. The Dirichlet function has no right or left limit for *any a* in $\mathbb{R}$!

   Because: If $a \in \mathbb{R}$ and $\delta > 0$, then there exist $x \in \mathbb{Q}$ and $y \in \mathbb{R} \setminus \mathbb{Q}$ (thus $f(x) = 1$ and $f(y) = 0$) with $a < x, y < a + \delta$. From this it follows that there is no right-sided limit at a. Similarly, one can see that there is no left-sided limit. In particular, this means that f is discontinuous at all points $a \in \mathbb{R}$.

As the function sgn shows, the images of intervals under functions with jump discontinuities are generally not intervals. The situation is different for continuous functions, as the following central theorem for analysis shows. The proof is based largely on the completeness of $\mathbb{R}$.

Theorem 2.1.20 (Intermediate Value Theorem.) *Let there be* $X \subset \mathbb{R}$ *and* $f : X \to \mathbb{R}$ *continuous. If* $I \subset X$ *is an interval, then* $f(I)$ *is also an interval.*

Proof We need to show: If $\eta \in \mathbb{R}$ with $\inf_I f < \eta < \sup_I f$, then there exists a $\xi \in I$ with $f(\xi) = \eta$, thus $\eta \in f(I)$.

First, by definition of the supremum and the infimum, there exist $u, v \in f(I)$ with $u < \eta < v$. Therefore, there exist $\alpha, \beta \in I$ with $f(\alpha) = u$ and $f(\beta) = v$. Without loss of generality, we can assume $\alpha < \beta$. We set

$$M := \{x \in [\alpha, \beta] : f(x) \le \eta\} .$$

Then $M \ne \emptyset$ (since $\alpha \in M$) and is bounded, so due to the completeness of $\mathbb{R}$

$$\xi := \sup M \in [\alpha, \beta].$$

Since I is an interval, it follows that $[\alpha, \beta] \subset I$, so $\xi \in I$. Furthermore, $\xi \in M' \cup M$ (Exercise 2.1.28.9). If $\xi \in M'$, then $f(x) \to f(\xi)$ $(x \to \xi, x \in M)$, since f is continuous at $\xi \in I$. From $f(x) \le \eta$ for all $x \in M$ it follows that $f(\xi) \le \eta$ with Remark 2.1.10. If $\xi \in M$, then $f(\xi) \le \eta$ by definition of M. Therefore, always

$$f(\xi) \le \eta.$$

In particular, $\xi < \beta$ and with $\eta < f(x) \to f(\xi)$ $(x \to \xi^+)$ thus also

$$\eta \leq f(\xi),$$

again according to Remark 2.1.10, so in total $f(\xi) = \eta$.   □

Remark 2.1.21 For $n \in \mathbb{N}$ we consider $f : \mathbb{R} \to \mathbb{R}$ with $f(x) := x^n$ and $I = [0, \infty)$. Since f is continuous, $f(I)$ is an interval according to the intermediate value theorem. Furthermore, $0 = f(0) \in f(I)$ and from $s \geq 1$ and $x > s$ it also follows that $f(x) = x^n \geq x > s$. Therefore, $f(I)$ is unbounded above and thus $[0, \infty) \subset f(I)$ (of course, $f(I) \subset [0, \infty)$ also holds). Therefore, for every $c \geq 0$, the equation $x^n = c$ has a solution in I. Thus, from the intermediate value theorem, we obtain once again—and now much more easily—the existence of n th roots.

Remark and Definition 2.1.22 Let $X \subset \mathbb{R}$ and $f : X \to \mathbb{C}$. We write $\infty \in X'$, if X is unbounded above and $-\infty \in X'$ if X is unbounded below. We also set

$$U_\rho(\pm\infty) := U_{\rho,X}(\pm\infty) := \{x \in X : \pm x > 1/\rho\}.$$

If $\infty \in X'$, we write

$$f(x) \to c \quad (x \to \infty) \quad \text{or} \quad f(\infty^-) := \lim_{x \to \infty} f(x) := c,$$

if for every $\varepsilon > 0$ there exists a $\delta = \delta_\varepsilon > 0$ such that $|f(x) - c| < \varepsilon$ for all $x \in U_\delta(\infty)$. This is exactly the case when $f(1/u) \to c$ $(u \to 0^+)$ holds. Correspondingly, in the case of $-\infty \in X'$ we write

$$f(x) \to c \quad (x \to -\infty) \quad \text{or} \quad f(-\infty^+) := \lim_{x \to -\infty} f(x) := c,$$

if for every $\varepsilon > 0$ there exists a $\delta = \delta_\varepsilon > 0$ such that $|f(x) - c| < \varepsilon$ for all $x \in U_\delta(-\infty)$. This is exactly the case when $f(1/u) \to c$ $(u \to 0^-)$ holds.

Definition 2.1.23 If $X \subset \mathbb{R}$ and $f : X \to \mathbb{R}$, then f is called

1. **(monotonically) increasing**, if $f(x_1) \leq f(x_2)$ for all $x_1, x_2 \in X$ with $x_1 < x_2$,
2. **strictly (monotonically) increasing**, if $f(x_1) < f(x_2)$ for all $x_1, x_2 \in X$ with $x_1 < x_2$,
3. **(monotonically) decreasing** or **strictly (monotonically) decreasing**, if $-f$ is increasing or strictly increasing.

If f is increasing or decreasing, we briefly say, f is **monotonic**.

The Dirichlet function shows that bounded functions generally do not have right or left-sided limits. The following theorem shows that *monotonic* functions always have right and left-sided limits. Again, the—simple—proof is based significantly on the existence of the supremum and the infimum of bounded sets, i.e., the completeness of $\mathbb{R}$.

Theorem 2.1.24 (Main theorem on monotonic functions).[2] *Let there be $X \subset \mathbb{R}$ and $f : X \to \mathbb{R}$ bounded and monotone.*

1. *If $a \in (X \cap (a, \infty))'$, then $f(a^+)$ exists and*

$$f(a^+) = \begin{cases} \inf\limits_{X \cap (a,\infty)} f, & \text{if } f \text{ is increasing} \\ \sup\limits_{X \cap (a,\infty)} f, & \text{if } f \text{ is decreasing} \end{cases}.$$

2. *If $b \in (X \cap (-\infty, b))'$, then $f(b^-)$ exists and*

$$f(b^-) = \begin{cases} \sup\limits_{X \cap (-\infty,b)} f, & \text{if } f \text{ is increasing} \\ \inf\limits_{X \cap (-\infty,b)} f, & \text{if } f \text{ is decreasing} \end{cases}.$$

Proof We only show 1. The statements in 2. result in an analogous way. Furthermore, without restriction, let f be increasing (otherwise consider $-f$).
   Since f is bounded, there exists

$$c := \inf_{X \cap (a,\infty)} f \in \mathbb{R}.$$

If $\varepsilon > 0$ is given, then there exists an $x_\varepsilon > a$ with $f(x_\varepsilon) < c + \varepsilon$. Since f is increasing, it also holds

$$c \le f(x) \le f(x_\varepsilon) < c + \varepsilon$$

for all x with $a < x < x_\varepsilon$. Thus, $f(a^+) = c$.   □

Remark and Definition 2.1.25 Let $X \subset \mathbb{K}$ and $a \in X'$ (in the case of $X \subset \mathbb{R}$ possibly also $a = \pm\infty$). If $f : X \to \mathbb{R}$ is bounded, then the functions $g^* : (0, \infty) \to \mathbb{R}$ and $g_* : (0, \infty) \to \mathbb{R}$, defined by

$$g^*(r) := \sup_{U_r(a)\backslash\{a\}} f \quad \text{and} \quad g_*(r) := \inf_{U_r(a)\backslash\{a\}} f,$$

are bounded and monotonic. The limit

$$\limsup_{x \to a} f(x) := \overline{\lim_{x \to a}} f(x) := \lim_{r \to 0^+} g^*(r)$$

existing according to Theorem 2.1.24 is called **Limes superior** of f at a. The **Limes inferior** of f at a is defined similarly with g_* instead of g^*. For this, we write $\liminf\limits_{x \to a} f(x)$ or $\underline{\lim\limits_{x \to a}} f(x)$.

---

[2] In the literature—or on the internet—you will not find a theorem under this name, but you will find the special case of the theorem for sequences under the name monotone convergence theorem.

Example 2.1.26 The Dirichlet function $f = \mathbb{1}_{\mathbb{Q}}$ is bounded, and thus the Limes superior and Limes inferior exist at all points $a \in \mathbb{R} \cup \{\pm\infty\}$. More precisely,

$$\limsup_{x \to a} f(x) = 1 \quad \text{and} \quad \liminf_{x \to a} f(x) = 0$$

holds for all a (since g^* is constant $= 1$ and g_* is constant $= 0$).

Definition 2.1.27 Let $X \subset \mathbb{K}$ and $a \in X'$ (again, $a = \pm\infty$ is allowed for $X \subset \mathbb{R}$). If $f : X \to \mathbb{R}$, we write

$$f(x) \to \pm\infty \qquad (x \to a),$$

if for all $\varepsilon > 0$ there exists a $\delta = \delta_\varepsilon > 0$ such that $f(x) \in U_\varepsilon(\pm\infty)$ for all $x \in U_\delta(a) \setminus \{a\}$. Under the conditions of Remark 2.1.18,

$$f(x) \to \pm\infty \qquad (x \to a, x \in M).$$

is defined as there.

Exercises 2.1.28
1. Let $\rho > 0$. Show:
   a) For $a \in \mathbb{R}, U_{\rho,\mathbb{R}}(a) = (a - \rho, a + \rho)$,
   b) For $a \in \mathbb{C}$ with normal form $a = \alpha + i\beta$,

$$U_{\rho,\mathbb{C}}(a) = \left\{ (s, t) \in \mathbb{R}^2 : (s - \alpha)^2 + (t - \beta)^2 < \rho^2 \right\}.$$

2. Examine the following functions $f : [0, \infty) \to \mathbb{R}$ for convergence as $x \to 0$ and for continuity at the point 0:
   a) $f(x) := \sqrt{x}$,
   b) $f(x) = \begin{cases} 1, & x > 0 \\ 0, & x = 0 \end{cases}$.

3. (**Reverse Triangle Inequality**) Show: For $z, w \in \mathbb{C}$

$$\big||z| - |w|\big| \leq |z - w|.$$

4. Let $X \subset \mathbb{C}, a \in X'$ and $f : X \to \mathbb{C}$ with $f(x) \to c$ $(x \to a)$. Prove:
   a) If $c' \in \mathbb{K}$ with $f(x) \to c'$ $(x \to a)$, then $c = c'$.
   b) From $f \geq 0$ follows $c \geq 0$.
   Does $f > 0$ always imply $c > 0$?

5. Show:
   a) For $a, x \geq 0$ is $(\sqrt{x} - \sqrt{a})(\sqrt{x} + \sqrt{a}) = x - a$.
   b) $x \mapsto \sqrt{x}$ is continuous on $[0, \infty)$.

6. Let $M = \{1/k : k \in \mathbb{N}\} \subset \mathbb{R}$ be. Show that the function $\mathbb{1}_M = \mathbb{1}_{M,\mathbb{R}}$ does not have a limit at the point $a = 0$.

7. Let $M = \{1/k : k \in \mathbb{N}\}$ and $\varphi := \mathbb{1}_M \mathrm{id}_{\mathbb{R}}$. Does the function $\mathbb{1}_{\mathbb{R}\setminus\{0\}} \circ \varphi$ have a limit at the point 0?

8. For $x \in \mathbb{R}$,

$$\lfloor x \rfloor := \max\{m \in \mathbb{Z} : m \leq x\}$$

is called the **Gauss bracket** of x. Examine the functions $\mathbb{R} \ni x \mapsto \lfloor x \rfloor \in \mathbb{R}$ and $\mathbb{R} \ni x \mapsto x - \lfloor x \rfloor \in \mathbb{R}$ for the existence of right-sided and left-sided limits and for continuity.

9. Prove: If $\emptyset \neq X \subset \mathbb{R}$ is bounded above, then

$$\sup X \in X \cup X'.$$

holds.

10. (**Bolzano's theorem on zeros**) Let $I \subset \mathbb{R}$ be an interval and $f : I \to \mathbb{R}$ continuous. Show: If there exist $\alpha, \beta \in I$ with $f(\alpha)<0<f(\beta)$, then f has a zero ξ with $\min\{\alpha, \beta\} < \xi < \max\{\alpha, \beta\}$.

11. Show: If $p : \mathbb{R} \to \mathbb{R}$ is a polynomial and $\deg(p)$ is odd, then $p(\mathbb{R}) = \mathbb{R}$, thus p is surjective.

12. Prove: For all $n \in \mathbb{N}$ the following holds
    a) (**Polynomial division**) If $p : \mathbb{K} \to \mathbb{C}$ is a polynomial of degree n and a is a root of p, then there exists a polynomial q of degree $n - 1$ with

$$p(x) = (x - a)q(x) \quad (x \in \mathbb{K}).$$

b) Every polynomial $p : \mathbb{K} \to \mathbb{C}$ of degree n has at most n roots.

2.2   Sequences and Series in K

Many methods in mathematics are based on recursive execution of a mapping $\varphi : X \to X$, where X denotes a non-empty set. If $a_0 \in X$ (the **starting point** of the iteration), then set

$$a_{n+1} = \varphi(a_n) \quad (n \in \mathbb{N}_0). \tag{2.2.1}$$

This recursively defines a sequence $(a_n)_{n \in \mathbb{N}_0}$ in X.

Typically, one is interested in the behavior of a_n for large n, where we generally say for an unbounded set $X \subset \mathbb{R}$ that a property holds for all **sufficiently large** $x \in X$, if there exists a $\rho > 0$ such that the property holds on $U_{\rho,X}(\infty)$ (i.e., for all $x \in X$ with $x > 1/\rho$).

In this section, we consider sequences $(a_n)_{n \in N}$ in $\mathbb{R}$ or $\mathbb{C}$, that is, sequences of real or complex numbers. Since sequences in $\mathbb{C}$ are special $\mathbb{C}$-valued functions with domain $N \subset \mathbb{N}_0 \subset \mathbb{R}$,[3] all the concepts and results of the previous section are available (where we again consider real sequences as complex if necessary).

---

[3] If desired, one can restrict sequences to the index set and thus the domain $\mathbb{N}_0$ (or $\mathbb{N}$). If $N \subset \mathbb{N}_0$ is infinite, there exists exactly one strictly increasing sequence $(n_k)_{k=0}^{\infty}$ with $N = \{n_k : k \in \mathbb{N}_0\}$, inductively defined by $n_0 := \min N$ and $n_{k+1} := \min(N \setminus \{n_0, \ldots, n_k\})$. The behavior of $(a_n)_{n \in N}$ for large n then corresponds to that of $(a_{n_k})_{k=0}^{\infty}$ for large k. Often the term sequence is actually introduced in this sense.

Remark and Definition 2.2.1 A sequence $(a_n)_{n \in N}$ in $\mathbb{K}$ is called **convergent**, if an $c \in \mathbb{K}$ exists with

$$a_n \to c \quad (n \to \infty).$$

We then also write $\lim a_n := \lim_{n \to \infty} a_n = c$. If $c = 0$, we also speak of a **null sequence**. A sequence that is not convergent is called **divergent**. According to Remark 2.1.22, $(a_n)_{n \in N}$ is convergent towards c if and only if for all $\varepsilon > 0$, $|a_n - c| < \varepsilon$ for all sufficiently large $n \in N$. Moreover, it follows from

$$|a_n| \leq |a_n - c| + |c| < 1 + |c|$$

for sufficiently large n that convergent sequences are necessarily bounded.

If a subsequence $(a_n)_{n \in J}$ of $(a_n)_{n \in N}$ converges to c, we write $a_n \to c$ $(n \to \infty$, $n \in J)$. From the definition it immediately follows: If a sequence is convergent, then every subsequence is also convergent, and with the same limit.

Example 2.2.2 An important family of sequences are **geometric sequences** $(a_n) = (q^n)$ for $q \in \mathbb{K}$, which recursively result from (2.2.1) for $\varphi : \mathbb{K} \to \mathbb{K}$ with $\varphi(x) = qx$ and $a_0 = 1$. The following applies

1. For $|q| > 1$, (q^n) is unbounded.
   Because: For $\delta := |q| - 1 > 0$, the Bernoulli inequality applies with

$$|q^n| = |q|^n = (1 + \delta)^n \geq 1 + n\delta > n\delta \qquad (n \in \mathbb{N}).$$

   Therefore, (q^n) is unbounded.

2. For $|q| < 1$, (q^n) is a null sequence, so $q^n \to 0$ $(n \to \infty)$.

   Because: For $q = 0$, the assertion is clear. If $0 < |q| < 1$, then $1/|q| = 1 + \delta$ with a $\delta > 0$ and thus as in 1.

$$|q^n| < 1/(n\delta) \qquad (n \in \mathbb{N}).$$

   From $1/n \to 0$ follows $q^n \to 0$ $(n \to \infty)$.

Remark 2.2.3 (Monotone convergence theorem) As a special case of the main theorem on monotone functions (Theorem 2.1.24), we obtain the following simple and important sufficient condition for the convergence of sequences of real numbers:

If (a_n) is monotonic and bounded then (a_n) is convergent.

Example 2.2.4 (Heron's method; Babylonian square root extraction) Let $c > 0$ be given and $\varphi : (0, \infty) \to (0, \infty)$ be defined by

$$\varphi(x) := \frac{1}{2}\left(x + \frac{c}{x}\right) \quad (x > 0).$$

Then, according to the inequality between the geometric and arithmetic mean (Exercise 1.4.27.8), one has $\sqrt{c} = \sqrt{x \cdot c/x} \leq \varphi(x)$. We consider the sequence (a_n) in $(0, \infty)$ with an arbitrary starting value a_0 and

$$a_{n+1} := \varphi(a_n) = \frac{1}{2}\left(a_n + \frac{c}{a_n}\right) \qquad (n \in \mathbb{N}_0) .$$

Using the monotone convergence theorem, it can be shown that (a_n) is convergent with

$$\lim_{n \to \infty} a_n = \sqrt{c}$$

(Exercise 2.2.22.2), that is, the a_n are approximations for $\sqrt{c}$. In the case of $c \in \mathbb{Q}$ and $a_0 \in \mathbb{Q}$, the a_n are always rational numbers.

What about the error when using a_n instead of $\sqrt{c}$? We estimate the error from above. For this, let

$$\varepsilon_n = \frac{a_n - \sqrt{c}}{\sqrt{c}} = \frac{a_n}{\sqrt{c}} - 1 \geq 0 \quad (n \in \mathbb{N})$$

be the relative error. Then

$$1 + \varepsilon_{n+1} = \frac{1}{\sqrt{c}} a_{n+1} = \frac{1}{2}\left(\frac{a_n}{\sqrt{c}} + \frac{\sqrt{c}}{a_n}\right) = \frac{1}{2}\left(1 + \varepsilon_n + \frac{1}{1 + \varepsilon_n}\right),$$

so

$$\varepsilon_{n+1} = \frac{1}{2}\frac{\varepsilon_n^2}{1 + \varepsilon_n} \leq \frac{1}{2}\varepsilon_n^2 .$$

If after n steps one has for a_n a relative error $\varepsilon_n \leq 10^{-m}$ for some $m \in \mathbb{N}$, then the relative error ε_{n+1} in the next step is $\leq \frac{1}{2}(10^{-m})^2 = \frac{1}{2}10^{-2m}$; the number of exact digits essentially doubles!

We now consider general bounded sequences.

Example 2.2.5 For $|q| = 1$, the geometric sequence (q^n) is bounded. Specifically, if $q = -1$, thus $q^n = (-1)^n$, then the subsequence $((-1)^n)_{n \in 2\mathbb{N}_0}$ is constant with value 1 (and thus converges to 1) and the subsequence $((-1)^n)_{n \in 2\mathbb{N}_0 + 1}$ is constant with value -1 (and thus converges to -1). Therefore, there exist two subsequences with different limits. In particular, the sequence $((-1)^n)$ is divergent.

Remark 2.2.6 If $(a_n)_{n \in N}$ is a bounded sequence in $\mathbb{R}$, then according to Remark 2.1.25 (and thus essentially due to the completeness of $\mathbb{R}$)

$$\limsup a_n := \lim \sup_{n \to \infty} a_n \qquad \text{und} \qquad \liminf a_n := \lim \inf_{n \to \infty} a_n.$$

We show: There exists a subsequence of $(a_n)_{n \in N}$ that converges to $\limsup a_n$.

Because: Let $c := \limsup a_n$. We set $n_0 := \min N$ and define a sequence $(n_k)_{k=0}^{\infty}$ in N inductively: If $n_0, \ldots, n_k$ are defined, then the set

$$N_k := \{n \in N : n > n_k, \, c - 1/k < a_n < c + 1/k\}$$

is not empty (Exercise 2.2.22.6). With that we set $n_{k+1} := \min N_k$. For the thus defined sequence (n_k) we have

$$|a_{n_k} - c| < 1/k \qquad (k \in \mathbb{N}).$$

If $J := \{n_k : k \in \mathbb{N}_0\}$, then it follows $a_n \to c$ $(n \to \infty, n \in J)$. Accordingly, one can show that there exists a subsequence that converges to $\liminf a_n$.

As a consequence, we obtain another central result of analysis:

Theorem 2.2.7 (Bolzano-Weierstrass) *Every bounded sequence $(a_n)_{n \in N}$ in $\mathbb{K}$ has a convergent subsequence.*

Proof

1. Let $\mathbb{K} = \mathbb{R}$ be. Then choose a subsequence as in Remark 2.2.6.
2. Let $\mathbb{K} = \mathbb{C}$ be. If

$$a_n = \alpha_n + i\beta_n$$

is the normal form of a_n, then the sequences $(\alpha_n)_{n \in N}$ and $(\beta_n)_{n \in N}$ in $\mathbb{R}$ are bounded (it holds $|\alpha_n| \leq |a_n|$ and $|\beta_n| \leq |a_n|$). According to 1., there exist a subsequence $(\alpha_n)_{n \in I}$ of $(\alpha_n)_{n \in N}$ and $\alpha \in \mathbb{R}$ with $\alpha_n \to \alpha$ $(n \to \infty, n \in I)$. Again according to 1., there also exist a subsequence $(\beta_n)_{n \in J}$ of $(\beta_n)_{n \in I}$ and $\beta \in \mathbb{R}$ with $\beta_n \to \beta$ $(n \to \infty, n \in J)$. With Theorem 2.1.9 it follows $a_n = \alpha_n + i\beta_n \to \alpha + i\beta (n \to \infty, n \in J)$. □

Remark and Definition 2.2.8 Let $(a_n)_{n \geq m}$ be a sequence in $\mathbb{K}$. The sequence $(s_n)_{n \geq m}$ of **partial sums** or **subtotals**

$$s_n := \sum_{\nu=m}^{n} a_\nu =: a_m + \cdots + a_n \qquad (n \geq m)$$

is called (the with the sequence (a_n) formed) **series**. The a_ν are then called **series terms**.

If the sequence (s_n) is convergent, then $\lim_{n \to \infty} s_n$ is called the **series value** and one writes

$$\sum_{\nu=m}^{\infty} a_\nu := \lim_{n \to \infty} s_n.$$

Traditionally, in addition to the series value, the sequence of partial sums (s_n) is denoted by $\sum\limits_{\nu=m}^{\infty} a_\nu$. This is quite practical because one can then briefly speak of convergence or divergence of $\sum\limits_{\nu=m}^{\infty} a_\nu$. However, note that the symbol $\sum\limits_{\nu=m}^{\infty} a_\nu$ has two meanings: First, it stands for the sequence (s_n) of partial sums and secondly (in the case of convergence!) for their limit.

If $k > m$, then $\sum\limits_{\nu=k}^{\infty} a_\nu$ is convergent if and only if $\sum\limits_{\nu=m}^{\infty} a_\nu$ converges, and in this case

$$\sum_{\nu=m}^{\infty} a_\nu = \sum_{\nu=m}^{k-1} a_\nu + \sum_{\nu=k}^{\infty} a_\nu .$$

For convergence investigations, it is therefore irrelevant what the lower summation limit looks like.

Example 2.2.9 (Geometric Series) Let $a_n = q^n$ for an $q \in \mathbb{K}$, $|q| < 1$. Then $\sum\limits_{\nu=0}^{\infty} q^\nu$ is convergent with

$$\sum_{\nu=0}^{\infty} q^\nu = \lim_{n\to\infty} \sum_{\nu=0}^{n} q^\nu = \lim_{n\to\infty} \frac{1 - q^{n+1}}{1-q} = \frac{1}{1-q} .$$

For $q = 1/2$, we get $\sum_{\nu=0}^{\infty} 1/2^\nu = 2$ and thus also $\sum_{\nu=1}^{\infty} 1/2^\nu = 1$. Fig. 2.1 illustrates the last series as the limit of the sequence of partial sums in the form of rectangular areas. The gray shaded area corresponds to the partial sum s_5.

Fig. 2.1 Illustration of the geometric series $\sum_{\nu=1}^{\infty} 1/2^\nu = 1$ as the limit of the sequence of partial sums

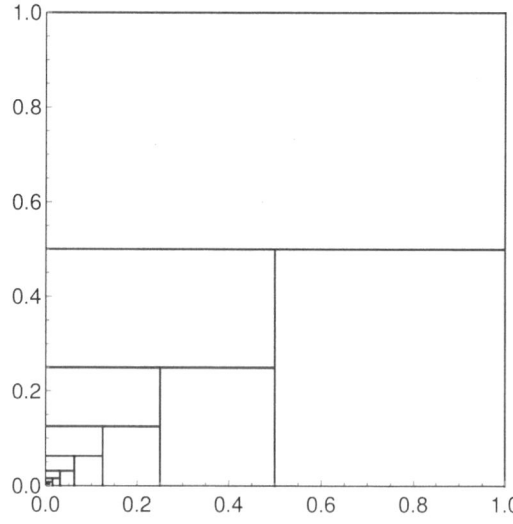

Remark 2.2.10 By applying Theorem 2.1.9 it is easy to see: If $\sum_{v=m}^{\infty} a_v$ and $\sum_{v=m}^{\infty} b_v$ are convergent series in $\mathbb{K}$ and if $\lambda \in \mathbb{K}$, then also $\sum_{v=m}^{\infty}(a_v + b_v)$ and $\sum_{v=m}^{\infty} \lambda a_v$ are convergent with

$$\sum_{v=m}^{\infty}(a_v + b_v) = \sum_{v=m}^{\infty} a_v + \sum_{v=m}^{\infty} b_v \quad \text{und} \quad \sum_{v=m}^{\infty} \lambda a_v = \lambda \sum_{v=m}^{\infty} a_v.$$

Example 2.2.11 With Example 2.2.9 and Remark 2.2.10,

$$\sum_{v=0}^{\infty} \frac{2 \cdot 3^v + 4}{5^v} = 2 \cdot \sum_{v=0}^{\infty} \frac{3^v}{5^v} + 4 \cdot \sum_{v=0}^{\infty} \frac{1}{5^v} = 2 \cdot \frac{1}{1 - 3/5} + 4 \cdot \frac{1}{1 - 1/5} = 10.$$

Remark 2.2.12 A *necessary* condition for the convergence of a series is that the series terms form a null sequence, that is, if $\sum_{v=m}^{\infty} a_v$ is convergent, then

$$a_n \to 0 \qquad (n \to \infty).$$

Because: With $s_n := \sum_{v=m}^{n} a_v$ and $s := \sum_{v=m}^{\infty} a_v$, $a_n = s_n - s_{n-1} \to s - s = 0 \ (n \to \infty)$.

Example 2.2.13

1. If $a_n = q^n$ with $|q| \geq 1$, then $|a_n| \geq 1$ $(n \in \mathbb{N})$, so $\sum_{v=0}^{\infty} q^v$ is certainly divergent. This results in the following for geometric series: $\sum_{v=0}^{\infty} q^v$ is convergent exactly when $|q| < 1$.

2. We consider the **harmonic series** $\sum_{v=1}^{\infty} 1/v$. Here, $(a_n) = (1/n)$ is a null sequence and

$$s_{2^k} = \sum_{v=1}^{2^k} \frac{1}{v} = 1 + \sum_{\ell=1}^{k} \sum_{v=2^{\ell-1}+1}^{2^\ell} \frac{1}{v} \geq 1 + \sum_{\ell=1}^{k} 2^{\ell-1} \frac{1}{2^\ell} = 1 + \frac{k}{2} \to \infty \quad (k \to \infty).$$

Therefore, (s_n) is unbounded, and thus the harmonic series is divergent. This example shows that the necessary condition $a_n \to 0 \ (n \to \infty)$ from Remark 2.2.12 is in general *not sufficient* for the convergence of the series formed with a_n.

In the case of series with non-negative terms, such as the harmonic series, only two fundamentally different situations can occur.

Remark 2.2.14 If (a_n) is a sequence in $[0, \infty)$, i.e., $a_n \geq 0$ for all n, then the partial sum sequence (s_n) is increasing. Thus,

- either (s_n) is bounded and then convergent according to the monotone convergence theorem

$$\sum_{\nu=m}^{\infty} a_\nu = \sup\{s_n : n \in \mathbb{N}\} < \infty$$

- or (s_n) is unbounded with $s_n \to \infty$ $(n \to \infty)$.

In the second case, we also write $\sum_{\nu=m}^{\infty} a_\nu = \infty$.

Remark and Definition 2.2.15 (Majorant criterion) Let there be sequences $(a_n)_{n \geq m}$ and $(b_n)_{n \geq k}$ with

$$0 \leq a_n \leq b_n$$

for all sufficiently large n. We then call $\sum_{\nu=k}^{\infty} b_\nu$ a **majorant** of $\sum_{\nu=m}^{\infty} a_\nu$. If $\sum_{\nu=k}^{\infty} b_\nu$ is convergent, then $\sum_{\nu=m}^{\infty} a_\nu$ is also convergent.

Because: Let $n_0 \in \mathbb{N}$ with $0 \leq a_n \leq b_n$ for $n \geq n_0$. From

$$\sum_{\nu=n_0}^{n} a_\nu \leq \sum_{\nu=n_0}^{n} b_\nu \leq \sum_{\nu=n_0}^{\infty} b_\nu \qquad (n \geq n_0)$$

the boundedness of the partial sums $s_n = \sum_{\nu=n_0}^{n} a_\nu$ follows. According to Remark 2.2.14, $\sum_{\nu=n_0}^{\infty} a_\nu$ is convergent and thus also $\sum_{\nu=m}^{\infty} a_\nu$.

Example 2.2.16 (General harmonic series) Let $d \in \mathbb{N}$. Then

$$\sum_{\nu=1}^{\infty} \frac{1}{\nu^d} \begin{cases} = \infty & \text{if } d = 1 \\ < \infty & \text{if } d > 1 \end{cases}.$$

Because: For $d = 1$, the assertion results from Example 2.2.13. For $d > 1$,

$$\frac{1}{\nu^d} \leq \frac{1}{\nu^2} \leq \frac{1}{(\nu-1)\nu} =: b_\nu .$$

Furthermore, $\sum_{\nu=2}^{n} b_\nu = 1 - 1/n$ is easily seen by induction, so $\sum_{\nu=2}^{\infty} b_\nu = 1$. Thus, $\sum_{\nu=2}^{\infty} b_\nu$ is a convergent majorant. The assertion follows from Remark 2.2.15.

If one chooses a geometric series as a special majorant, further convergence criteria are obtained:

Theorem 2.2.17 (Root criterion and Ratio criterion) *Let $(a_n)_{n \geq m}$ be a sequence in $[0, \infty)$. Then $\sum_{\nu=m}^{\infty} a_\nu$ is convergent if one of the following conditions is met*

1. *There exists a $q < 1$ with $\sqrt[n]{a_n} \leq q$ for all sufficiently large n.*
2. *There exists a $q < 1$ with $a_n > 0$ and $a_{n+1}/a_n \leq q$ for all sufficiently large n.*

Proof

1. By assumption, $0 \leq a_n \leq q^n$ for all sufficiently large n. From the convergence of the geometric series $\sum_{\nu=0}^{\infty} q^\nu$ follows the convergence of $\sum_{\nu=m}^{\infty} a_\nu$ with Remark 2.2.15.
2. Let $n_0 \in \mathbb{N}$ be such that $a_n > 0$ and $a_{n+1}/a_n \leq q$ for $n \geq n_0$. Inductively, with $\lambda := a_{n_0} q^{-n_0}$

$$a_n \leq q^{n-n_0} a_{n_0} = \lambda q^n \qquad (n \geq n_0).$$

As in 1., the assertion follows from the convergence of $\sum_{\nu=0}^{\infty} q^\nu$.   □

Remark 2.2.18 If $(a_n)_{n \geq m}$ is a sequence in $[0, \infty)$ and $r < 1$ with $\sqrt[n]{a_n} \to r$ $(n \to \infty)$, then condition 1. from Theorem 2.2.17 is fulfilled for every q with $r < q < 1$. Similarly, in the case $a_{n+1}/a_n \to r$ $(n \to \infty)$, condition 2. is fulfilled for every q with $r < q < 1$. Thus, in both cases $\sum_{\nu=m}^{\infty} a_\nu$ is convergent.

Example 2.2.19 (Geometric versus polynomial growth) Let $d \in \mathbb{N}$ and $0 < r < 1$. Then $\sum_{\nu=1}^{\infty} \nu^d r^\nu$ is convergent.

Because: From $\sqrt[n]{n} \to 1$ for $n \to \infty$ (Exercise 2.2.22.5) follows $\sqrt[n]{n^d r^n} = r(\sqrt[n]{n})^d \to r$ $(n \to \infty)$. Thus, the assertion results with Remark 2.2.18. In particular, it follows from the convergence of the series with Remark 2.2.12, that $(n^d r^n)$ is a null sequence. More precisely, the above consideration shows that the sequence $(n^d r^n)$ for each q with $r < q < 1$ even decays as fast as the geometric sequence (q^n). A realization that is of fundamental importance for all of mathematics.

Finally, we consider series of the form $\sum_{\nu=m}^{\infty} (-1)^\nu a_\nu$ with $a_n \geq 0$. Such series are called **alternating**.

Theorem 2.2.20 *Let $(a_n)_{n \geq 0}$ be a decreasing sequence in $[0, \infty)$.*

1. *For $s_n := \sum_{\nu=0}^{n} (-1)^\nu a_\nu$ we have $s_n \geq 0$ with a decreasing subsequence $(s_n)_{n \in 2\mathbb{N}_0}$ and an increasing subsequence $(s_n)_{n \in 2\mathbb{N}_0 + 1}$.*
2. **(Leibniz Criterion)** *If (a_n) is a null sequence, then $\sum_{\nu=0}^{\infty} (-1)^\nu a_\nu$ is convergent.*

Proof

1. For odd $n \in \mathbb{N}$ is

$$s_n = \sum_{\nu=0}^{n} (-1)^\nu a_\nu = (a_0 - a_1) + (a_2 - a_3) + \cdots + (a_{n-1} - a_n) \geq 0$$

and thus also $s_{n+1} = s_n + a_{n+1} \geq 0$. Furthermore, for $n \geq 2$

$$s_n - s_{n-2} = (-1)^{n-1}(a_{n-1} - a_n) \begin{cases} \leq 0 & \text{for } n \text{ even} \\ \geq 0 & \text{for } n \text{ odd} \end{cases}.$$

Thus, $(s_n)_{n \in 2\mathbb{N}_0}$ is decreasing and $(s_n)_{n \in 2\mathbb{N}_0+1}$ is increasing.

2. By 1. and the monotone convergence theorem, there exists $s \in [0, \infty)$ with $s_n \to s$ ($n \to \infty$, $n \in 2\mathbb{N}_0$). Then also $s_{n+1} = s_n - a_{n+1} \to s$ ($n \to \infty$, $n \in 2\mathbb{N}_0$). Together, this results in $s_n \to s$ for $n \to \infty$.   □

Example 2.2.21 (Alternating harmonic series). While the harmonic series $\sum_{\nu=1}^{\infty} 1/\nu$ (according to Example 2.2.13) diverges, $\sum_{\nu=1}^{\infty} (-1)^\nu/\nu$ converges according to Theorem 2.2.20.

Exercises 2.2.22

1. Let $X \subset \mathbb{R}$ be bounded from below and $\varphi : X \to X$ with $\varphi(x) \leq x$ ($x \in X$). Furthermore, let $a_0 \in X$ and $a_{n+1} = \varphi(a_n)$ ($n \in \mathbb{N}_0$). Show:
   a) (a_n) is convergent.
   b) If $a^* := \lim_{n \to \infty} a_n \in X$ and φ is continuous at a^*, then $\varphi(a^*) = a^*$.

2. **(Heron's method)** Let $c > 0$ and $\varphi : [\sqrt{c}, \infty) \to [\sqrt{c}, \infty)$ be defined by

$$\varphi(x) := \frac{1}{2}\left(x + \frac{c}{x}\right) \qquad (x \geq \sqrt{c}).$$

   Furthermore, let $a_0 \geq \sqrt{c}$. Show that the sequence (a_n), defined by

$$a_{n+1} = \varphi(a_n) \qquad (n \in \mathbb{N}_0),$$

   converges to $\sqrt{c}$.
   Hint: You can use the previous exercise.

3. Let $X \subset \mathbb{K}$ and $a \in \mathbb{C}$. Consider that $a \in X'$ holds exactly when a sequence (x_n) in $X \setminus \{a\}$ exists with $x_n \to a$ ($n \to \infty$).

4. Let $X \subset \mathbb{C}$ and $f : X \to \mathbb{C}$. If $a \in X$, the function f is called **sequentially continuous** at a, if for all sequences (x_n) in X with $x_n \to a$ ($n \to \infty$) also $f(x_n) \to f(a)$ holds. Prove: f is continuous at a if and only if f is sequentially continuous at a.

5. Show:
   a) For all $c \geq 1$, $\sqrt[n]{c} \to 1$ ($n \to \infty$) holds.
   b) $\sqrt[n]{n} \to 1$ ($n \to \infty$).

6. Let $(a_n)_{n \in N}$ be a bounded sequence in $\mathbb{R}$ and $c \in \mathbb{R}$. Show:
   a) It is $c = \limsup a_n$ if and only if for every $\varepsilon > 0$

$$a_n \begin{cases} < c + \varepsilon & \text{for all sufficiently large } n \in N \\ > c - \varepsilon & \text{for infinitely many } n \in N \end{cases}.$$

b) It is $c = \liminf a_n$ if and only if for every $\varepsilon > 0$

$$a_n \begin{cases} > c - \varepsilon & \text{for all sufficiently large } n \in N \\ < c + \varepsilon & \text{for infinitely many } n \in N \end{cases}.$$

c) $a_n \to c \ (n \to \infty)$ holds if and only if $\limsup a_n = \liminf a_n = c$.

7. Examine the following series for convergence and determine the series value if applicable

$$\text{a)} \quad \sum_{\nu=0}^{\infty} \frac{i^\nu}{2^\nu}, \qquad \text{b)} \quad \sum_{\nu=0}^{\infty} i^\nu.$$

8. Examine the following series for convergence

$$\sum_{\nu=1}^{\infty} \frac{1}{\sqrt[\nu]{\nu}}, \quad \sum_{\nu=1}^{\infty} \frac{1}{\sqrt{\nu}}, \quad \sum_{\nu=1}^{\infty} \frac{(-1)^\nu}{\sqrt{\nu}}, \quad \sum_{\nu=1}^{\infty} \frac{1}{\sqrt{\nu!}}, \quad \sum_{\nu=1}^{\infty} \nu^{123} \left(\frac{123}{124} \right)^\nu.$$

9. Let $q \in \mathbb{N}$, $q \geq 2$ and $A := \{0, 1, \ldots, q - 1\}$. Show:
a) If $(a_n)_{n \in \mathbb{N}}$ is a sequence in A, then

$$0 \leq \sum_{j=1}^{\infty} a_j q^{-j} \leq 1$$

with equality in the right inequality in the case $a_j = q - 1$ for all $j \in \mathbb{N}$.

b) (q-**adic representation**[4]) If $x \in [0, 1]$, then there exists a sequence $(a_n)_{n \in \mathbb{N}}$ in A with

$$0 \leq x - \sum_{j=1}^{n} a_j q^{-j} \leq q^{-n} \quad (n \in \mathbb{N}),$$

so in particular $x = \sum_{j=1}^{\infty} a_j q^{-j}$.

10. Let $(a_n)_{n \geq 1}$ be a decreasing sequence in $[0, \infty)$. Show: For all $k \in \mathbb{N}_0$

$$2 \sum_{\nu=1}^{2^k} a_\nu \geq a_1 + \sum_{\mu=0}^{k} 2^\mu a_{2^\mu} \quad \text{and} \quad \sum_{\nu=1}^{2^{k+1}-1} a_\nu \leq \sum_{\mu=0}^{k} 2^\mu a_{2^\mu}.$$

11. (**Cauchy's Condensation Test**) Let $(a_n)_{n \geq 1}$ be a decreasing sequence in $[0, \infty)$. Show: $\sum_{\nu=1}^{\infty} a_\nu$ converges if and only if $\sum_{\mu=0}^{\infty} 2^\mu a_{2^\mu}$ converges.

---

[4] The binary expansion of x derived here for $q = 2$ results solely from the fact that $\mathbb{R}$ is a completely ordered field and is thus independent of the approach to the real numbers as binary expansions chosen in Remark 1.6.5. In addition, the statement in a) makes it clear that the division into equivalence classes in Remark 1.6.5 is somewhat compelling.

12. Show:

a) For $k \in \mathbb{N}_0$

$$\sum_{\nu=1}^{2^{k+1}-1} \frac{1}{\nu} \le k+1 \le 2 \sum_{\nu=1}^{2^k} \frac{1}{\nu}.$$

b) The series $\sum_{\nu=1}^{\infty} \frac{1}{\nu\sqrt{\nu}}$ is convergent

Hint: Cauchy's Condensation Test.

2.3   Cauchy Criterion and Elementary Functions

Definition 2.3.1 A sequence $(a_n)_{n \in N}$ in $\mathbb{K}$ is called a **Cauchy sequence**, if for every $\varepsilon > 0$ there exists an $R = R_\varepsilon > 0$ such that

$$|a_n - a_{n'}| < \varepsilon$$

holds for all $n, n' \in N$ with $n, n' > R$, in short, if for every $\varepsilon > 0$ the inequality $|a_n - a_{n'}| < \varepsilon$ holds for sufficiently large $n, n' \in N$.

Remark 2.3.2 Let $(a_n)_{n \in N}$ be a Cauchy sequence. Then the following holds

1. (a_n) is bounded.

   Because: For $\varepsilon = 1$ there exists an $n_0 \in N$ such that $|a_n - a_{n'}| < 1$ for all $n, n' \in N$ with n, $n' \ge n_0$, and also

   $$|a_n| = |a_n - a_{n_0} + a_{n_0}| \le |a_n - a_{n_0}| + |a_{n_0}| < |a_{n_0}| + 1$$

   for all $n \in N$ with $n \ge n_0$. Thus, $|a_n| \le \max\{|a_k| + 1 : k \in N, k \le n_0\}$ for all $n \in N$.

2. If $(a_n)_{n \in N}$ has a convergent subsequence, then $(a_n)_{n \in N}$ is convergent.

   Because: Let $(a_n)_{n \in J}$ be a subsequence with $a_n \to c$ $(n \to \infty, n \in J)$. We show: $a_n \to c$ $(n \to \infty)$. For this, let $\varepsilon > 0$ be given. Then there exists an $R > 0$ with

   $$|a_n - a_{n'}| < \varepsilon/2 \qquad (n, n' > R).$$

   Furthermore, there exists a $j \in J$ such that $j > R$ and $|a_j - c| < \varepsilon/2$. Thus,

   $$|a_n - c| \le |a_n - a_j| + |a_j - c| < \varepsilon \qquad (n > R).$$

Another central building block of analysis is

Theorem 2.3.3 (Cauchy Criterion for Sequences) *A sequence $(a_n)_{n \in N}$ in $\mathbb{K}$ is convergent if and only if it is a Cauchy sequence.*

Proof $\Rightarrow$: Let $c \in \mathbb{K}$ with $a_n \to c$ $(n \to \infty)$. Then for every $\varepsilon > 0$ there exists an $R > 0$ such that $|a_n - c| < \varepsilon/2$ for $n > R$. Therefore,

$$|a_n - a_{n'}| \le |a_n - c| + |c - a_{n'}| < \varepsilon \qquad (n, n' > R).$$

$\Leftarrow$: Let $(a_n)_{n \in N}$ be a Cauchy sequence. Then $(a_n)_{n \in N}$ is bounded according to Remark 2.3.2.1. Therefore, by the Bolzano-Weierstrass theorem, $(a_n)_{n \in N}$ has a convergent subsequence. According to Remark 2.3.2.2, $(a_n)_{n \in N}$ is convergent. □

We draw first conclusions from the Cauchy criterion.

Theorem 2.3.4 *Let there be* $X \subset \mathbb{K}$ *and* $a \in X'$. *If* $f : X \to \mathbb{C}$, *then the following statements are equivalent:*

a) *f has a limit at the point a.*
b) *For all sequences* (x_n) *in* $X \setminus \{a\}$ *with* $x_n \to a$, *the sequence* $(f(x_n))$ *is convergent.*
c) *For all sequences* (x_n) *in* $X \setminus \{a\}$ *with* $x_n \to a$, *the sequence* $(f(x_n))$ *is a Cauchy sequence.*

Proof The equivalence of b) and c) results from the Cauchy criterion. The implication a) $\Rightarrow$ b) follows from Remark 2.1.15 with (x_n) instead of φ.

b) $\Rightarrow$ a): Let (x_n) be a sequence in $X \setminus \{a\}$ with $x_n \to a$. Then there exists an $c \in \mathbb{C}$ with $f(x_n) \to c$ $(n \to \infty)$. Suppose it does not hold that $f(x) \to c$ $(x \to a)$. Then there exist an $\varepsilon > 0$ and a sequence (y_n) in $X \setminus \{a\}$ with $y_n \to a$ $(n \to \infty)$ and $|f(y_n) - c| \geq \varepsilon$ for all n. For the sequence (z_n) in $X \setminus \{a\}$ with

$$z_n := \begin{cases} x_n, & \text{if } n \text{ even} \\ y_n, & \text{if } n \text{ odd} \end{cases}$$

it holds that $z_n \to a$ $(n \to \infty)$, but the sequence $(f(z_n))$ is not convergent, since the subsequence $(f(z_n))_{n \in 2N_0}$ converges to c, but the subsequence $(f(z_n))_{n \in 2N_0+1}$ does not. This contradicts the assumption b). Therefore, $f(x) \to c$ $(x \to a)$. □

Theorem 2.3.5 (Cauchy Criterion for Series) *Let* $(a_n)_{n \geq m}$ *be a sequence in* $\mathbb{K}$. *Then* $\sum_{v=m}^{\infty} a_v$ *is convergent if and only if for all* $\varepsilon > 0$ *there exists an* $R > 0$ *such that*

$$\left| \sum_{v=n'+1}^{n} a_v \right| < \varepsilon \qquad (n > n' > R).$$

Proof If $s_n = \sum_{v=m}^{n} a_v$, then for $n > n' \geq m$

$$|s_{n'} - s_n| = |s_n - s_{n'}| = \left| \sum_{v=n'+1}^{n} a_v \right|.$$

Thus, the claim immediately follows from the Cauchy Criterion for sequences. □

Theorem 2.3.6 *Let* $(a_n)_{n \geq m}$ *be a sequence in* $\mathbb{K}$. *If* $\sum_{v=m}^{\infty} |a_v|$ *is convergent, then* $\sum_{v=m}^{\infty} a_v$ *is also convergent.*

Proof Given $\varepsilon > 0$, according to Theorem 2.3.5 there exists an $R > 0$ such that

$$\sum_{\nu=n'+1}^{n} |a_\nu| < \varepsilon \qquad (n > n' > R).$$

From the triangle inequality follows

$$\left| \sum_{\nu=n'+1}^{n} a_\nu \right| \le \sum_{\nu=n'+1}^{n} |a_\nu| < \varepsilon \qquad (n > n' > R).$$

Again, according to Theorem 2.3.5, $\sum_{\nu=m}^{\infty} a_\nu$ is convergent.   □

Remark and Definition 2.3.7 Let $(a_n)_{n \ge m}$ be a sequence in $\mathbb{K}$. The series $\sum_{\nu=m}^{\infty} a_\nu$ is called **absolutely convergent**, if $\sum_{\nu=m}^{\infty} |a_\nu|$ converges. According to theorem 2.3.6, every absolutely convergent series is also convergent. Moreover, then (Exercise 2.3.32.3)

$$\left| \sum_{\nu=m}^{\infty} a_\nu \right| \le \sum_{\nu=m}^{\infty} |a_\nu|.$$

If $\sum_{\nu=m}^{\infty} a_\nu$ is convergent and $\sum_{\nu=m}^{\infty} |a_\nu|$ is divergent, then $\sum_{\nu=m}^{\infty} a_\nu$ is called **conditionally convergent**.

Example 2.3.8

1. For $|q| < 1$, the geometric series $\sum_{\nu=0}^{\infty} q^\nu$ is absolutely convergent (since $\sum_{\nu=0}^{\infty} |q|^\nu$ converges).

2. Let $d \in \mathbb{N}$. The series $\sum_{\nu=1}^{\infty} (-1)^\nu / \nu^d$ is conditionally convergent for $d = 1$ and absolutely convergent for $d \ge 2$ (Example 2.2.21 and Example 2.2.16).

Remark and Definition 2.3.9 Let $z \in \mathbb{C}$ and $a_n := z^n/n!$ for $n \in \mathbb{N}_0$. Then for $z \ne 0$,

$$\frac{|a_{n+1}|}{|a_n|} = \frac{|z|^{n+1}}{(n+1)!} \frac{n!}{|z|^n} = \frac{|z|}{n+1} \to 0 \quad (n \to \infty).$$

Thus, the series $\sum_{\nu=0}^{\infty} z^\nu / \nu!$ is absolutely convergent for $z \ne 0$ according to the quotient criterion. For $z = 0$, the series is also convergent (all partial sums are $= 1$). The function $\exp : \mathbb{C} \to \mathbb{C}$, given by

$$\exp(z) := \sum_{\nu=0}^{\infty} \frac{z^{\nu}}{\nu!} \qquad (z \in \mathbb{C}) \,,$$

is called **(complex) exponential function**. By definition, $\exp(0) = 1$ and also $\exp(\mathbb{R}) \subset \mathbb{R}$. More generally, for all $z \in \mathbb{C}$ (Exercise 2.3.32.7)

$$\exp(\bar{z}) = \overline{\exp(z)}.$$

We want to derive properties of the exponential function that are of fundamental importance for mathematics.

Theorem 2.3.10 *For all $z, w \in \mathbb{C}$*

$$\exp(z + w) = \exp(z) \cdot \exp(w)$$

and $\exp(-z) = 1/\exp(z)$.

Proof

1. For $n \in \mathbb{N}$ we set $L_n := \{(\mu, \nu) \in \{0, \ldots, n\}^2 : \mu + \nu \le n\}$ and

$$J_n := \{(\mu, \nu) \in \{0, \ldots, n\}^2 : \mu + \nu > n\}.$$

Then for $z, w \in \mathbb{C}$

$$\left(\sum_{\mu=0}^{n} \frac{z^{\mu}}{\mu!} \right) \left(\sum_{\nu=0}^{n} \frac{w^{\nu}}{\nu!} \right) = \sum_{(\mu,\nu) \in L_n} \frac{z^{\mu} w^{\nu}}{\mu! \nu!} + \sum_{(\mu,\nu) \in J_n} \frac{z^{\mu} w^{\nu}}{\mu! \nu!} =: s_n + \varepsilon_n \,.$$

The left side converges for $n \to \infty$ towards $\exp(z)\exp(w)$, and also with the binomial theorem

$$s_n = \sum_{k=0}^{n} \sum_{\mu=0}^{k} \frac{z^{\mu} w^{k-\mu}}{\mu!(k-\mu)!} = \sum_{k=0}^{n} \frac{1}{k!}(z+w)^k \to \exp(z+w) \quad (n \to \infty).$$

Therefore, it is sufficient to show that a subsequence of (ε_n) is a null sequence. We set $r := \max\{|z|, |w|, 1\}$. From $\#(J_n) = n(n-1)/2$ it follows for $n = 2m \in 2\mathbb{N}$ with $\max\{\mu, \nu\} \ge m+1$ for $(\mu, \nu) \in J_{2m}$

$$|\varepsilon_{2m}| \le \frac{1}{(m+1)!} \sum_{(\mu,\nu) \in J_{2m}} r^{\mu+\nu} \le \frac{r^{4m}}{(m+1)!} m(2m-1) \le 2r^4 \frac{(r^4)^{m-1}}{(m-1)!} \to 0$$

for $m \to \infty$. Thus, $\varepsilon_n \to 0$ $(n \to \infty, n \in 2\mathbb{N})$.
2. From 1. follows $1 = \exp(0) = \exp(z - z) = \exp(z) \cdot \exp(-z)$. □

Remark and Definition 2.3.11 From Theorem 2.3.10 follows

$$\exp(mz) = (\exp(z))^m \qquad (z \in \mathbb{C}, m \in \mathbb{Z}).$$

The number $e := \exp(1)$ is called **Euler's number**. Therefore, it holds that $\exp(m) = e^m$ for all $m \in \mathbb{Z}$, and hence the notation e^z instead of $\exp(z)$ for general $z \in \mathbb{C}$ is consistent with the definition of integer powers. We will mostly use this in the following.

Theorem 2.3.12 *The function* exp *is continuous.*

Proof For $|h| \leq 1$, with Remark 2.3.7,

$$|e^h - 1| = \left| \sum_{\nu=1}^{\infty} \frac{h^\nu}{\nu!} \right| \leq \sum_{\nu=1}^{\infty} \frac{|h|^\nu}{\nu!} = |h| \sum_{\nu=1}^{\infty} \frac{|h|^{\nu-1}}{\nu!} \leq |h| \sum_{\nu=1}^{\infty} \frac{1}{\nu!} = |h|(e - 1).$$

From this follows $e^h - 1 \to 0 \ (h \to 0)$. For $a, h \in \mathbb{C}$ follows

$$e^{h+a} - e^a = e^a \cdot (e^h - 1) \to 0 \quad (h \to 0),$$

thus $e^{h+a} \to e^a \ (h \to 0)$ and therefore $e^z \to e^a \ (z \to a)$. $\qquad \square$

Theorem 2.3.13 (Exponential Growth) *The function* $\exp|_{\mathbb{R}}$ *is strictly increasing with* $\exp(\mathbb{R}) = (0, \infty)$. *Moreover, for all* $n \in \mathbb{N}$

$$e^t / t^n \to \infty \qquad (t \to \infty).$$

Proof From

$$e^x = 1 + x + \sum_{\nu=2}^{\infty} \frac{x^\nu}{\nu!} \geq 1 + x$$

for $x \geq 0$ and $e^{-x} = 1/e^x$ it follows that $\exp(\mathbb{R}) \subset (0, \infty)$ and also $e^t \to \infty$ $(t \to \infty)$ as well as $e^t = 1/e^{-t} \to 0 \ (t \to -\infty)$. According to the intermediate value theorem, $\exp(\mathbb{R}) = (0, \infty)$. For $s < t$ one has $e^t/e^s = e^{t-s} \geq 1 + (t- s) > 1$ and thus $e^t > e^s$. Therefore, $\exp|_{\mathbb{R}}$ is strictly increasing. Finally, for $n \in \mathbb{N}$ and $t > 0$ also $e^t \geq t^{n+1}/(n+1)!$ and thus it follows

$$e^t / t^n \geq t/(n+1)! \to \infty \qquad (t \to \infty). \qquad \square$$

Definition 2.3.14 The function $\cos : \mathbb{C} \to \mathbb{C}$, given by

$$\cos z := \cos(z) := \frac{1}{2}(e^{iz} + e^{-iz}) \qquad (z \in \mathbb{C}),$$

is called **(complex) cosine function**. The function $\sin : \mathbb{C} \to \mathbb{C}$, given by

$$\sin z := \sin(z) := \frac{1}{2i}(e^{iz} - e^{-iz}) \qquad (z \in \mathbb{C}),$$

is called **(complex) sine function**. Therefore, the **Euler's formula**

$$e^{iz} = \cos z + i \sin z \qquad (z \in \mathbb{C})$$

holds.

Remark 2.3.15 With Remark 2.1.11 and the continuity of exp, the continuity of cos and sin can be easily derived. Furthermore, from the respective definition, $\cos(0) = 1$, $\sin(0) = 0$ and $\cos(-z) = \cos z$ as well as $\sin(-z) = -\sin z$ follow immediately. Finally,

$$\cos(z) = \sum_{k=0}^{\infty} \frac{(-1)^k}{(2k)!} z^{2k}$$

and

$$\sin(z) = \sum_{k=0}^{\infty} \frac{(-1)^k}{(2k+1)!} z^{2k+1}$$

hold with absolute convergence of the series for all $z \in \mathbb{C}$ (Exercise 2.3.32.5). This particularly implies $\sin(\mathbb{R}) \subset \mathbb{R}$ and $\cos(\mathbb{R}) \subset \mathbb{R}$. More precisely, for $t \in \mathbb{R}$, with $e^{-it} = \overline{e^{it}}$,

$$\cos t = \frac{1}{2}(e^{it} + \overline{e^{it}}) = \mathrm{Re}(e^{it}), \qquad \sin t = \frac{1}{2i}(e^{it} - \overline{e^{it}}) = \mathrm{Im}(e^{it})$$

and

$$\cos^2 t + \sin^2 t = |e^{it}|^2 = e^{it} e^{-it} = 1.$$

Fig. 2.2 shows the real part of the complex exponential function. The oscillation behavior of the real cosine function is reflected in the behavior of the real part of the exponential function along the imaginary axis.

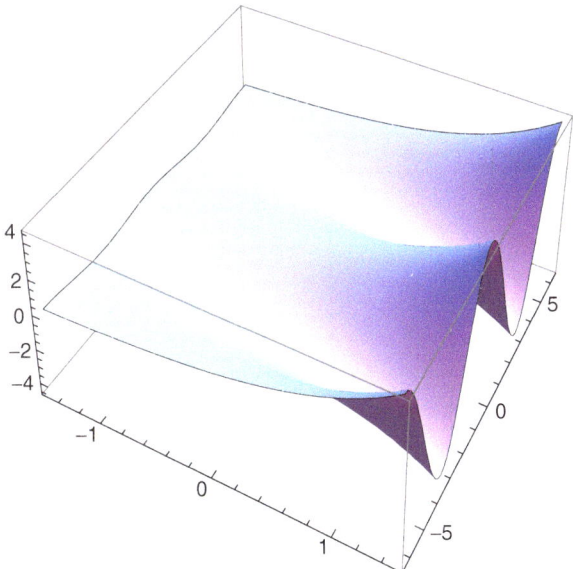

Fig. 2.2 $s + it \mapsto \mathrm{Re}(e^{s+it}) = e^s \cos(t)$

Theorem 2.3.16 (Addition Theorems) *For all* $z, w \in \mathbb{C}$ *the following holds*

$$\cos(z + w) = \cos z \cos w - \sin z \sin w$$

and

$$\sin(z + w) = \sin z \cos w + \cos z \sin w.$$

Proof For arbitrary $u, v \in \mathbb{C}^*$ the following holds

$$(u \pm u^{-1})(v + v^{-1}) + (u \mp u^{-1})(v - v^{-1}) = 2(uv \pm u^{-1}v^{-1}).$$

With $u = e^{iz}$ and $v = e^{iw}$ we get

$$\begin{aligned} 2\cos(z)2\cos(w) + 2i\sin(z)2i\sin(w) &= 2\left(e^{iz}e^{iw} + e^{-iz}e^{-iw}\right) \\ &= 2\left(e^{i(z+w)} + e^{-i(z+w)}\right) = 4\cos(z+w). \end{aligned}$$

After division by 4, we get 1. Accordingly, we see 2. $\qquad\square$

We use the intermediate value theorem and once again the completeness of $\mathbb{R}$, to define the circle number π. To do this, we first prove

Theorem 2.3.17 *The following holds:* $\sin |_{[0,2]} \geq 0$, *and there exists a* $t \in (0, 2)$ *with* $\cos t = 0$.

Proof

1. We set for $t \in (0, 2]$

$$a_n := \frac{t^{2n+1}}{(2n+1)!} \qquad (n \in \mathbb{N}_0).$$

Then

$$\frac{a_n}{a_{n-1}} = \frac{t^2}{(2n+1)(2n)} \leq \frac{4}{6} < 1 \qquad (n \in \mathbb{N}),$$

so (a_n) is decreasing. From Theorem 2.2.20.1 it follows that

$$\sin t = \sum_{v=0}^{\infty} (-1)^v \frac{t^{2v+1}}{(2v+1)!} = \lim_{n\to\infty} s_n \geq 0.$$

2. Note that

$$\cos(2) = 1 - \frac{2^2}{2!} + \sum_{k=2}^{\infty} \frac{(-1)^k 2^{2k}}{(2k)!} = -1 + \sum_{v=0}^{\infty} (-1)^v \frac{4^{v+2}}{(2v+4)!}.$$

We set

$$a_n := \frac{4^{n+2}}{(2n+4)!} \qquad (n \in \mathbb{N}_0),$$

so

$$\frac{a_n}{a_{n-1}} = \frac{4}{(2n+3)(2n+4)} < 1 \qquad (n \in \mathbb{N}).$$

Hence, (a_n) is decreasing. For $s_n := \sum_{\nu=0}^{n}(-1)^{\nu}a_{\nu}$ it follows from Theorem 2.2.20.1 that

$$s_n \le s_0 = a_0 = \frac{4^2}{4!} = \frac{2}{3} \qquad (n \in 2\mathbb{N}_0).$$

Consequently, $\sum_{\nu=0}^{\infty}(-1)^{\nu}a_{\nu} \le 2/3$ and thus $\cos(2) \le -1 + 2/3 < 0$. Since cos is continuous, there exists, according to the intermediate value theorem, a $t \in (0, 2)$ with $\cos t = 0$. □

Remark and Definition 2.3.18
1. Let $(X, +)$ be an additive group and $a \in X$. If Y is a set and $f : X \to Y$, then f is called an a-**periodic function**, if $f(x+a) = f(x)$ for all $x \in X$.
2. According to Theorem 2.3.17, the set $M := \{t > 0 : \cos(t) = 0\}$ is not empty, so $s := \inf M$ exists. From the continuity of cos follows $\cos(s) = 0$ and with $\cos(0) = 1$, $s > 0$. According to the intermediate value theorem and $\cos(t) = \cos(-t)$, $\cos(t) > 0$ also holds for $t \in (-s, s)$. We thus define the **circle number** π as $\pi := 2s$. Then, with Remark 2.3.15 and Theorem 2.3.17,

$$\cos(\pi/2) = 0, \qquad \sin(\pi/2) = 1$$

and thus also

$$e^{i\pi/2} = i.$$

From this follows $e^{\pi i} = i^2 = -1$ and $e^{2\pi i} = 1$, so $e^{z + 2k\pi i} = e^z$ for all $k \in \mathbb{Z}$, that is exp is $2\pi i$-periodic.[5]

By utilizing the addition theorems, periodicity properties of the trigonometric functions are obtained.

Theorem 2.3.19 *For all $z \in \mathbb{C}$ holds*

1. $\cos(z + \pi/2) = -\sin z$ *and* $\sin(z + \pi/2) = \cos z$,
2. $\cos(z + \pi) = -\cos z$ *and* $\sin(z + \pi) = -\sin z$,
3. $\cos(z + 2\pi) = \cos z$ *and* $\sin(z + 2\pi) = \sin z$.

---

[5] The formula $e^{\pi i} + 1 = 0$ combines the real numbers 0, 1, e and π in a simple way via the imaginary unit i.

Proof 1. With Theorem 2.3.16 we obtain

$$\cos(z + \pi/2) = \cos(z)\cos(\pi/2) - \sin(z)\sin(\pi/2) = -\sin z$$

and

$$\sin(z + \pi/2) = \cos(z)\sin(\pi/2) + \sin(z)\cos(\pi/2) = \cos z .$$

The statements in 2. result from applying the first twice and those in 3. from applying 2. twice. □

Theorem 2.3.20 *It holds*

1. $\sin |_{[-\pi/2,\pi/2]}$ *is strictly increasing with* $\sin([-\pi/2, \pi/2]) = [-1, 1]$.
2. $\cos |_{[0,\pi]}$ *is strictly decreasing with* $\cos([0, \pi]) = [-1, 1]$.

Proof

1. From the addition theorems, it follows for $z, w \in \mathbb{C}$

$$\sin(z + w) - \sin(z - w) = 2\cos(z)\sin(w) .$$

Therefore, it follows for $s, t \in \mathbb{R}$

$$\sin t - \sin s = \sin\left(\frac{t + s}{2} + \frac{t - s}{2}\right) - \sin\left(\frac{t + s}{2} - \frac{t - s}{2}\right)$$
$$= 2 \cdot \cos\left(\frac{t + s}{2}\right) \cdot \sin\left(\frac{t - s}{2}\right) .$$

If $-\pi/2 \leq s < t \leq \pi/2$, then

$$(t + s)/2 \in (-\pi/2, \pi/2) \quad \text{and} \quad (t - s)/2 \in (0, \pi/2].$$

According to Remark 2.3.18, $\cos((t + s)/2) > 0$ is true, and with Theorem 2.3.19, $\sin((t - s)/2) > 0$ also follows. Therefore, $\sin s < \sin t$ is true. From $\sin(\pi/2) = 1$ and $\sin(-\pi/2) = -\sin(\pi/2) = -1$, the intermediate value theorem yields

$$\sin([-\pi/2, \pi/2]) = [-1, 1].$$

2. The second statement follows from 1. and Theorem 2.3.19. □

We have already seen that the complex exponential function is without zeros. We now show that every complex number $w \neq 0$ is taken as a function value.

Theorem 2.3.21 *It holds* $\exp(i\mathbb{R}) = \mathbb{S}$ *and* $\exp(\mathbb{C}) = \mathbb{C}^*$.

Proof

1. According to Remark 2.3.15, $\exp(i\mathbb{R}) \subset \mathbb{S}$. We show $\supset$. For this, let $w \in \mathbb{S}$ be with normal form $u + iv$. Without loss of generality, we can assume $v \geq 0$ (if $e^{it} = u + iv$, then $e^{-it} = u - iv$). According to Theorem 2.3.20, there exists a $t \in [0, \pi]$ with $u = \cos t$. Then $\sin t \geq 0$ and

$$v^2 = 1 - u^2 = 1 - \cos^2 t = \sin^2 t.$$

Therefore, $v = \sin t$ holds and thus $e^{it} = \cos t + i \sin t = w$.

2. With 1., according to Remark 1.5.7 and Theorem 2.3.13,

$$\mathbb{C}^* = (0, \infty) \cdot \mathbb{S} = \exp(\mathbb{R}) \cdot \exp(i\mathbb{R}) = \exp(\mathbb{R} + i\mathbb{R}). \qquad \square$$

Remark and Definition 2.3.22 As already indicated, in the field $\mathbb{C}$ equations of the form $z^n = c$ always, that is, for all $c \in \mathbb{C}$ and $n \in \mathbb{N}$, are solvable. We have not yet provided a proof. Using Theorem 2.3.21 the assertion resuts quite easily: Without restriction let $c \neq 0$. Then $c = e^w$ for a $w \in \mathbb{C}$. For $z := e^{w/n}$ we get

$$z^n = (e^{w/n})^n = c.$$

Due to the $2\pi i$-periodicity of $\exp$ then also $(ze^{2k\pi i/n})^n = c$ for $k \in \mathbb{Z}$, where the n numbers

$$z_k := ze^{2k\pi i/n} \qquad (k = 0, \ldots, n-1)$$

are pairwise different and thus represent all solutions of the equation. We call $z_0, \ldots, z_{n-1}$ the n-th **roots** of c. In the case $c = 1$ we also speak of the n-th **unit roots**. So, for example, ± 1 are the second unit roots and $\pm i, \pm 1$ are the fourth unit roots. In general, the n-th unit roots are given by $z_k = e^{2\pi i k/n}$ for $k = 0, \ldots, n - 1$. Fig. 2.3 shows the tenth unit roots.[6]

The following theorem provides information about the zeros of the trigonometric functions, also in the complex plane.

Theorem 2.3.23 *It holds*

1. $Z(\exp -1) = 2\pi i\mathbb{Z}$.
2. $Z(\sin) = \pi\mathbb{Z}$ *and* $Z(\cos) = \pi(\mathbb{Z} + 1/2)$.

---

[6] Geometrically interpreted, the n-th unit roots are the corners of the regular n-gon inscribed in the unit circle.

Fig. 2.3 Tenth unit roots

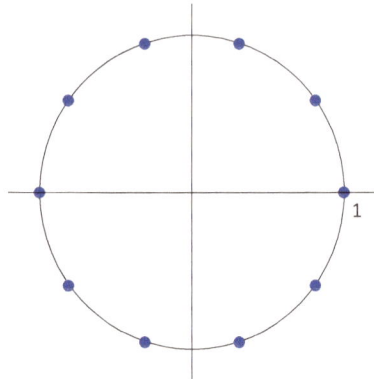

Proof

1. From the $2\pi i$-periodicity of exp and $e^0 = 1$ follows $Z(\exp -1) \supset 2\pi i\mathbb{Z}$.
   $\subset$: Since cos is strictly decreasing on $[0, \pi]$ and $\cos(2\pi - t) = \cos(-t) = \cos t$
   holds for all $t \in \mathbb{R}$, $\cos(t) < 1$ is true for all $t \in \mathbb{R} \setminus 2\pi\mathbb{Z}$, thus $e^{it} \neq 1$ for
   $t \in \mathbb{R} \setminus 2\pi\mathbb{Z}$. If now $z = s + it$ with $e^z = 1$, then $1 = |e^z| = e^s|e^{it}| = e^s$ and conse-
   quently $s = 0$. Thus $e^{it} = 1$, so $t \in 2\pi\mathbb{Z}$, that is $z = it \in 2\pi i\mathbb{Z}$.
2. $0 = 2i \sin z = e^{iz} - e^{-iz}$ holds if and only if $e^{2iz} - 1 = 0$. From 1. it follows that
   $Z(\sin) = \pi\mathbb{Z}$ and with Theorem 2.3.19 also $Z(\cos) = \pi(\mathbb{Z} + 1/2)$.   □

Example 2.3.24 Let $f, g : \mathbb{R}^* \to \mathbb{R}$ be defined by

$$f(x) := \cos(\pi/x) \quad \text{and} \quad g(x) := xf(x) = x\cos(\pi/x).$$

Then $Z(f) = Z(g) = \{(n + 1/2)^{-1} : n \in \mathbb{Z}\}$. In particular, f and g have infinitely
many zeros in every neighborhood of 0 (Fig. 2.4 and 2.5).

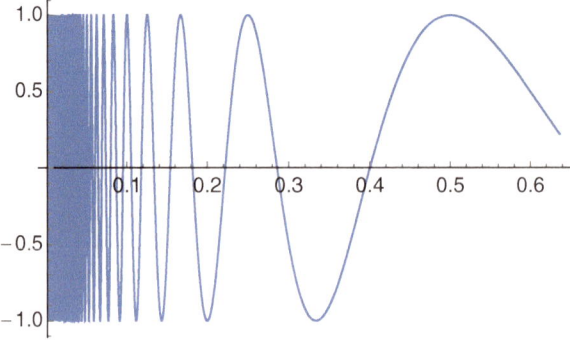

Fig. 2.4 $x \mapsto \cos(\pi/x)$

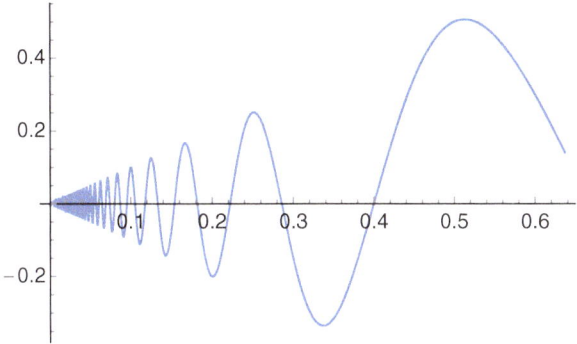

Fig. 2.5 $x \mapsto x \cos(\pi/x)$

Since $\cos|_{\mathbb{R}}$ is bounded, g is decaying at the point 0 according to Remark 2.1.6, so $g(x) \to 0$ $(x \to 0)$. The function f has neither a right-sided nor a left-sided limit at 0.

Remark and Definition 2.3.25 The **tangent function** $\tan : \mathbb{C} \setminus \pi(\mathbb{Z} + 1/2) \to \mathbb{C}$ and the **cotangent function** $\cot : \mathbb{C} \setminus \pi\mathbb{Z} \to \mathbb{C}$ are defined by

$$\tan z := \frac{\sin z}{\cos z}, \qquad \cot z := \frac{\cos z}{\sin z}.$$

According to Remark 2.1.11 and Remark 2.3.15, tan and cot are continuous on their respective domains.

Remark and Definition 2.3.26 (Polar Coordinates) From Theorem 2.3.21 another representation of complex numbers emerges, which proves to be appropriate for many purposes. If $z \in \mathbb{C}^*$ is in polar form $z = r\zeta$, then there exists $\theta \in \mathbb{R}$ with $\zeta = e^{i\theta}$, thus

$$z = re^{i\theta}.$$

If one fixes $\alpha \in \mathbb{R}$ and restricts θ to the interval $(\alpha - \pi, \alpha + \pi]$, the representation, according to Theorem 2.3.23, is unique. If $z = (s, t) \in \mathbb{R}^2$, then also

$$(s, t) = (r \cos\theta, r \sin\theta).$$

We then call $r > 0$ and $\theta \in (\alpha - \pi, \alpha + \pi]$ the **polar coordinates** of (s, t) with respect to α. Usually, one chooses $\alpha = 0$ or $\alpha = \pi$. Using polar coordinates, the multiplication of complex numbers becomes very natural: If $z = re^{i\theta}$ and $w = \rho e^{i\varphi}$, then $zw = r\rho e^{i(\theta+\varphi)}$.

We now turn to the reversibility of elementary functions. To this end, we first prove the following general result.

Theorem 2.3.27 *Let $I \subset \mathbb{R}$ be an interval and let $f : I \to W(f) \subset \mathbb{R}$ be strictly increasing (or decreasing). Then f is bijective and it holds that*

1. f^{-1} *is strictly increasing (or decreasing).*
2. f^{-1} *is continuous.*

Proof We set $J := W(f)$. From the strict monotony, it follows that $f : I \to J$ is injective (and therefore also bijective), which means $f^{-1} : J \to I$ exists.

1. Without loss of generality, let f be strictly increasing. Suppose there exist u, $v \in J$ with $u<v$ and $s:=f^{-1}(u) \geq f^{-1}(v) =: t$. Then $u = f(s) \geq f(t) = v$, since f is (strictly) increasing. Contradiction! Therefore, $f^{-1}: J \to I$ is strictly increasing.
2. Let $u \in J$ and $\varepsilon > 0$ be given. We set $t := f^{-1}(u)$. If $t \neq \sup I$, then there exists a $h = h_\varepsilon \in (0, \varepsilon)$ with $t + h \in I$. We set

$$\delta^+ := \delta_\varepsilon^+ := f(t + h) - f(t) .$$

Then $\delta^+ > 0$ and for all $v \in J$ with $u \leq v<u + \delta^+ = f(t+ h)$ it follows

$$0 \leq f^{-1}(v) - f^{-1}(u) < f^{-1}(u + \delta^+) - f^{-1}(u) = h < \varepsilon .$$

If $t \neq \inf I$, then one can see accordingly: There exists a $\delta^- > 0$ such that

$$0 \leq f^{-1}(u) - f^{-1}(v) < \varepsilon$$

for all $v \in J$ with $u - \delta^-<v \leq u$. This results in $| f^{-1}(v) - f^{-1}(u)| <\varepsilon$ for all $v \in J$ with $|v - u| < \delta := \min\{\delta^+, \delta^-\}$.   □

Remark and Definition 2.3.28 According to Theorem 2.3.13 and Theorem 2.3.27, the inverse function of exp exists on the interval $(0, \infty)$ and is continuous and strictly increasing there. This function is called the (natural) **logarithm function** and is written as ln or also log. From the corresponding properties of the exponential function, it easily follows (Exercise 2.3.32.15):

1. For all $s, t > 0$, $\ln(st) = \ln(s) + \ln(t)$ is true.
2. For all $t > 0$ and all $m \in \mathbb{Z}$, $\ln(t^m) = m \ln(t)$ is true.

Definition 2.3.29 For $a > 0$ and $m \in \mathbb{Z}$,

$$a^m = e^{\ln(a^m)} = e^{m \ln a}$$

is true according to Remark 2.3.28. We set for general $z \in \mathbb{C}$

$$a^z := \exp(z \cdot \ln a) = e^{z \cdot \ln a} .$$

From the calculation rules for ln and exp, one obtains $a^{1/n} = \sqrt[n]{a}$ for $a > 0$ and the following power laws.

Theorem 2.3.30 *Let $a, b > 0$, $c \in \mathbb{R}$ and $z, w \in \mathbb{C}$. Then the following holds*

$$a^z a^w = a^{z+w}, \quad a^z b^z = (ab)^z \quad and \quad (a^c)^z = a^{cz}.$$

Proof We have

$$a^z a^w = e^{z \ln a} e^{w \ln a} = e^{z \ln a + w \ln a} = e^{(z+w) \ln a} = a^{z+w}$$

and

$$a^z b^z = e^{z \ln a} e^{z \ln b} = e^{z(\ln a + \ln b)} = e^{z \ln(ab)} = (ab)^z.$$

For $c \in \mathbb{R}$ we have $a^c = e^{c \ln a} > 0$ and thus

$$(a^c)^z = e^{z \ln(e^{c \ln a})} = e^{cz \ln a} = a^{cz} . \qquad \square$$

Remark and Definition 2.3.31 The inverse function of sin that exists on $[-1, 1]$ according to Theorem 2.3.27 and Theorem 2.3.20, and is strictly increasing and continuous there, is called **Arcsine** (short arcsin). Similarly, the inverse function of cos that exists on $[-1, 1]$, and is strictly decreasing and continuous there, is referred to as **Arccosine** (short arccos).

Furthermore, it holds that tan is strictly increasing in $(-\pi/2, \pi/2)$ and cot is strictly decreasing in $(0, \pi)$ with $W(\tan |_{(-\pi/2, \pi/2)}) = W(\cot |_{(0,\pi)}) = \mathbb{R}$. Therefore, the inverse functions, called **Arctangent** (short arctan) and **Arccotangent** (short arccot), with corresponding monotonicity properties, exist on $\mathbb{R}$ and are continuous there.

Exercises 2.3.32

1. Examine the following series for convergence and absolute convergence:

   a) $\sum_{\nu=1}^{\infty} \frac{x^\nu}{\nu}$ $(x \in \mathbb{R})$,   b) $\sum_{\nu=1}^{\infty} \frac{x^\nu}{\sqrt{\nu}}$ $(x \in \mathbb{R})$,   c) $\sum_{\nu=1}^{\infty} \frac{x^\nu}{\nu^2}$ $(x \in \mathbb{C})$.

2. a) (**Dirichlet Criterion**)

   Let $(a_n)_{n \geq 1}$ be a decreasing sequence in $[0, \infty)$ and $(b_n)_{n \geq 1}$ be a sequence in $\mathbb{K}$. Show: If (a_n) is a null sequence and the series $\sum_{\nu=1}^{\infty} b_\nu$ (i.e., the sequence of partial sums) is bounded, then the series $\sum_{\nu=1}^{\infty} a_\nu b_\nu$ is convergent.

   Hint: Use Abel's partial summation.

   b) Examine the series from Exercise 1 a) and b) for $x \in \mathbb{S}$ for convergence.

3. a) Let (s_n) be a sequence in $\mathbb{K}$. Show: From $s_n \to c$ $(n \to \infty)$ it follows that

   $$|s_n| \to |c| \qquad (n \to \infty).$$

   b) Let $\sum_{\nu=m}^{\infty} a_\nu$ be absolutely convergent. Show: $\left| \sum_{\nu=m}^{\infty} a_\nu \right| \leq \sum_{\nu=m}^{\infty} |a_\nu|$.

4. Show: For all $n \in \mathbb{N}$, $\left(1 + \frac{1}{n}\right)^n \leq \sum_{\nu=0}^{n} \frac{1}{\nu!}$ is true.

5. Prove: For all $z \in \mathbb{C}$

$$\cos z = \sum_{k=0}^{\infty} \frac{(-1)^k}{(2k)!} z^{2k} \quad \text{and} \quad \sin z = \sum_{k=0}^{\infty} \frac{(-1)^k}{(2k+1)!} z^{2k+1}$$

with absolute convergence.

6. Show: For $z, w \in \mathbb{C}$ is

$$\sin z \cos w + \cos z \sin w = \sin(z + w).$$

7. Let $(c_n)_{n \geq 0}$ be a sequence in $\mathbb{K}^*$ with $|c_{n+1}|/|c_n| \to 0$ $(n \to \infty)$. Show:
   a) The series $\sum_{\nu=0}^{\infty} c_\nu z^\nu$ is absolutely convergent for all $z \in \mathbb{C}$.
   b) If $\mathbb{K} = \mathbb{R}$ and $f(z) := \sum_{\nu=0}^{\infty} c_\nu z^\nu$ for $z \in \mathbb{C}$, then $\overline{f(z)} = f(\bar{z})$.

8. Prove:
   a) For all $n \in \mathbb{N}$ is $0 < e - \sum_{\nu=0}^{n-1} \frac{1}{\nu!} < \frac{e}{n!}$.
   b) e is irrational.

9. (de Moivre Formulas) Show: For all $t \in \mathbb{R}$ and $n \in \mathbb{N}$ applies

$$\cos(nt) = \sum_{\substack{\nu=0 \\ \nu \text{ gerade}}}^{n} \binom{n}{\nu}(-1)^{\nu/2} \sin^\nu(t) \cos^{n-\nu}(t),$$

$$\sin(nt) = \sum_{\substack{\nu=0 \\ \nu \text{ ungerade}}}^{n} \binom{n}{\nu}(-1)^{(\nu-1)/2} \sin^\nu(t) \cos^{n-\nu}(t).$$

   Hint: $\cos(nt) + i \sin(nt) = e^{int} = (\cos(t) + i \sin(t))^n$.

10. Show:
    a) For $x > -1$, $e^x \geq 1 + x$.
    b) For $x < 1$, $e^x \leq \frac{1}{1-x}$.

11. For $\alpha \in \mathbb{R}$, let $f_\alpha : (0, \infty) \to \mathbb{R}$ be defined by $f_\alpha(x) = x^\alpha \sin(\pi/x)$. Investigate for which α the function f_α has a (right-sided) limit at 0.

12. Let $f : \mathbb{R} \to \mathbb{R}$ be defined by $f(x) = e^x + x$ for $x \in \mathbb{R}$. Show:
    a) f is strictly increasing.
    b) For all $c \in \mathbb{R}$, there exists exactly one $x \in \mathbb{R}$ with $f(x) = c$.

13. The function $j : \mathbb{C}^* \to \mathbb{C}$ with

$$j(z) := \frac{1}{2}\left(z + \frac{1}{z}\right) \qquad (z \in \mathbb{C}^*)$$

is called **Joukowski mapping**. Show that j is surjective and that $j^{-1}([-1, 1]) = \mathbb{S}$.

14. Let $E_n := \{z \in \mathbb{S} : z^n = 1\}$ be the set of n th roots of unity. Show that $(E_n, \cdot, 1)$ is an abelian group.

15. Let $t, s > 0$ and $n \in \mathbb{N}$. Prove:
    a) $\ln(t \cdot s) = \ln t + \ln s$.
    b) $\sqrt[n]{t} = t^{1/n}$.

16. Show: $x \ln x \to 0$ $(x \to 0^+)$ and $x^x \to 1$ $(x \to 0^+)$.

17. (**Hyperbolic cosine**) The function $\cosh : \mathbb{C} \to \mathbb{C}$ is defined by

$$\cosh(z) := \frac{1}{2} \left(e^{z} + e^{-z} \right) \qquad (z \in \mathbb{C}).$$

Prove:

a) For all $z \in \mathbb{C}$, $\cosh(z) = \sum_{k=0}^{\infty} \frac{z^{2k}}{(2k)!}$ with absolute convergence.

b) $\cosh(\mathbb{R}) = [1, \infty)$.

c) $\cosh(\mathbb{C}) = \mathbb{C}$ and $\cos(\mathbb{C}) = \mathbb{C}$.

Hint: $\cosh = j \circ \exp$, where j denotes the Joukowski mapping.

18. (**Hyperbolic sine**) The function $\sinh : \mathbb{C} \to \mathbb{C}$ is defined by

$$\sinh(z) := \frac{1}{2} \left(e^{z} - e^{-z} \right) \qquad (z \in \mathbb{C}).$$

a) Show: For all $z \in \mathbb{C}$, $\sinh(z) = \sum_{k=0}^{\infty} \frac{z^{2k+1}}{(2k+1)!}$ with absolute convergence.

b) Show that $\sinh|_{\mathbb{R}}$ is strictly increasing with $\sinh(\mathbb{R}) = \mathbb{R}$.

c) Determine the inverse function of $\sinh|_{\mathbb{R}}$, called **Area hyperbolic sine**(short arsinh).

2.4 Metric Spaces

As we have already seen in the previous sections, the concept of limits plays a central role in analysis. It is essential to be able to speak of distances between two elements in a set.

Definition 2.4.1 Let $X \neq \emptyset$ be a set. A mapping $d : X \times X \to \mathbb{R}$ is called a **metric** (or **distance**) on X, if the following conditions are met:

(d1) (Definiteness) For all $x, y \in X$, $d(x, x) = 0$ and $d(x, y) > 0$, if $x \neq y$.

(d2) (Symmetry) For all $x, y \in X$, $d(x, y) = d(y, x)$.

(d3) (Triangle inequality) For all $x, y, z \in X$, $d(x, y) \leq d(x, z) + d(z, y)$. The pair (X, d) is then called a **metric space**.

Remark and Definition 2.4.2

1. If $X \neq \emptyset$ is an arbitrary set, then

$$\delta(x, y) := \begin{cases} 0, & \text{if } x = y \\ 1, & \text{if } x \neq y \end{cases}$$

defines a metric on X, the so-called **discrete metric**. In particular, every non-empty set can thus be provided with a metric.

2. If (X, d) is a metric space and $M \subset X$ is non-empty, then a metric on M is given by $d_M := d|_{M \times M}$.

3. If (X_1, d_1), ..., (X_m, d_m) are metric spaces, then

$$d_\infty(x, y) := \max\{d_j(x_j, y_j) : j = 1, \ldots, m\} \qquad (x = (x_1, \ldots, x_m), y = (y_1, \ldots, y_m))$$

and

$$d_{\text{sum}}(x,y) := \sum_{j=1}^{m} d_j(x_j, y_j) \qquad (x = (x_1, \ldots, x_m), y = (y_1, \ldots, y_m))$$

define metrics on $X_1 \times \ldots \times X_m$ with (Exercise 2.4.33.1)

$$d_\infty \le d_{\text{sum}} \le m \cdot d_\infty . \tag{2.4.1}$$

We consider a class of spaces that possess a linear structure and a metric structure.

Definition 2.4.3

1. Let K be a field and $V \ne \emptyset$ a set. Further, let $+$ be an operation on V and $s :$ $K \times V \to V$. Then $V = (V, +, s)$ is called a **vector space** (over K) or K-**vector space** or also $(K\text{-})$**linear space**, if the following holds

   (V1) $(V, +)$ is an abelian group.
   (V2) For all $\lambda, \mu \in K$, $v \in V$, $s(\lambda, s(\mu, v)) = s(\lambda\mu, v)$.
   (V3) For all $v \in V$, $s(1, v) = v$.
   (V4) (Distributive laws) For all $\lambda, \mu \in K$, $u, v \in V$, the following

   $$s(\lambda, u + v) = s(\lambda, u) + s(\lambda, v) \quad \text{and} \quad s(\lambda + \mu, v) = s(\lambda, v) + s(\mu, v).$$

   holds.
   The elements of V are called **vectors**, the elements from K **scalars** and s **scalar multiplication**. We also write λv instead of $s(\lambda, v)$.
2. If V and W are two K-vector spaces and $T : V \to W$, then T is called **linear**, if for all $u, v \in V$ and all $\lambda, \mu \in K$ the following

   $$T(\lambda u + \mu v) = \lambda T(u) + \mu T(v)$$

   holds.

Remark and Definition 2.4.4 If V is a K-vector space and $M \subset V$, then

$$\text{span}(M) := \Big\{ \sum_{v \in E} \lambda_v v : E \subset M \text{ finite}, (\lambda_v)_{v \in E} \in K^E \Big\}$$

is called a **linear span** or simply **span** of M. Furthermore, M is called a **vector subspace** or **linear subspace**, if $\text{span}(M) = M$ applies. It can easily be shown: A non-empty set $U \subset V$ is a subspace if and only if for all $u, v \in U$, $\lambda \in K$ also $u + v \in U$ and $\lambda u \in U$ holds. Furthermore, then $(U, +, s|_{K \times U})$ is a vector space.

Definition 2.4.5 If $(V, +, s)$ is a K-vector space and $(V, +, \cdot)$ is a ring, then $(V, +, s, \cdot)$ is called a **unitary $(K\text{-})$Algebra**, if for $\lambda \in K$ and $u, v \in V$

$$\lambda(u \cdot v) = (\lambda u) \cdot v = u \cdot (\lambda v)$$

applies. If one dispenses with the requirement for the existence of a unit element in $(V, +, \cdot)$, one speaks of a $(K\text{-})$**Algebra**.

Remark 2.4.6 Let K be a field, $X \neq \emptyset$ and $(K^X, +, \cdot)$ the commutative ring from Remark 1.2.12. By identifying λ with $\lambda 1_{K^X}$, $(K^X, +, s, \cdot)$ (with $s(\lambda, f) := \lambda \cdot f$) becomes a unitary K-Algebra.

Definition 2.4.7 Let $V = (V, +, \cdot)$ be a vector space over $\mathbb{K}$. A mapping $||\cdot|| : V \to \mathbb{R}$ is called a **norm** (on V), if the following conditions are met

- (N1) (Definiteness) It holds that $||0|| = 0$ and $||v|| > 0$ for all $v \neq 0$.
- (N2) (Homogeneity) For all $v \in V$ and all $\lambda \in \mathbb{K}$ is $||\lambda v|| = |\lambda| \cdot ||v||$.
- (N3) (Triangle inequality) For all $u, v \in V$ it holds that $||u + v|| \leq ||u|| + ||v||$. The pair $(V, ||\cdot||)$ is then called a **normed space** (over $\mathbb{K}$).

Remark and Definition 2.4.8 We set

$$||x||_2 := \sqrt{\sum_{j=1}^{m} |x_j|^2} \qquad (x = (x_1, \ldots, x_m) \in \mathbb{K}^m).$$

Then for $x, y \in \mathbb{K}^m \setminus \{0\}$ according to the inequality between geometric and arithmetic mean (Exercise 1.4.27.8)

$$\sum_{j=1}^{m} \frac{|x_j|}{||x||_2} \cdot \frac{|y_j|}{||y||_2} \leq \frac{1}{2} \sum_{j=1}^{m} \left(\frac{|x_j|^2}{||x||_2^2} + \frac{|y_j|^2}{||y||_2^2} \right) = 1.$$

Therefore, (also for $x = 0$ or $y = 0$) the **Cauchy-Schwarz inequality**

$$\sum_{j=1}^{m} |x_j y_j| \leq ||x||_2 ||y||_2$$

holds. From this it follows that

$$||x + y||_2^2 \leq \sum_{j=1}^{m} |x_j| \cdot |x_j + y_j| + \sum_{j=1}^{m} |y_j| \cdot |x_j + y_j| \leq (||x||_2 + ||y||_2) \cdot ||x + y||_2,$$

thus also $||x + y||_2 \leq ||x||_2 + ||y||_2$. Thus, (N3) is fulfilled for $||\cdot||_2$. Since (N1) and (N2) are also fulfilled, $||\cdot||_2$ is a norm on $\mathbb{K}^m$. In the case of $\mathbb{K} = \mathbb{R}$, $||x||_2$ is called the **Euclidean length** of x.

As one can easily see that further norms on $\mathbb{K}^m$ are given by

$$||x||_1 := \sum_{j=1}^{m} |x_j| \qquad (x = (x_1, \ldots, x_m) \in \mathbb{K}^m)$$

and

$$||x||_\infty := \max\{|x_j| : j = 1, \ldots, m\} \qquad (x = (x_1, \ldots, x_m) \in \mathbb{K}^m)$$

From

$$|x_k|^2 \le \sum_{j=1}^{m} |x_j|^2 \le \sum_{j,\ell=1}^{m} |x_j x_\ell| = \left(\sum_{j=1}^{m} |x_j| \right)^2$$

for all $x \in \mathbb{K}^m$ and $k \in \{1, \ldots, m\}$ follows

$$||x||_\infty \le ||x||_2 \le ||x||_1 \ .$$

Unless otherwise specified, $\mathbb{K}^m$ should always be provided with the norm $|| \cdot ||_2$. We also write briefly $| \cdot | := || \cdot ||_2$.

Remark and Definition 2.4.9 If $(V, || \cdot ||)$ is a normed space, then by

$$d_{||\cdot||}(u, v) := ||u - v|| \qquad (u, v \in V)$$

a metric is given on V, the so-called **induced metric**. In particular, by

$$d(x, y) := |x - y| = ||x - y||_2 \qquad (x, y \in \mathbb{K}^m)$$

a metric is defined on $\mathbb{K}^m$ (cf. Remark 2.4.8). In the case of $m = 1$, we speak of the **absolute value metric** and in the case of $\mathbb{K} = \mathbb{R}$ also of the **Euclidean metric**. When we speak of $\mathbb{K}^m$ as a metric space, this metric should always be meant, unless otherwise stated.

Furthermore, here $d_{||\cdot||_\infty} = d_\infty$ as well as $d_{||\cdot||_1} = d_{\text{sum}}$ and with Remark 2.4.8

$$d_\infty(x, y) \le d(x, y) \le d_{\text{sum}}(x, y) \qquad (x, y \in \mathbb{K}^d). \tag{2.4.2}$$

We now consider sequences in metric spaces.

Remark and Definition 2.4.10 Let (X, d) be a metric space. A sequence $(x_n)_{n \in N}$ in X is called $(d\text{-})$**convergent**, if $(d(x_n, c))_{n \in N}$ for a $c \in X$ is a null sequence (in $\mathbb{R}$). Then c is uniquely determined. We call c the $(d\text{-})$**limit**, and we then also write $x_n \to c$ $(n \to \infty)$ and

$$\lim x_n := \lim_{n \to \infty} x_n := c.$$

In the case $d = d_{||\cdot||}$ we also write briefly $|| \cdot ||$ instead of $d_{||\cdot||}$, so for example $|| \cdot ||$-convergent.

Remark 2.4.11 Let X be a set and δ the discrete metric on X. A sequence $(x_n)_n$ in X is δ-convergent with limit c exactly when $x_n = c$ for all sufficiently large n (Exercise 2.4.33.4). If $X = \mathbb{R}$, then $(1/n)$ is *not* δ-convergent (although $(1/n)$ is of course $| \cdot |$-convergent).

Remark and Definition 2.4.12 If (X, d) is a metric space, then a sequence $(x_n)_{n \in N}$ in X is called a $(d\text{-})$**Cauchy sequence**, if for every $\varepsilon > 0$ there exists an $R > 0$ with

$$d(x_n, x_{n'}) < \varepsilon \qquad (n, n' > R) \, .$$

In general (with the same proof as in the case $(\mathbb{K}, d_{|\cdot|})$):

1. Every convergent sequence is a Cauchy sequence.
2. Every Cauchy sequence that has a convergent subsequence is convergent.

Remark 2.4.13 If (X, d) is a metric space and (x_n) is a sequence in X with $x_n \neq c$ for all n and $x_n \to c$ $(n \to \infty)$, then with $M := X \setminus \{c\}$ the sequence (x_n) is a d_M-Cauchy sequence, but not d_M-convergent. In general, therefore, *not every Cauchy sequence is convergent*!

Definition 2.4.14 A metric d on X or the metric space (X, d) is called **sequentially complete** or briefly **complete**, if every Cauchy sequence converges. If V is a normed space with the property that $d_{\|\cdot\|}$ is complete, then $(V, \|\cdot\|)$ is called a **Banach space** (over $\mathbb{K}$).

Remark 2.4.15 1. Let $(X_1, d_1), \ldots, (X_m, d_m)$ be metric spaces and let

$$(x_n)_{n \in N} = (x_{1,n}, \ldots, x_{m,n})_{n \in N}$$

be a sequence in $X_1 \times \cdots \times X_m$. From the estimates (2.4.1) it follows directly that the following statements are equivalent:

a) $(x_n)_{n \in N}$ is d_∞-convergent (or a d_∞-Cauchy sequence).
b) $(x_n)_{n \in N}$ is d_{sum}-convergent (or a d_{sum}-Cauchy sequence).
c) Every **component sequence**$(x_{j,n})_{n \in N}$ is d_j-convergent (or a d_j-Cauchy sequence).

Moreover, in the case of convergence

$$\lim_{n \to \infty} x_n = \Big(\lim_{n \to \infty} x_{1,n}, \ldots, \lim_{n \to \infty} x_{m,n} \Big).$$

holds. Specifically, if $X_1 \times \cdots \times X_m = \mathbb{K}^m$, then with (2.4.2), a)-c) are also equivalent to the sequence being $|\cdot|$-convergent (or a $|\cdot|$-Cauchy sequence).

Remark 2.4.16 Using the Cauchy criterion for sequences in $\mathbb{K}$, together with Remark 2.4.15, the completeness of $(\mathbb{K}^m, d_{|\cdot|})$ is obtained. Thus, $(\mathbb{K}^m, |\cdot|)$ is a Banach space.

Definition 2.4.17 Let (X, d) be a metric space.

1. For $a \in X$ and $0 \leq \rho \leq \infty$ we set

$$U_\rho(a) := U_{\rho,d}(a) := U_{\rho,X}(a) := \{x \in X : d(x, a) < \rho\}.$$

For $\rho > 0$, $U_\rho(a)$ is called the ρ-**neighborhood** of a. Furthermore, we set

$$B_\rho(a) := B_{\rho,d}(a) := B_{\rho,X}(a) := \{x \in X : d(x,a) \le \rho\}$$

and

$$K_\rho(a) := K_{\rho,d}(a) := K_{\rho,X}(a) := \{x \in X : d(x,a) = \rho\}$$

as well as

$$U_\rho^*(a) := U_{\rho,d}^*(a) := U_{\rho,X}^*(a) := \{x \in X : 0 < d(x,a) < \rho\}.$$

2. If $M \subset X$, then $a \in M$ is an **interior point** of M (in (X, d)), if a $\rho > 0$ exists with $U_\rho(a) \subset M$. In this case, M is also called a **neighborhood** of a (in (X, d)).

Definition 2.4.18 Let (X, d) be a metric space and $M \subset X$. Then M is called

1. **open** (in (X, d)), if every point $x \in M$ is an interior point of M,
2. **closed** (in (X, d)), if $M^c = X \setminus M$ is open.

Example 2.4.19

1. In every metric space (X, d), X and $\emptyset$ are open and closed. Moreover, it easily follows from the triangle inequality that $U_\rho(a)$ is open for any $a \in X$ and $\rho > 0$.
2. For $a, b \in \mathbb{R} \cup \{\pm\infty\}$ with $a < b$, the interval (a, b) is open in $\mathbb{R}$. If $a, b \in \mathbb{R}$ holds, then the intervals $[a, b]$, $(-\infty, b]$ and $[a, \infty)$ are closed in $\mathbb{R}$. Intervals of the form $(a, b]$ or $[a, b)$ are neither open nor closed in $\mathbb{R}$.

The following theorem provides a characterization of closedness using sequences.

Theorem 2.4.20 *If (X, d) is a metric space and $M \subset X$, then M is closed if and only if for all sequences (x_n) in M with $x_n \to a \in X$ $(n \to \infty)$ it follows that $a \in M$.*

Proof $\Rightarrow$: Let (x_n) be a sequence in M with $x_n \to a$ $(n \to \infty)$. For any $\varepsilon > 0$, then $M \cap U_\varepsilon(a) \ne \emptyset$ (since $x_n \in U_\varepsilon(a)$ for all sufficiently large n). By assumption, M^c is open. Therefore, $a \notin M^c$, that is, $a \in M$.

$\Leftarrow$: Suppose, M^c is not open. Then there exists an $a \in M^c$ with $U_{1/n}(a) \cap M \ne \emptyset$ for all $n \in \mathbb{N}$. If $x_n \in U_{1/n}(a) \cap M$, then $x_n \to a$ $(n \to \infty)$. Contradiction. $\square$

Theorem 2.4.21 *Let (X, d) be a metric space. If $(M_\alpha)_{\alpha \in I}$ is a family of sets in X, then*

1. *If all M_α are open, then $\bigcup_{\alpha \in I} M_\alpha$ and for finite I also $\bigcap_{\alpha \in I} M_\alpha$ is open.*
2. *If all M_α are closed, then $\bigcap_{\alpha \in I} M_\alpha$ and for finite I also $\bigcup_{\alpha \in I} M_\alpha$ is closed.*

Proof

1. If $x \in M_\alpha$, then by assumption there exists an $\varepsilon_\alpha = \varepsilon_{\alpha,x} > 0$ with $U_{\varepsilon_\alpha}(x) \subset M_\alpha$. So if $x \in \bigcup_{\alpha \in I} M_\alpha$, then there exists a $\beta \in I$ with $x \in M_\beta$ and thus $U_{\varepsilon_\beta}(x) \subset M_\beta$. If I is finite, $x \in \bigcap_{\alpha \in I} M_\alpha$ and $\varepsilon := \min\{\varepsilon_\alpha : \alpha \in I\}$, then $U_\varepsilon(x) \subset M_\alpha$ for all $\alpha \in I$.
2. If all M_α are closed, then all M_α^c are open and thus by De Morgan's laws and 1. also

$$\left(\bigcap_{\alpha \in I} M_\alpha \right)^c = \bigcup_{\alpha \in I} M_\alpha^c.$$

The assertion for $\bigcup_{\alpha \in I} M_\alpha$ in the case of a finite index set I follows accordingly. $\quad\square$

Remark 2.4.22 In general, infinite intersections of open sets are no longer open: If for example $X = \mathbb{R}$, then for the open sets

$$M_n := \left(-1/n, 1/n \right) \qquad (n \in \mathbb{N})$$

the countable intersection

$$\bigcap_{n \in \mathbb{N}} M_n = \{0\}$$

is no longer open. By taking the complement, one can see that, in general, infinite unions of closed sets are also no longer closed.

Remark and Definition 2.4.23 Let (X, d_X), (Y, d_Y) be metric spaces and $f : X \to Y$. The continuity of f at a point $a \in X$ can be defined exactly as in the case of $X \subset \mathbb{K}$ and $Y = \mathbb{C}$, where one simply replaces $|x - a|$ with $d_X(x, a)$ and $|f(x) - f(a)|$ with $d_Y(f(x), f(a))$. That is, f is said to be **continuous at the point** a, if for every $\varepsilon > 0$ there exists a $\delta = \delta_\varepsilon > 0$ such that $d_Y(f(x), f(a)) < \varepsilon$ for all $x \in X$ with $d_X(x, a) < \delta$. This is exactly the case when for every open neighborhood V of $f(a)$ there exists an (open) neighborhood U of a with

$$f(U) \subset V.$$

Furthermore, f is continuous at a if and only if f is **sequentially continuous** at a, that is, if for all sequences (x_n) in X with $x_n \to a$, $f(x_n) \to f(a)$ also holds (see Exercise 2.2.22.4).

Note that continuity depends significantly on the metrics on X and Y. Therefore, we also write $f: (X, d_X) \to (Y, d_Y)$. As usual, f is called **continuous on** $M \subset X$, if f is continuous at every point in M and in the case $M = X$ simply **continuous**. We set

$$C(X, Y) := \{f : X \to Y : f \text{ continuous}\}$$

and $C(X) := C(X, \mathbb{C})$.

Remark and Definition 2.4.24 Let (X, d) be a metric space, Y a set and $f: X \to Y$.

1. We say that a property is local at the point $a \in X$ or in short, that the property holds around a, if a neighborhood U of a exists such that the property is fulfilled for $f|_U$. For example, we say f is **locally constant at** a or in short **constant around** $a \in X$, if a neighborhood U of a exists such that $f|_U$ is constant. If f is constant around every point $a \in X$, then f is called **locally constant**. If d_Y is any metric on Y, it immediately follows from the definition of continuity that any function constant around a is also continuous at a.
2. With the equivalence of continuity and sequential continuity, it follows from Remark 2.4.15: If $f = (f_1, \ldots, f_m) : X \to \mathbb{C}^m$, then f is continuous at the point $a \in X$ if and only if every **component function** $f_j : X \to \mathbb{C}$ is continuous at a.

Remark 2.4.25 Let (X, d_X), (Y, d_Y) and (Z, d_Z) be metric spaces.

1. As in Remark 2.1.15, it is easy to see: If $f : X \to Y$ is continuous at a and $g: Y \to Z$ is continuous at $f(a)$, then $g \circ f$ is also continuous at a.
2. The limit statements from Remark 2.1.11 show that the mappings $(x, y) \mapsto x \pm y$ and $(x, y) \mapsto xy$ are (sequential-)continuous on $\mathbb{K}^2$. The same applies to $(x, y) \mapsto x/y$ on $\mathbb{K} \times \mathbb{K}^*$. With 1. and Remark 2.4.24.2, it follows: If $f, g : X \to \mathbb{C}$ are continuous at a, then $f \pm g, f \cdot g$ and (if defined) f/g are also continuous at a. In particular, $C(X, \mathbb{K})$ is a unitary $\mathbb{K}$-algebra.

Theorem 2.4.26 *Let (X, d_X) and (Y, d_Y) be metric spaces, and let $f : X \to Y$. Then the following are equivalent:*

a) *f is continuous.*
b) *For all open sets $V \subset Y$, $f^{-1}(V) \subset X$ is open.*
c) *For all closed sets $B \subset Y$, $f^{-1}(B) \subset X$ is closed.*

Proof a) $\Rightarrow$ b): Let $V \subset Y$ be open. If $a \in f^{-1}(V)$, then V is an open neighborhood of $f(a)$. Since f is continuous at a, there exists a neighborhood U of a with $f(U) \subset V$, thus $U \subset f^{-1}(f(U)) \subset f^{-1}(V)$. Therefore, $f^{-1}(V)$ is open.

b) $\Rightarrow$ a): Let $a \in X$ and V be an open neighborhood of $f(a)$. By assumption, $U := f^{-1}(V)$ is open in X. Since $a \in U$, U is a neighborhood of a with $f(U) = f(f^{-1}(V)) \subset V$. Therefore, f is continuous at a.

The equivalence of b) and c) results from taking the complement (note that $(f^{-1}(M))^c = f^{-1}(M^c)$ for any set $M \subset Y$).   □

Definition 2.4.27 Let (X, d) be a metric space and $M \subset X$.

1. A point $a \in X$ is called an **accumulation point** of M (in (X, d)), if for every neighborhood U of a the set $M \cap (U \setminus \{a\})$ is not empty (or equivalently, if a sequence in M exists with $a \neq x_n \to a$). We write $M' := M'_{(X,d)}$ for the set of accumulation points of M. If $M' = \emptyset$, then M is called **discrete** (in (X, d)).

2. The set $M^° = M^°_{(X,d)}$ of inner points of M is called the **interior** of M. Furthermore, the set $\overline{M} = \overline{M}_{(X,d)} := M \cup M'$ is called the **closure** of M and $\partial M := \partial_{(X,d)} M := \overline{M} \setminus M^°$ the **boundary** of M.

3. M is called **dense** (in (X, d)), if $\overline{M} = X$ holds. Furthermore, (X, d) is called **separable**, if a countable dense subset exists.

4. With $\mathrm{diam}(\emptyset) := 0$, $\mathrm{diam}(M) := \sup\{d(x, y) : x, y \in M\}$ is called the **diameter** of M.

Remark 2.4.28 Let (X, d) be a metric space. From Definition 2.4.27 it immediately follows that for any set $M \subset X$ the interior $M^°$ is open and that M is open if and only if $M = M^°$ holds. Accordingly, the closure $\overline{M}$ is closed and M is closed if and only if $M = \overline{M}$ holds (Exercise 2.4.33.5). From $\partial M = \overline{M} \cap (M^°)^c$ it follows with Theorem 2.4.21, that also ∂M is closed.

Example 2.4.29

1. Let $X = \mathbb{C}$. We write

$$\mathbb{D} := U_1(0) = \{z \in \mathbb{C} : |z| < 1\}$$

for the open **unit disk** in $\mathbb{C}$. Here $\overline{\mathbb{D}} = \{z : |z| \leq 1\} = B_1(0)$ and $\partial\mathbb{D} = \{z : |z| = 1\} = K_1(0) = \mathbb{S}$.

2. In $X = \mathbb{R}$ is $\mathbb{Q}^° = \emptyset$ and $\overline{\mathbb{Q}} = \mathbb{Q}' = \partial\mathbb{Q} = \mathbb{R}$. In particular, $\mathbb{R}$ is separable.

We now want to show that in complete metric spaces, open dense sets are in a certain sense "large".

Theorem 2.4.30 (Baire) *Let (X, d) be a complete metric space. If $(U_n)_{n \in \mathbb{N}}$ is a sequence of open and dense sets in X, then $\bigcap_{n \in \mathbb{N}} U_n$ is also dense in X.*

Proof Let $U \subset X$ be open and non-empty. It is to be shown that: $U \cap \bigcap_{n \in \mathbb{N}} U_n$ is not empty.

Since U_1 is open and dense in X, $U \cap U_1$ is non-empty and open. Therefore, there exists a closed set $A_1 \subset U \cap U_1$ with $A_1^° \neq \emptyset$ and $\mathrm{diam}(A_1) \leq 1$ (if $x \in U \cap U_1$ and $0 < \delta \leq 1/2$ with $B_\delta(x) \subset U \cap U_1$, then $A_1 := B_\delta(x)$ is suitable).

Remark and Definition 2.4.31 Let (X, d) be a metric space. A set $M \subset X$ is called a G_δ-**set**, if M is a countable intersection of open sets. From Baire's theorem, it easily follows: If (X, d) is complete, then countable intersections of dense G_δ-sets are dense in X (Exercise 2.4.33.7).

Since U_2 is open and dense, $A_1^° \cap U_2$ is non-empty and open. As before, there exists a closed set $A_2 \subset A_1^° \cap U_2$ with $A_2^° \neq \emptyset$ and $\mathrm{diam}(A_2) \leq 1/2$. Inductively, one obtains a sequence (A_n) of closed sets with $A_n^° \neq \emptyset$ and $\mathrm{diam}(A_n) \leq 1/n$ as well as

$$A_n \subset A_{n-1} \cap U_n.$$

If $a_n \in A_n$, then (a_n) is a Cauchy sequence in X. Since X is complete, $c := \lim a_n \in X$ exists. If $n \in \mathbb{N}$, then $a_j \in A_j \subset A_n$ for all $j \geq n$, so $c \in A_n \subset U_n$, since A_n is closed. From $A_1 \subset U$ it also follows that $c \in U$. $\qquad\square$

Remark and Definition 2.4.32 A metric space (X, d) is called **perfect**, if no isolated points exist, i.e., if every point is an accumulation point. If X is complete and perfect, then X is locally uncountable, meaning, every open, non-empty subset is uncountable (Exercise 2.4.33.8). As a special case, we obtain—once again—the local uncountability of $\mathbb{R}$.

Exercises 2.4.33
1. Show that for the metrics defined in Remark 2.4.2,

$$d_\infty \leq d_{\mathrm{sum}} \leq m \cdot d_\infty$$

   holds.
2. Let $d_\mathbb{Q}$ be the absolute value metric restricted to $\mathbb{Q}$. Show that $d_\mathbb{Q}$ is not complete.
3. Let (X, d) be a metric space. Show: For $a \in X$ and $\rho > 0$, $U_\rho(a)$ is open.
4. Let $X \neq \emptyset$ be a set and δ the discrete metric on X. Show:
   a) If (x_n) is a sequence in X, then (x_n) is convergent if and only if there exists a $c \in X$ with $x_n = c$ for all sufficiently large n.
   b) Every set $M \subset X$ is open in (X, δ).
5. Let (X, d) be a metric space and $M \subset X$. Prove:
   a) $(\overline{M})^c = (M^c)^\circ$.
   b) $\overline{M}$ is closed.
   c) It holds $M = \overline{M}$ if and only if M is closed.
6. Let $(V, \|\cdot\|)$ be a Banach space and $U \subset V$ be a subspace. Show: The closure $\overline{U}$ of U is a subspace and thus $\overline{U}$ (with the correspondingly restricted norm) is also a Banach space.
7. Prove: If (X, d) is a complete metric space and (M_n) is a sequence of dense G_δ sets in X, then $\bigcap_{n \in \mathbb{N}} M_n$ is also a dense G_δ set.
8. Let X be a complete and perfect metric space. Prove that every non-empty G_δ set $A \subset X$ is locally uncountable, that is, for all open sets $U \subset X$, $A \cap U$ is uncountable.
   Hint: Use Baire's theorem.
9. A number $x \in \mathbb{R}$ is called **algebraic**, if x is a root of a polynomial $p : \mathbb{R} \to \mathbb{R}$, $p \neq 0$, with integer coefficients. If $x \in \mathbb{R}$ is not algebraic, then x is called **transcendental**. Show: For $\mathbb{A} := \{x \in \mathbb{R} : x\ \text{algebraisch}\}$ the following holds
   a) $\mathbb{A}$ is countable.
      Hint: First prove that the set of polynomials with integer coefficients is countable.
   b) The set of transcendental numbers $\mathbb{R} \setminus \mathbb{A}$ is a dense G_δ-set in $\mathbb{R}$.
   c) $\mathbb{R} \setminus \mathbb{A}$ is locally uncountable.

2.5   Compact Spaces

Definition 2.5.1 A metric space (X, d) is called **(sequentially-)compact**, if every sequence in X has a convergent subsequence. A set $M \subset X$ is called **compact**, if (M, d_M) is compact (or $M = \emptyset$), i.e., if every sequence in M has a convergent subsequence with limit in M. Furthermore, M is called **relatively compact** (in (X, d)), if every sequence in M has a (in X) convergent subsequence.

Remark 2.5.2 If (X, d) is a metric space, then $M \subset X$ is compact if and only if M is relatively compact and closed.

> Because: Let M be compact. Then M is also relatively compact. If (x_n) is a sequence in M with $x_n \to a$, then $a \in M$, since every subsequence also converges to a. According to Theorem 2.4.20, M is closed. Conversely, if M is relatively compact and closed, then every sequence in M has a convergent subsequence and according to Theorem 2.4.20, the limit is in M.

Remark 2.5.3 According to the Bolzano-Weierstrass theorem, bounded sets in $\mathbb{K}$ are relatively compact. With Remark 2.5.2, it follows that intervals of the form $[a, b]$ are compact.

Theorem 2.5.4 *Let* (X_1, d_1), ..., (X_m, d_m) *be compact metric spaces. Then* $(X_1 \times \cdots \times X_m, d_\infty)$ *is also compact.*

Proof We prove the assertion by induction on m.

For $m = 1$ there is nothing to show.

$m \to m+1$: We set $X := X_1 \times \cdots \times X_m$ and define $p : X \times X_{m+1} \to X$, $q : X \times X_{m+1} \to X_{m+1}$ by

$$p(x) := (x_1, \ldots, x_m), \quad q(x) := x_{m+1} \quad (x = (x_1, \ldots, x_m, x_{m+1}) \in X \times X_{m+1}).$$

If $(x_n)_{n \in N}$ is a sequence in $X_1 \times \cdots \times X_{m+1}$, then $(p(x_n))_{n \in N}$ is a sequence in X. Therefore, by the induction hypothesis, there exists a d_∞-convergent subsequence $(p(x_n))_{n \in I}$. Furthermore, $(q(x_n))_{n \in I}$ is a sequence in X_{m+1}. Therefore, there exists a convergent subsequence $(q(x_n))_{n \in J}$. From

$$d_\infty(x_n, c) = \max\{d_\infty(p(x_n), p(c)), d_{m+1}(q(x_n), q(c))\}$$

for $c \in X \times X_{m+1}$ by applying Remark 2.4.15, the d_∞-convergence of $(x_n)_{n \in J} = (p(x_n), q(x_n))_{n \in J}$ is obtained. $\square$

Example 2.5.5 According to Remark 2.5.3 and Theorem 2.5.4, sets of the form $[a_1, b_1] \times \ldots \times [a_m, b_m]$, where $a_j, b_j \in \mathbb{R}$ with $a_j \leq b_j$ for $j = 1, \ldots, m$, are compact.

Remark and Definition 2.5.6 Let $(V, \| \cdot \|)$ be a normed space. A set $M \subset V$ is called **bounded**, if there exists a $\rho > 0$ such that $M \subset U_\rho(0)$, thus $\|x\| < \rho$ for all

$x \in M$. If $M \subset V$ is relatively compact (thus relatively compact in the metric space $(V, d_{\|\cdot\|})$), then M is bounded (otherwise, a sequence (x_n) in M would exist with $\|x_n\| \to \infty$. This sequence would not have a convergent subsequence).

For subsets M of $\mathbb{K}^m$ the converse also holds: If M is bounded, then M is already relatively compact.

> Because: There exists a $\rho > 0$ such that $M \subset U_\rho(0) \subset [-\rho, \rho]^m$. Therefore, M is a subset of a compact set according to Example 2.5.5 and thus relatively compact.

Another central result of analysis is almost proven with the above considerations:

Theorem 2.5.7 (Heine-Borel) *Subsets of $\mathbb{K}^m$ are compact if and only if they are bounded and closed.*

Proof According to Remark 2.5.6, for sets $M \subset \mathbb{K}^m$ boundedness is equivalent to relative compactness. With Remark 2.5.2, this results in the assertion.   □

Remark 2.5.8 If $M \subset \mathbb{R}$ is bounded, then $\sup M \in \overline{M}$ and $\inf M \in \overline{M}$. If M is compact and thus additionally closed, then $\max M$ and $\min M$ exist.

Many mathematical questions involve maximizing or minimizing real-valued functions. We want to introduce the appropriate terms first.

Definition 2.5.9 Let $X \neq \emptyset$ be a set and $f : X \to \mathbb{R}$.

1. We say that f becomes **maximal** (or f has a **maximum**), if

$$\max_X f := \max_{x \in X} f(x) := \max f(X)$$

exists. If $x_0 \in X$ such that $f(x_0) = \max_X f$ holds, that is, if $f(x) \leq f(x_0)$ for all $x \in X$, then we say that f becomes maximal at x_0 (or that f takes its maximum at x_0).
2. We say that f becomes **minimal** (or f has a **minimum**), if

$$\min_X f := \min_{x \in X} f(x) := \min f(X)$$

exists. If $x_0 \in X$ such that $f(x_0) = \min_X f$ holds, that is, if $f(x) \geq f(x_0)$ for all $x \in X$, then we say that f becomes minimal at x_0 (or that f takes its minimum at x_0).

If $M \subset X$, we say that f becomes **maximal on M**, if $f|_M$ has a maximum. Similarly, we say that f becomes **minimal on M**, if $f|_M$ has a minimum.

Bounded and continuous real-valued functions generally have neither a maximum nor a minimum, as shown by $\arctan : \mathbb{R} \to \mathbb{R}$ (the range

$$\arctan(\mathbb{R}) = (-\pi/2, \pi/2)$$

is open). The situation is more favorable for continuous functions on compact sets.

Theorem 2.5.10 *Let* (X, d_X), (Y, d_Y) *be metric spaces and* $f : X \to Y$ *continuous. Then the following holds:*

1. *If* $M \subset X$ *is relatively compact, then* $f(M) \subset Y$ *is also relatively compact.*
2. *If* X *is compact, then* $f(X)$ *is also compact.*
3. *If* X *is compact and* $Y = \mathbb{R}$, *then* f *becomes both maximal and minimal.*

Proof

1. Let $(y_n)_{n \in \mathbb{N}}$ be a sequence in $f(M)$. We choose $x_n \in M$ with $y_n = f(x_n)$. By assumption, there exist an $a \in X$ and a subsequence $(x_n)_{n \in I}$ of $(x_n)_{n \in \mathbb{N}}$ with $x_n \to a$ ($n \to \infty, n \in I$). Since f is continuous, it follows

$$y_n = f(x_n) \to f(a) \qquad (n \to \infty, n \in I) .$$

Thus, $f(M)$ is relatively compact.
2. If $X = M$ in 1., then of course $f(a) \in f(X)$.
3. By 2. and Remark 2.5.8 there exist $\max\limits_X f$ and $\min\limits_X f$. $\qquad\qquad\square$

Remark and Definition 2.5.11 If $(E, | \cdot |_E)$ is a normed space, we set

$$B(X, E) := \{f : X \to E : |f|_E \text{ bounded}\}$$

and $B(X) := B(X, \mathbb{C})$. As one can easily see, $B(X, E)$ is a subspace of E^X and thus a vector space. From $\sup(\alpha A) = \alpha \sup(A)$ and $\sup(A + B) \leq \sup A + \sup B$ for $\alpha \geq 0$ and bounded sets $A, B \subset [0, \infty)$ (equality actually also holds in the second case; Exercise 1.4.27.5) it follows that

$$\|f\|_\infty := \|f\|_{\infty,X} := \|f\|_{\infty,X,E} := \sup_X |f|_E \qquad (f \in B(X, E))$$

defines a norm on $B(X, E)$. We call $\| \cdot \|_\infty$ the **supremum norm** (with respect to X and $| \cdot |_E$). In the case of a *compact* metric space (X, d), $C(X, E) \subset B(X, E)$ holds according to Theorem 2.5.10 and $\|f\|_\infty = \max\limits_X |f|_E$.

In special cases, the existence of maxima or minima in non-compact situations can be traced back to Theorem 2.5.10:

Remark 2.5.12

1. Let $I = (a, b)$ be an open interval and $f : I \to \mathbb{R}$ continuous. If the limits $f(a^+)$ and $f(b^-)$ exist and $f(a^+) = f(b^-)$, then f becomes maximal *or* minimal.

   Because: If f is constant, the assertion is clear. If f is not constant, there exists an $x_0 \in I$ with $f(x_0) \neq f(a^+)(=f(b^-))$. If $f(x_0) > f(a^+)$, there exists a compact interval $J \subset I$ with $f(x_0) > f(x)$ for all $x \in I \setminus J$. According to Theorem 2.5.10, f has a maximum on J. Therefore, $\max\limits_J f = \max\limits_I f$. If $f(x_0) < f(a^+)$, then correspondingly f has a minimum.

2. If $p : \mathbb{K} \to \mathbb{C}$ is a polynomial, then $|p|$ becomes minimal.

Because: According to Example 2.1.17, there exists an $R > 0$ such that $|p(0)| < |p(x)|$ for $|x| > R$. Moreover, according to Theorem 2.5.10, $|p|$ has a minimum on the compact set $B_R(0)$. Then also $\min\limits_{B_R(0)} |p| = \min\limits_{\mathbb{K}} |p|$.

In Remark 2.3.22 we have proven the existence of complex roots. The existence of roots means that for $d \in \mathbb{N}$ polynomials $p : \mathbb{C} \to \mathbb{C}$ of the form $p(z) = z^d - c$ always have roots. We now show more generally:

Theorem 2.5.13 (Fundamental Theorem of Algebra) *Every non-constant polynomial $p : \mathbb{C} \to \mathbb{C}$ has a root.*

Proof According to Remark 2.5.12 there exists a $z_0 \in \mathbb{C}$ with $|p(z_0)| \leq |p|$. Assume, p has no root and therefore in particular $p(z_0) \neq 0$. We can assume without loss of generality $z_0 = 0$ and $p(z_0) = 1$ (otherwise consider $p(z + z_0)/p(z_0)$). Then there exist an $m \in \{1, \ldots, d\}$, a $c_m \neq 0$ and a polynomial q with $q(0) = 0$ and

$$p(z) = 1 + c_m z^m + z^m q(z).$$

According to Remark 2.3.22 there exists an m-th root $\zeta \in \mathbb{S}$ of $-|c_m|/c_m$. If one chooses $r > 0$ so small that both $|q(r\zeta)| < |c_m|$ and $r^m |c_m| < 1$, then it follows

$$|p(r\zeta)| \leq \left|1 - r^m |c_m|\right| + r^m |q(r\zeta))| = 1 - r^m |c_m| + r^m |q(r\zeta)| < 1,$$

in contradiction to $|p| \geq 1$. $\square$

A continuous, bijective function between metric spaces, whose inverse is also continuous, is called a **homeomorphism**. In general, the inverse function of a continuous, bijective function is not continuous (Exercise 2.5.18.8). However, as an application of Theorem 2.5.10, we obtain:

Theorem 2.5.14 *Let (X, d_X) and (Y, d_Y) be metric spaces. If (X, d_X) is compact and $f : X \to Y$ is bijective and continuous, then Y is also compact and f is a homeomorphism.*

Proof Firstly, $Y = f(X)$ is compact according to Theorem 2.5.10. Let $A \subset X$ be closed. To prove the continuity of f^{-1}, it is sufficient according to Theorem 2.4.26 to show that

$$(f^{-1})^{-1}(A) = f(A) \subset Y$$

is closed. Since subsets of relatively compact sets are again relatively compact and since A is closed, A is compact according to Remark 2.5.2. Therefore, $f(A) \subset Y$ is compact according to Theorem 2.5.10 and thus in particular closed (again according to Remark 2.5.2). $\square$

Another important property of continuous functions on compact sets is the uniform continuity:

Definition 2.5.15 Let (X, d_X), (Y, d_Y) be metric spaces and $f : X \to Y$. Then f is called **uniformly continuous**, if for all $\varepsilon > 0$ a $\delta = \delta_\varepsilon > 0$ exists such that

$$d_Y(f(x), f(x')) < \varepsilon$$

for all $x, x' \in X$ with $d_X(x, x') < \delta$.

Furthermore, f is called an **isometry**, if $d_Y(f(x), f(x')) = d_X(x, x')$ for all $x, x' \in X$.

From the definition, it immediately follows that every uniformly continuous function is also continuous and that every isometry is uniformly continuous. The following example shows that continuity in general does not imply uniform continuity.

Example 2.5.16 Let it be $f : \mathbb{R} \to \mathbb{R}$ with $f(x) = x^2$ for $x \in \mathbb{R}$. Then f is continuous on $\mathbb{R}$, but not uniformly continuous.

Because: Let $\varepsilon = 1$ and $\delta > 0$ be arbitrary. We choose $x = 1/\delta$ and $x' = 1/\delta + \delta/2$. Then $|x - x'| = \delta/2 < \delta$, but

$$|f(x) - f(x')| = |x + x'| \cdot |x - x'| > 2/\delta \cdot \delta/2 = 1 = \varepsilon \, .$$

Consequently, f is not uniformly continuous on $\mathbb{R}$.

Theorem 2.5.17 *Let (X, d_X) and (Y, d_Y) be metric spaces. If (X, d_X) is compact and $f : X \to Y$ is continuous, then f is uniformly continuous.*

Proof Assume the contrary. Then there exists an $\varepsilon > 0$ such that for all $n \in \mathbb{N}$ two points $x_n, x_n' \in X$ exist with

$$d_X(x_n, x_n') < 1/n \qquad \text{and} \qquad d_Y(f(x_n), f(x_n')) \geq \varepsilon \, .$$

Since X is compact, the sequence (x_n) has a subsequence $(x_n)_{n \in I}$ with $x_n \to a \in X$ $(n \to \infty, n \in I)$. This also implies

$$d_X(a, x_n') \leq d_X(a, x_n) + d_X(x_n, x_n') \to 0 \qquad (n \to \infty, n \in I),$$

thus $x_n' \to a$ $(n \to \infty, n \in I)$. Due to the (sequence) continuity of f at the point a, we get

$$\varepsilon \leq d_Y(f(x_n), f(x_n')) \leq d_Y(f(x_n), f(a)) + d_Y(f(a), f(x_n')) \to 0 \quad (n \to \infty, n \in I) \, .$$

Contradiction! □

Exercises 2.5.18

1. We set

$$\mathbb{H} := \{z \in \mathbb{C} : z = s + it : t > 0\}.$$

Which of the sets $\mathbb{D}$, $\mathbb{S}$, $\mathbb{H}$, $\mathbb{H} \cup \mathbb{R}$, $\mathbb{H} \cap \mathbb{S}$ are open or closed or (relatively) compact in $\mathbb{C}$?

2. Show that every compact metric space is complete.
3. Let $X \neq \emptyset$ be a set and δ the discrete metric on X. Prove: A set $M \subset X$ is compact if and only if it is finite.
4. Let (X, d) be a metric space and $M \subset X$. Show: M is discrete if and only if $K \cap M$ is finite for all compact sets $K \subset X$.
5. Prove that the set $\{\mathbb{1}_{\{n\}} : n \in \mathbb{N}\}$ is bounded in $(B(\mathbb{R}), \|\cdot\|_\infty)$, but not relatively compact.
6. Let (X, d_X), (Y, d_Y) be metric spaces and $f : X \to Y$ continuous. Show: If $B \subset Y$ is compact, then $f^{-1}(B) \subset X$ is closed. Is $f^{-1}(B)$ always compact?
7. Find a bounded function $f : [0, 1] \to \mathbb{R}$ that becomes neither maximal nor minimal.
8. Let $f : (-\pi, \pi] \to \mathbb{S}$ be defined by
$$f(\theta) = e^{i\theta} \qquad (\theta \in (-\pi, \pi]).$$
Show that f is continuous and bijective and that f^{-1} is continuous at all points $\zeta \in \mathbb{S} \setminus \{-1\}$, but not continuous at the point -1.
9. Examine the following functions $f : (0, 1] \to \mathbb{R}$ for uniform continuity:
   a) $f(x) = \sqrt{x} \qquad (x \in (0, 1])$,
   b) $f(x) = 1/x \qquad (x \in (0, 1])$.
10. Let (X, d) be a metric space and $A \subset X$ non-empty. Then
$$\text{dist}(x, A) := \inf_{a \in A} d(x, a)$$
is called the **distance** between x and A. Show:
   a) If A is closed and $x \in X$ with $\text{dist}(x, A) = 0$, then $x \in A$.
   b) For all $x, y \in X$,
$$|\text{dist}(x, A) - \text{dist}(y, A)| \leq d(x, y).$$
   c) (**Reverse triangle inequality**) For all $x, y, a \in X$,
$$|d(x, a) - d(y, a)| \leq d(x, y).$$
   d) If $K \subset X$ is compact, then $\min_{x \in K} \text{dist}(x, A)$ exists.
11. Prove the following version of the Fundamental Theorem of Algebra: Every polynomial p of degree $d \in \mathbb{N}$ decomposes into linear factors, that is, there exist $z_1, \ldots, z_d \in \mathbb{C}$ and $c \in \mathbb{C}^*$ with
$$p(z) = c \prod_{j=1}^{d} (z - z_j) \quad (z \in \mathbb{C}).$$

Hint: Use polynomial division.

2.6   Sequences of Functions and Series of Functions

We have already seen that important functions such as the exponential function are defined as certain limits. The goal now is to make general structural statements about functions that result from the limits of so-called sequences of functions or series of functions.

Definition 2.6.1 Let $X \neq \emptyset$ be a set and (Y, d) a metric space. A sequence $(f_n)_{n \in N}$ in $\mathrm{Map}(X, Y)$ is called a **sequence of functions**. The sequence of functions $(f_n)_{n \in N}$ is said to be **pointwise convergent** on the set $M \subset X$, if for all $x \in M$ the sequence $(f_n(x))_{n \in N}$ in Y converges. The function $f : M \to Y$ with $f(x) := \lim_{n \to \infty} f_n(x)$ is called the **limit function** of the sequence $(f_n)_{n \in N}$ (on M). We then also write

$$f_n \to f \ (n \to \infty) \quad \text{pointwise on } M.$$

Example 2.6.2 We consider the functions $f_n : \mathbb{R} \to \mathbb{R}$ with

$$f_n(x) := x^n \qquad (x \in \mathbb{R}, n \in \mathbb{N}) .$$

Then

$$f_n(x) \to \begin{cases} 0 \text{ , if } x \in (-1, 1) \\ 1 \text{ , if } x = 1 \end{cases} ,$$

that is, (f_n) converges pointwise on $(-1, 1]$ with the limit function $f = \mathbb{1}_{\{1\},(-1,1]}$. In addition, the sequence $(f_n(x))$ diverges for all other x. The example particularly shows that the limit function is discontinuous (at the point 1), although all sequence terms f_n are continuous functions on $\mathbb{R}$.

We now introduce a stricter concept of convergence for sequences of functions, which has the decisive advantage that continuity is transferred to the limit function.

Remark and Definition 2.6.3 Let $X \neq \emptyset$ be a set and (Y, d) a metric space.

1. For $f, g : X \to Y$ with the property that $X \ni x \mapsto d(f(x), g(x)) \in \mathbb{R}$ is bounded, we set

$$d_\infty(f, g) := d_{\infty,X}(f, g) := d_{\infty,X,Y}(f, g) := \sup_{x \in X} d(f(x), g(x)).$$

Furthermore, if $h : X \to Y$ has the property that $x \mapsto d(g(x), h(x))$ is bounded, then the triangle inequality

$$d_\infty(f, h) \leq d_\infty(f, g) + d_\infty(g, h)$$

holds. If $(E, |\cdot|_E)$ is a normed space and $(Y, d) = (E, d_{|\cdot|_E})$, then

$$d_\infty(f, g) = \|f - g\|_\infty \qquad (f, g \in B(X, E)),$$

is namely d_∞ the metric induced by the supremum norm on $B(X, E)$.

2. A sequence $(f_n)_{n \in N}$ in Map(X, Y) is called **uniformly convergent** on the set $M \subset X$ towards the limit function $f : M \to Y$, if $(d_{\infty,M}(f, f_n))_{n \in N}$ exists and is a null sequence. We then write

$$f_n \to f \ (n \to \infty) \quad \text{uniformly on } M$$

or also

$$f_n(x) \to f(x) \ (n \to \infty) \quad \text{uniformly on } M \ .$$

If $(E, |\cdot|_E)$ is a normed space, then for $f_n, f \in B(X, E)$ *uniform convergence on X is the same as* $\|\cdot\|_\infty$*-convergence.*

3. If $f_n \to f$ converges uniformly on M, then $f_n(x) \to f(x)$ $(n \to \infty)$ for all $x \in M$, since $d(f_n(x), f(x)) \leq d_{\infty,M}(f_n, f)$ holds. In other words: uniform convergence implies pointwise convergence.

Example 2.6.4 We consider again f_n and f from Example 2.6.2. If $M = [-1/2, 1/2]$, then

$$d_{\infty,M}(f_n, f) = \|f_n - f\|_{\infty,M} = \max_{x \in M} |x^n| = 1/2^n \to 0 \qquad (n \to \infty) \ ,$$

so $f_n \to 0 \ (= f|_M) \ (n \to \infty)$ uniformly on $[-1/2, 1/2]$. For $M = [0, 1)$ is

$$d_{\infty,M}(f_n, f) = \|f_n - f\|_{\infty,M} = \sup_{x \in [0,1)} x^n = 1 \quad (n \in \mathbb{N}) \ .$$

So (f_n) is not uniformly convergent on $[0, 1)$.

Theorem 2.6.5 *Let* (Y, d) *be a metric and* (X, d_X) *a compact metric space. Then*

$$d_\infty(f, g) = \max_{x \in X} d\big(f(x), g(x)\big) \qquad (f, g \in C(X, Y))$$

and d_∞ *is a metric on* $C(X, Y)$.

Proof From the triangle inequality and the reverse triangle inequality (Exercise 2.5.18.10), it follows for $u, v, u', v' \in Y$

$$\big|d(u, v) - d(u', v')\big| \leq \big|d(u, v) - d(u', v)\big| + \big|d(u', v) - d(u', v')\big| \leq d(u, u') + d(v, v')$$

and thus

$$\big|d\big(f(x), g(x)\big) - d\big(f(x'), g(x')\big)\big| \leq d(f(x), f(x')) + d(g(x), g(x'))$$

for all $f, g \in C(X, Y)$ and $x, x' \in X$. Since X is compact, f and g are uniformly continuous according to Theorem 2.5.17. Therefore, $x \mapsto d(f(x), g(x))$ is also uniformly continuous and becomes maximal according to Theorem 2.5.10. As already mentioned in Remark 2.6.3, the triangle inequality applies. Definiteness and symmetry of d_∞ immediately follow from the corresponding properties of d. $\qquad \square$

We now come to the already indicated result about the inheritance of continuity to the limit function.

Theorem 2.6.6 *Let (X, d_X), (Y, d) be metric spaces and a $\in X$. Furthermore, let $(f_n)_{n \in N}$ be a sequence in* $\mathrm{Map}(X, Y)$*, which converges uniformly on a neighborhood U of a to f. If the functions f_n are continuous at a, then f is also continuous at a.*

Proof Let $\varepsilon > 0$ be given. Due to the uniform convergence of (f_n) towards f on U, there exists an $n = n_\varepsilon \in N$ with

$$\sup_{x \in U} d(f(x), f_n(x)) < \varepsilon/3 \ .$$

Since f_n is continuous at the point a, there exists a $\delta = \delta_\varepsilon > 0$ such that $U_\delta(a) \subset U$ and

$$d(f_n(x), f_n(a)) < \varepsilon/3 \qquad (x \in U_\delta(a)) \ .$$

Therefore, for $x \in U_\delta(a)$

$$d(f(x), f(a)) \leq d(f(x), f_n(x)) + d(f_n(x), f_n(a)) + d(f_n(a), f(a)) < \varepsilon \ .$$

$\square$

A necessary and sufficient criterion for uniform convergence is provided by

Theorem 2.6.7 (Cauchy criterion for uniform convergence) *Let $X \neq \emptyset$ be a set, (Y, d) a complete metric space and $(f_n)_{n \in N}$ a sequence in* $\mathrm{Map}(X, Y)$*. If $M \subset X$, then $(f_n)_{n \in N}$ is uniformly convergent on M if and only if, for every $\varepsilon > 0$, there exists an $R > 0$ with*

$$d_{\infty, M}(f_n, f_{n'}) < \varepsilon \qquad (n, n' > R).$$

Proof $\Rightarrow$: Let $f_n \to f$ converge uniformly on M. If $\varepsilon > 0$ is given, there exists an $R > 0$ with

$$d_{\infty, M}(f, f_n) < \varepsilon/2 \qquad (n > R) \ .$$

For all $n, n' > R$, the triangle inequality applies (see Remark 2.6.3)

$$d_{\infty, M}(f_n, f_{n'}) \leq d_{\infty, M}(f_n, f) + d_{\infty, M}(f, f_{n'}) < \varepsilon.$$

$\Leftarrow$: By assumption, in particular for each fixed $x \in M$, the sequence $(f_n(x))_{n \in N}$ is a d-Cauchy sequence. Since (Y, d) is complete, $(f_n(x))_{n \in N}$ is convergent. We define $f : M \to Y$ by $f(x) := \lim_{n \to \infty} f_n(x)$ $(x \in M)$ and show that (f_n) converges uniformly on M to f.

To do this, let $\varepsilon > 0$ be given. Then there exists an $R = R_\varepsilon > 0$ with $d_{\infty, M}(f_n, f_{n'}) < \varepsilon$ for $n, n' > R$. Now let $n > R$ be fixed. If $x \in M$, the mapping $y \mapsto d(f_n(x), y)$ is continuous according to the reverse triangle inequality (Exercise 2.5.18.10). Therefore,

$$d(f_n(x), f_m(x)) \to d(f_n(x), f(x)) \qquad (m \to \infty) \ .$$

From $d(f_n(x), f_m(x)) < \varepsilon$ for $m > R$, it follows that $d(f_n(x), f(x)) \leq \varepsilon$, and since $x \in M$ was arbitrary, $d_{\infty.M}(f, f_n) \leq \varepsilon$ is also true.   □

Remark 2.6.8 If $X \neq \emptyset$ is a set and $(E, |\cdot|_E)$ is a Banach space, then $(B(X, E),$ $\|\cdot\|_\infty)$ is also a Banach space.

Because: Let $(f_n)_{n \in N}$ be a $\|\cdot\|_\infty$-Cauchy sequence. According to Theorem 2.6.7, there exists a function $f : X \to E$ with $f_n \to f$ uniformly on X. According to Remark 2.6.3, it suffices to show that $|f|_E$ is bounded. For this, we choose an $n \in N$ with $\|f - f_n\|_\infty < 1$. Then for all $x \in X$

$$|f(x)|_E \leq |f(x) - f_n(x)|_E + |f_n(x)|_E \leq 1 + \|f_n\|_\infty.$$

So, $|f|_E$ is bounded.

Remark 2.6.9 Let (X, d_X) be *compact* and (Y, d) be *complete*. Then, according to Theorem 2.6.7 and Theorem 2.6.6, the metric space $(C(X, Y), d_\infty)$ is complete.

Now let (Y, d) be *compact*. If X is *finite*, then $C(X, Y) = \text{Map}(X, Y)$ and with Theorem 2.5.4 one can see that then also $(\text{Map}(X, Y), d_\infty)$ is compact. However, in general, the space $(C(X, Y), d_\infty)$ is not compact: If, for example, $X = Y = [0, 1]$ (each with the absolute value metric), then it follows from Example 2.6.4 that the sequence (f_n) in $C(X, Y)$ with

$$f_n(x) = x^n \qquad \left(x \in [0, 1]\right)$$

has no d_∞-convergent (thus uniformly convergent on $[0, 1]$) subsequence.

Definition 2.6.10 If $X \neq \emptyset$ is a set, $(E, |\cdot|_E)$ is a normed space, and $(f_n)_{n \geq m}$ is a sequence in $\text{Map}(X, E)$, then the sequence of functions (s_n) with

$$s_n(x) := \sum_{\nu=m}^{n} f_\nu(x) \qquad (x \in X, \, n \geq m)$$

is called a **function series**. We write again $\sum_{\nu=m}^{\infty} f_\nu$ instead of $(s_n)_{n \geq m}$. The function series $\sum_{\nu=m}^{\infty} f_\nu$ is called **pointwise convergent** on $M \subset X$, if the sequence of functions (s_n) converges pointwise on M. Correspondingly, the function series is called **uniformly convergent** on M, if (s_n) converges uniformly on M. The symbol $\sum_{\nu=m}^{\infty} f_\nu$ is then also used again for the limit function.

The question arises as to sufficient conditions for uniform convergence. Analogous to Theorem 2.3.6, we obtain

Theorem 2.6.11 (Weierstrass Criterion) *Let $X \neq \emptyset$ be a set, $(E, |\cdot|_E)$ a Banach space and $(f_n)_{n \geq m}$ a sequence in $B(X, E)$. If the series $\sum_{\nu=m}^{\infty} \|f_\nu\|_\infty$ converges, then*

$\sum\limits_{\nu=m}^{\infty} f_\nu$ *converges uniformly on X. This is the case if a sequence* (b_n) *exists with*

$\sum\limits_{\nu=m}^{\infty} b_\nu < \infty$ *and* $|f_n(x)|_E \le b_n$ *for* $x \in X$ *and* $n \ge m$.

Proof Firstly, $s_n \in B(X, E)$, since $B(X, E)$ is a linear space. If $\varepsilon > 0$ is given, then according to Theorem 2.3.5 there exists a $R > 0$ with

$$\sum_{\nu=n'+1}^{n} \|f_\nu\|_\infty < \varepsilon \qquad (n > n' > R).$$

Therefore, for $n > n' > R$

$$\|s_n - s_{n'}\|_\infty = \left\| \sum_{\nu=n'+1}^{n} f_\nu \right\|_\infty \le \sum_{\nu=n'+1}^{n} \|f_\nu\|_\infty < \varepsilon.$$

Thus, (s_n) is a $\|\cdot\|_\infty$-Cauchy sequence. According to Remark 2.6.8, (s_n) is also $\|\cdot\|_\infty$-convergent, thus uniformly convergent on X. The additional assertion results from Remark 2.2.15. $\qquad\qquad\square$

Remark and Definition 2.6.12 Let $a \in \mathbb{K}$ and $(c_n)_{n=0}^{\infty}$ be a sequence in $\mathbb{C}$. Then the function series $\sum_{\nu=0}^{\infty} f_\nu$ with $f_n : \mathbb{K} \to \mathbb{C}$, defined by

$$f_n(x) := c_n(x - a)^n \qquad (x \in \mathbb{K}, \ n \in \mathbb{N}_0),$$

is called power series with **center of development** a and **sequence of coefficients** (c_n). Furthermore,

$$R := \sup\{|h| : \sum_{\nu=0}^{\infty} c_\nu h^\nu \text{ converges}\},$$

is called the **radius of convergence** and $U_R(a)$ the **circle of convergence** (in the case of $\mathbb{K} = \mathbb{R}$ usually the **interval of convergence**) of the power series. In the case $R > 0$, the power series is uniformly convergent on $B_r(a)$ for all $r < R$.

Because: Let $h \in \mathbb{K}$ with $|h| > r$ such that $\sum_{\nu=0}^{\infty} c_\nu h^\nu$ converges. Then $(c_n h^n)$ is a null sequence, so there exists an n_0 with $|c_n h^n| \le 1$ for $n \ge n_0$. This results in $q := r/|h| < 1$ for $x \in B_r(a)$ and

$$|f_n(x)| = |c_n(x - a)^n| \le |x - a|^n/|h|^n \le q^n \quad (n \ge n_0).$$

With Theorem 2.6.11 the assertion follows. If $y \in U_R(a)$, then $B_r(a)$ for $|y| < r < R$ is a neighborhood of y. According to Theorem 2.6.6, the function is $f : U_R(a) \to \mathbb{C}$ with

$$f(x) := \sum_{\nu=0}^{\infty} c_\nu(x - a)^\nu \qquad (x \in U_R(a))$$

is continuous at the point y. Since $y \in B_R(a)$ was arbitrary, f is continuous on the circle of convergence.

Example 2.6.13

1. The geometric series $\sum\limits_{v=0}^{\infty} z^v$ is a power series with $a = 0$ and $c_n = 1$ for $n \in \mathbb{N}_0$.

   Here, $R = 1$ and according to Remark 2.6.12,

   $$\sum_{v=0}^{\infty} z^v = \frac{1}{1-z} \quad (|z| < 1)$$

   holds with uniform convergence on $B_r(0)$ for all $r<1$.

2. The **exponential series** $\sum\limits_{v=0}^{\infty} z^v/v!$ is a power series with $a = 0$ and $c_n = 1/n!$

   for $n \in \mathbb{N}_0$. Since the series converges for all $z \in \mathbb{C}$, $R = \infty$ and according to Remark 2.6.12,

   $$\sum_{v=0}^{\infty} \frac{z^v}{v!} = e^z \quad (z \in \mathbb{C})$$

   holds with uniform convergence on $B_r(0)$ for all $r<\infty$.

Example 2.6.14 For $\alpha > 1$ let $S_\alpha := \{z \in \mathbb{C} : \operatorname{Re} z \geq \alpha\}$. Then the function series $\sum\limits_{v=1}^{\infty} v^{-z}$ is uniformly convergent on S_α.

  Because: For all $z \in S_\alpha$ and all $n \in \mathbb{N}$ is

  $$|n^{-z}| = |e^{-z \ln n}| = e^{-\operatorname{Re}(z) \ln n} \leq e^{-\alpha \ln n} = n^{-\alpha}.$$

Since the series $\sum\limits_{v=1}^{\infty} v^{-\alpha}$ converges (Exercise 2.6.15.3), the assertion results from Theorem 2.6.11. With $S := \{z \in \mathbb{C} : \operatorname{Re} z > 1\}$ the function $\zeta : S \to \mathbb{C}$, defined by

$$\zeta(z) := \sum_{v=1}^{\infty} v^{-z} \quad (z \in S),$$

is called **(Riemann) zeta function**.

  Let $\operatorname{Re}(a) > 1$ and $1 <\alpha<\operatorname{Re}(a)$. Since $z \mapsto n^{-z}$ for all $n \in \mathbb{N}$ is continuous on $\mathbb{C}$ and S_α is a neighborhood of a, the continuity of ζ at the point a follows from Theorem 2.6.6. Since $a \in S$ was arbitrary, the Zeta function is continuous on S.

Exercises 2.6.15

1. (**Cauchy-Hadamard Theorem**) Let $\sum\limits_{\nu=0}^{\infty} c_\nu(x-a)^\nu$ be a power series with radius of convergence R. Show: with $1/\infty := 0$ and $1/0 := \infty$ the following holds

$$1/R = \limsup_{n\to\infty} \sqrt[n]{|c_n|}.$$

2. Examine the power series $\sum\limits_{\nu=1}^{\infty} z^\nu/\nu$ and $\sum\limits_{\nu=1}^{\infty} z^\nu/\nu^2$ with radius of convergence 1 for pointwise and uniform convergence on the unit circle $\mathbb{S}$.

3. Let $\alpha > 0$. Show: $\sum\limits_{\nu=1}^{\infty} \nu^{-\alpha}$ is convergent if and only if $\alpha > 1$.

   Hint: Use the Cauchy condensation theorem.

4. Let $X \subset \mathbb{K}$, $a \in X \cap X'$ and $f_n : X \to \mathbb{C}$ with $f_n \to f \ (n \to \infty)$ uniformly on X. Show: If f is continuous at the point a and if (x_n) is a sequence in X with $x_n \to a$ $(n \to \infty)$, then also

$$f_n(x_n) \to f(a) \qquad (n \to \infty).$$

5. The functions $f_n : [0,1] \to \mathbb{R}$ are defined by

$$f_n(x) := nx(1-x)^{n-1} \quad (x \in [0,1], n \in \mathbb{N}).$$

   a) Show that (f_n) converges pointwise on $[0,1]$ and determine the limit function.

   b) Show: $f_n(1/n) \geq 1/e$.

   Hint: Use Exercise 2.3.32.4

   c) Is (f_n) uniformly convergent on $[0,1]$?

2.7 Concepts I: Completeness and Compactness

The two introductory chapters of the book essentially provide an introduction to those structures of mathematics that are of fundamental importance for analysis. The focus is on the concept of completeness, first introduced in the form of order completeness and later as sequence completeness within the framework of metric spaces. It turns out that the demand for order completeness in ordered fields almost inevitably leads to real numbers. In this sense, there is no alternative to real numbers if one wants to combine the arithmetic structure of a field with the indispensable existence of supremum and infimum (of bounded sets) for meaningful analysis. The *Intermediate Value Theorem* proves to be the central result in this context, which is essentially behind every statement about the solvability of nonlinear equations in $\mathbb{R}$ and thus the invertibility of nonlinear functions on intervals. If one wants to perform analysis on more general structures, it turns out that one can much more easily dispense with algebraic structures or an order structure than with the demand for completeness.

The mere abandonment of the existence of an order, through suitable extension of $\mathbb{R}$, leads to the field of complex numbers $\mathbb{C}$, which has the decisive advantage of algebraic closedness: According to the *Fundamental Theorem of Algebra*, all equations of the form

$$c_d x^d + c_{d-1} x^{d-1} + \ldots + c_0 = 0$$

with $d \in \mathbb{N}$ and $c_d \neq 0$ are solvable in $\mathbb{C}$!

In fact, the step from real to complex analysis proves to be a step into a world where one can recognize a multitude of relationships that remain hidden in the real world. For example, it turns out that the complex exponential function not only exhibits the "exponential growth" known from the real, but reveals a fundamentally different periodic behavior when the viewing angle is rotated by $\pi/2$. The trigonometric functions, which appear completely different from the real exponential function, prove to be essentially suitable means of the complex exponential function.

The existence of the exponential function and, consequently, all known elementary functions is by no means a matter of course, but arises from the completeness "inherited" from the real line to the complex plane, now in the form of sequential completeness, that is, the property that Cauchy sequences converge, thus contracting to points in space. In one form or another, most objects appearing in analysis, optimized in a certain way, arise from suitable limit processes. In this sense, one can say that we typically owe the realization of ideal states to completeness.

If one also gives up arithmetic structures in addition to the order structure, one arrives at the concept of a complete metric space.[7] Complete metric spaces in which meaningful analysis can be conducted are necessarily large in a certain sense. From Baire's theorem it follows that a complete perfect space is locally uncountable, that is, in the vicinity of every point there are uncountably many other points of the space. Local infinity already enforces local uncountability.

As mentioned several times, the completeness of metric spaces is the essential basis for conducting analysis. If one can even ensure compactness beyond completeness, the conditions become much more favorable. Compactness typically allows global statements such as the existence of maxima and minima in real-valued functions. This fact is naturally of fundamental importance for questions about the solvability of optimization problems.

In many respects, compact sets show similar properties to finite sets. The concept of compactness is so flexible that it captures enough relevant and interesting cases. In $\mathbb{K}^d$ or more generally in finite-dimensional normed spaces, according to the *Heine-Borel theorem*, all sets that are bounded and closed are already compact—a fact that can hardly be overestimated in its importance. Unfortunately,

---

[7] One can even consider topological spaces, more generally.

the situation in the case of infinite-dimensional normed spaces is not so favorable. Closed balls are not compact, and thus a Heine-Borel statement is no longer valid. In general, possibly also complete, metric spaces, the situation regarding the existence of compact sets can become arbitrarily unfavorable. In the case of the—not particularly appealing for analysis—discrete metric, only finite sets are compact.

Against this background, one learns to appreciate the statements of the *Montel theorems*, which we will prove in Chap. 6. A typical—and elegant—application example of the Montel theorem we will get to know in connection with the existence of conformal mappings. The proof of the *Riemann mapping theorem* is based on the fact that the existence of conformal mappings can be traced back to the solvability of a suitable optimization problem. A compactness argument based on the Montel theorem provides the existence of a solution.

Of central importance for the derivation of the Montel theorems is the *Arzelà-Ascoli theorem*, which shows that in the case of spaces of continuous functions between compact sets, equicontinuity leads to (relative) compactness. The appropriate selection of successive subsequences (i.e., subsequences of subsequences of...) plays a decisive role. This principle appears in our proof of Theorem 6.1.4, in which essentially relative compactness is linked with precompactness.

Derivatives and integrals are *the* tools of analysis. The chapter starts with an introduction to the differential calculus of complex-valued functions, which are defined on suitable subsets of $\mathbb{K}$. From the very beginning, a common path is taken into both the real and the complex world of differentiable functions. The central position of the complex exponential function is once again clearly demonstrated, and central results such as the mean value inequality theorem appear in a corresponding guise. There is also a first impression of power series and their significance as germs of analytic functions.

As an introduction to integral calculus, the approach via the Riemann integral is chosen. The main theorem on integral functions for the first time clarifies the close relationship between derivatives and integrals: integrals can be interpreted as "antiderivatives" in the sense of an operation that is (right-)inverse to the derivative.

In the following, compact intervals as integration areas are extended in two respects: on the one hand, general intervals and thus improper integrals are examined, and on the other hand—again with the aim of orienting towards complex analysis—paths in the complex plane and corresponding path integrals. In this context, an important result about the differentiation of parameter integrals is also proven. The result finds immediate application, for example, in the extension of the real logarithm function and the real arctangent to suitable subsets of the complex plane.

3.1 Differential Calculus

We again examine functions $f : X \to \mathbb{C}$, where $X \subset \mathbb{K}$, i.e., functions of a real or complex variable. In order to be able to investigate the finer structure of the behavior of such functions, we need a concept of smoothness that goes beyond

© The Author(s), under exclusive license to Springer-Verlag GmbH, DE, part of
Springer Nature 2025
J. Müller, *Concepts of Function Theory*, Mathematics Study Resources 12,
https://doi.org/10.1007/978-3-662-69115-1_3

continuity. Roughly speaking, we want to define functions that can be approximated very well locally by affine-linear mappings.

Definition 3.1.1 Let $X \subset \mathbb{K}$ and $f : X \to \mathbb{C}$.

1. f is called **differentiable at the point** $a \in X$, if a is an accumulation point of X (thus $a \in X \cap X'$) and the limit

$$f'(a) := \lim_{x \to a} \frac{f(x) - f(a)}{x - a}$$

exists. $f'(a)$ is referred to as the **derivative** of f at the point a.
2. f is called **differentiable** (on X), if f is differentiable at every point $x \in X$. In this case, the function $f' : X \to \mathbb{C}$ is called the **derivative** of f. Other notations are Df or df or also (df/dx).[1] If f' is continuous, then f is called **continuously differentiable** (on X).

Remark 3.1.2 Let us again consider $X \subset \mathbb{K}$, $a \in X \cap X'$ and $f : X \to \mathbb{C}$. If we define the function $\tau_a f : (X - a) \to \mathbb{C}$ by

$$(\tau_a f)(h) := f(a + h) - f(a) \qquad (h \in X - a),$$

then $\tau_a f(0) = 0$ and f is differentiable at a according to Remark 2.1.15 if and only if

$$\lim_{h \to 0} \frac{1}{h} \tau_a f(h) = \lim_{h \to 0} \frac{f(a + h) - f(a)}{h}$$

exists, i.e., $\tau_a f$ is differentiable at 0, and in this case

$$f'(a) = (\tau_a f)'(0)$$

is valid. Therefore, if necessary, when examining derivatives, one can always revert to the case $a = f(a) = 0$.

Example 3.1.3
1. If $f(x) := cx + b$ for $x \in \mathbb{K}$ with constants $b, c \in \mathbb{C}$, then f is differentiable on $\mathbb{K}$ and

$$f'(x) = \lim_{h \to 0} \frac{f(x + h) - f(x)}{h} = \lim_{h \to 0} \frac{ch}{h} = c \qquad (x \in \mathbb{K}) .$$

2. If $f(x) := x^2$ for $x \in \mathbb{K}$, then

$$f'(x) = \lim_{h \to 0} \frac{(x + h)^2 - x^2}{h} = \lim_{h \to 0} \frac{2xh + h^2}{h} = \lim_{h \to 0} (2x + h) = 2x \qquad (x \in \mathbb{K}) .$$

---

[1] This notation is suggestive and practical in some situations, but problematic, as it silently assumes that the variable is called x. Moreover, one encounters certain formal problems when one wants to consider the derivative at the point x, something like $(df/dx)(x)$. We will therefore refrain from using this notation in the future.

More generally, using Theorem 1.3.1 or the binomial theorem (Exercise 3.1.34.1), it can be shown: If $n \in \mathbb{N}$ and $f(x) = x^n$ for $x \in \mathbb{K}$, then

$$f'(x) = nx^{n-1} \qquad (x \in \mathbb{K}) .$$

3. If $f(x) := |x|$ for $x \in \mathbb{R}$, then f is continuous on $\mathbb{R}$, but not differentiable at the point $a = 0$, since $f(h)/h = 1$ for $h > 0$ and $f(h)/h = -h /h = -1$ for $h < 0$. Therefore, $h \mapsto f(h)/h$ has no (two-sided) limit at 0. This example shows that continuity at a point does not generally imply differentiability at that point.

Another proof of the outstanding importance of the exponential function is

Theorem 3.1.4 *The function* exp *is differentiable on* $\mathbb{C}$ *with*

$$\exp{'} = \exp .$$

Proof. According to Remark 2.6.12 $\varepsilon : \mathbb{C} \to \mathbb{C}$ with

$$\varepsilon(h) := \sum_{\mu=0}^{\infty} \frac{h^{\mu}}{(\mu + 1)!}$$

is continuous on $\mathbb{C}$ with $\varepsilon(0) = 1$. Therefore, for $h \in \mathbb{C}^*$

$$\frac{1}{h}(e^h - 1) = \sum_{\nu=1}^{\infty} \frac{h^{\nu-1}}{\nu!} = \varepsilon(h) \to 1 \qquad (h \to 0).$$

For any $a \in \mathbb{C}$ one has

$$(e^{a+h} - e^a)/h = e^a \cdot (e^h - 1)/h \to e^a \cdot 1 \quad (h \to 0).$$

$\square$

The first part of the following theorem shows that the differentiability of a function f at a point a means that f can be approximated locally at a with a certain accuracy by an affine-linear function of the form

$$h \mapsto f(a) + c \cdot h$$

The second part shows that differentiability implies continuity. We write $X_a := (X - a) \setminus \{0\}$.

Theorem 3.1.5 *Let* $X \subset \mathbb{K}$, $a \in X \cap X''$ *and* $f : X \to \mathbb{C}$. *Then the following holds*

1. **(Decomposition formula, Affine-linear approximation)** *f is differentiable at a if and only if there exist an $c \in \mathbb{C}$ and a function $\varepsilon = \varepsilon_{f,a} : X_a \to \mathbb{C}$ decaying at 0 with*

$$f(a + h) = f(a) + c \cdot h + \varepsilon(h) \cdot h \qquad (h \in X_a).$$

   In this case, $f'(a) = c$.
2. *If f is differentiable at a, then f is also continuous at a.*

Proof.

1. $\Rightarrow$: If we set for $h \in X_a$

$$\varepsilon(h) := (\tau_a f)(h)/h - f'(a),$$

then ε decays at 0 due to the differentiability of f at a and it holds

$$f(a + h) = f(a) + f'(a) \cdot h + \varepsilon(h) \cdot h \qquad (h \in X_a),$$

thus $c = f'(a)$.

$\Leftarrow$: It holds $(\tau_a f)(h)/h = c + \varepsilon(h) \to c$ for $h \to 0$.

2. The statement follows from the decomposition formula for $h \to 0$. $\square$

Theorem 3.1.6 (Sum Rule, Product Rule and Quotient Rule) *Let $X \subset \mathbb{K}$ and $f, g : X \to \mathbb{C}$ be differentiable at the point $a \in X$. Then the following holds*

1. *$f + g$ is differentiable at a with*

$$(f + g)'(a) = f'(a) + g'(a).$$

2. *$f \cdot g$ is differentiable at a with*

$$(f \cdot g)'(a) = f'(a) \cdot g(a) + f(a) \cdot g'(a).$$

3. *If g is free of zeros, then f/g is differentiable at a with*

$$\left(\frac{f}{g}\right)'(a) = \frac{f'(a)g(a) - f(a)g'(a)}{g^2(a)} .$$

Proof. As one can easily verify, for $h \in X - a$

$$\tau_a(f + g)(h) = \tau_a f(h) + \tau_a g(h)$$

and

$$\tau_a(f \cdot g)(h) = g(a + h) \cdot \tau_a f(h) + f(a) \cdot \tau_a g(h).$$

According to Theorem 3.1.5, g is continuous at a. Thus, the sum rule and the product rule each result from division by h and taking limits for $h \to 0$. If additionally $Z(g) = \emptyset$, then also

$$\tau_a(1/g)(h) = \frac{-\tau_a g(h)}{g(a + h)g(a)} .$$

Again, after division by h and taking limit for $h \to 0$, the quotient rule for $f = 1$ is obtained. The statement for general f follows with the product rule. $\square$

Example *3.1.7* If $p(x) = \sum_{\nu=0}^{d} c_\nu x^\nu (x \in \mathbb{K})$ is a polynomial, then from Theorem 3.1.6 and Example 3.1.3 follows

$$p'(x) = \sum_{\nu=1}^{d} \nu \cdot c_\nu x^{\nu-1} \qquad (x \in \mathbb{K}).$$

Theorem 3.1.8 (Chain Rule) *Let $X, Y \subset \mathbb{K}$ and let f be: $X \to Y$. Furthermore, let $g : Y \to \mathbb{C}$. If f is differentiable at $a \in X$ and g is differentiable at $f(a)$, then $g \circ f$ is differentiable at a with*
$$(g \circ f)'(a) = g'\big(f(a)\big)f'(a) .$$

Proof. We first consider the special case $a = 0$ and $f(0) = g(0) = 0$. If $\varepsilon := \varepsilon_{g,0}$ as in the decomposition formula and $\varepsilon(0) := 0$, then $\varepsilon \circ f$ is decreasing at 0 and thus

$$\begin{aligned}
\tfrac{1}{h}(g \circ f)(h) &= \frac{1}{h}\big(g(f(h)) - g'(0)f(h)\big) + g'(0)\frac{f(h)}{h} \\
&= \varepsilon(f(h))\frac{f(h)}{h} + g'(0)\frac{f(h)}{h} \to g'(0)f'(0) \quad (h \to 0).
\end{aligned}$$

Therefore, $g \circ f$ is differentiable at 0 with $(g \circ f)'(0) = g'(0)f'(0)$.

If now a as well as $f(a)$ and $g(f(a))$ are arbitrary, then

$$\tau_a(g \circ f) = \tau_{f(a)}g \circ \tau_a f,$$

and thus the assertion results from applying the special case to the right side. □

Example 3.1.9 For $p : \mathbb{K} \to \mathbb{C}$ with

$$p(x) = (x^3 + 2x + 1)^5 \quad (x \in \mathbb{K})$$

one has $p = g \circ f$ with $f(x) = x^3 + 2x + 1$ and $g(y) = y^5$. Thus, from the chain rule results

$$p'(x) = g'(f(x))f'(x) = 5(x^3 + 2x + 1)^4(3x^2 + 2) \quad (x \in \mathbb{K}).$$

Theorem 3.1.10 *The functions* $\sin$ *and* $\cos$ *are differentiable on* $\mathbb{C}$ *with*

$$\sin' = \cos \quad and \quad \cos' = -\sin .$$

Proof. For $z \in \mathbb{C}$ according to the chain rule

$$\sin'(z) = \frac{1}{2i}(ie^{iz} + ie^{-iz}) = \frac{1}{2}(e^{iz} + e^{-iz}) = \cos z .$$

Accordingly, $\cos' = -\sin$ results. □

Example 3.1.11
1. From Theorem 3.1.10 and the quotient rule, it follows

$$\tan' = \frac{1}{\cos^2}\big(\cos^2 + \sin^2\big) = \frac{1}{\cos^2} = 1 + \tan^2$$

   on $\mathbb{C} \setminus \pi(\mathbb{Z} + 1/2)$ and correspondingly on $\mathbb{C} \setminus \pi\mathbb{Z}$

$$\cot' = -\sin^{-2} = -1 - \cot^2.$$

2. If $a > 0$ is fixed and $f(z) := a^z = e^{z \ln a}$ for $z \in \mathbb{C}$, then from the chain rule follows

$$f'(z) = a^z \ln a \qquad (z \in \mathbb{C}).$$

Theorem 3.1.12 (Inversion Rule) *Let $X \subset \mathbb{K}$ and $f : X \to Y \subset \mathbb{C}$ be bijective. If f is differentiable at the point $a \in X$ with $f'(a) \neq 0$ and the inverse function f^{-1} is continuous at $c := f(a)$, then f^{-1} is differentiable at c with*

$$(f^{-1})'(c) = 1/f'(a) = 1/(f'(f^{-1}(c))) .$$

Proof. If (x_n) is a sequence in X with $a \neq x_n \to a \ (n \to \infty)$, then due to the continuity of f at a and the injectivity,

$$f(a) \neq f(x_n) \to c \qquad (n \to \infty) .$$

Thus, c is an accumulation point of Y. Let's first assume $a = c = 0$. Then f^{-1} is continuous at 0. Therefore, $f^{-1}(u) \to 0 \ (u \to 0)$ and consequently, with Remark 2.1.15 and $f^{-1}(u) \neq 0$ for $u \neq 0$

$$\frac{f^{-1}(u)}{u} = \frac{f^{-1}(u)}{f(f^{-1}(u))} \to \frac{1}{f'(0)} \qquad (u \to 0) .$$

The general case results from Remark 3.1.2 and $\tau_c f^{-1} = (\tau_a f)^{-1}$. $\qquad\qquad$ □

Remark 3.1.13 Let $X \subset \mathbb{K}$ and $f : X \to Y \subset \mathbb{C}$ be bijective with continuous inverse function. If $g : Y \to \mathbb{C}$ is free of zeros with $f' = g \circ f$ on X, then according to the inversion rule

$$(f^{-1})' = 1/(f' \circ f^{-1}) = 1/g.$$

Example 3.1.14
1. It holds

$$\ln'(t) = 1/t \qquad (t > 0).$$

Because: According to Remark 3.1.13, applied to $g(t) := t$ for $t > 0$ and $f(s) = e^s$ for $s \in \mathbb{R}$, is $\ln' = (f^{-1})' = 1/g$. This results in the differentiability of $t \mapsto t^\alpha = e^{\alpha \ln t}$ on $(0, \infty)$ with derivative $t \mapsto \alpha t^{\alpha-1}$ for fixed $\alpha \in \mathbb{C}$ using the chain rule.

2. It holds

$$\arctan'(t) = \frac{1}{1 + t^2}, \qquad \operatorname{arccot}'(t) = -\frac{1}{1 + t^2} \qquad (t \in \mathbb{R}).$$

Because: According to Example 3.1.11 and Remark 3.1.13, applied to $g(t) := 1 + t^2$ for $t \in \mathbb{R}$ and $f(s) = \tan(s)$ for $s \in (-\pi/2, \pi/2)$, is $\arctan' = (f^{-1})' = 1/g$. The same applies to arccot.

3. It holds (Exercise 3.1.34.3)

$$\arcsin'(t) = \frac{1}{\sqrt{1 - t^2}}, \qquad \arccos'(t) = -\frac{1}{\sqrt{1 - t^2}} \qquad (t \in (-1, 1)).$$

Remark 3.1.15 If $X = (-\pi, \pi]$ and $f : X \to \mathbb{S}$ with $f(x) = e^{ix}$ $(x \in X)$, then f is bijective and differentiable with $f'(x) = ie^{ix} \neq 0$ for all x. However, the inverse function is not continuous at the point -1 (Exercise 2.5.18.7) and therefore not differentiable. Thus, it can be seen that the continuity requirement for f^{-1} in Theorem 3.1.12 cannot generally be dispensed with.

The inverse function f^{-1} is "automatically" continuous if X is an interval and f is strictly monotonic (Theorem 2.3.27), or if X is compact and f is continuous (Theorem 2.5.14). If f^{-1} is continuous and f is continuously differentiable with a derivative f' that has no zeros, then according to the inversion rule, f^{-1} is also continuously differentiable, since $(f^{-1})' = 1/(f' \circ f^{-1})$ is then continuous.

We now turn to how we can use differential calculus to examine the local behavior of differentiable functions in more detail. To do this, we first define what we mean by extreme points.

Definition 3.1.16 Let (X, d) be a metric space and $f : X \to \mathbb{R}$.

1. A point $x_{\max} \in X$ is called a **maximum point** (of f), if a neighborhood U of $x_{\max}$ exists with

$$f(x) \leq f(x_{\max}) \qquad (x \in U),$$

   that is, if $f|_U$ at the point $x_{\max}$ becomes maximal. In this case, the function value $f(x_{\max})$ is called a **local maximum.** If $<$ instead of $\leq$ holds for $x \neq x_{\max}$, we speak of a **strict** local maximum.
2. A point $x_{\min} \in X$ is called a **minimum point** (of f), if a neighborhood U of $x_{\min}$ exists with

$$f(x) \geq f(x_{\min}) \qquad (x \in U),$$

   that is, if $f|_U$ at the point $x_{\min}$ becomes minimal. Then $f(x_{\min})$ is called a **local minimum.** If $>$ instead of $\geq$ holds for $x \neq x_{\min}$, we speak of a **strict** local minimum.

If a is a maximum or a minimum point of f, then a is also called an **extreme point** of f.

We will in the sequel demonstrate that differential calculus provides an efficient toolset for determining extreme points. If $f : X \to \mathbb{C}$ is differentiable, a zero of f' is also called a **critical point** of f.

Theorem 3.1.17 *Let $X \subset \mathbb{R}$ and a be an interior point of X. If $f : X \to \mathbb{R}$ is differentiable at a and if a is an extreme point of f, then a is a critical point, thus $f'(a) = 0$.*

Proof. Without loss of generality, let a be a minimum point. Then there exists a $\delta > 0$ with $\tau_a f(h) \geq 0$ for $|h| < \delta$. Since f is differentiable at a,

$$0 \le \lim_{h \to 0^+} (\tau_a f)(h)/h = f'(a) = \lim_{h \to 0^-} (\tau_a f)(h)/h \le 0.$$

□

Remark and Definition 3.1.18

1. The vanishing of f' at a point a is merely a *necessary* condition for a to be a maximum or minimum point. For example, for $f : \mathbb{R} \to \mathbb{R}$ with $f(x) = x^3$, the zero point is a critical point, but not an extreme point. Therefore, Theorem 3.1.17 is typically used to exclude the extremality of f at the interior points a that are not critical. Critical points are the only "candidates" for extreme points within X.

2. If a is an extreme point, but not an interior point of X, then a is not necessarily a critical point. For example, ± 1 for $f : [-1, 1] \to \mathbb{R}$ with $f(x) = |x|$ are maximum points, but not critical points. Also, 0 is a minimum point, but also not a critical point (since f is not differentiable at 0).

Theorem 3.1.19 (Rolle) *Let $I = (\alpha, \beta)$ be an open interval and $f : I \to \mathbb{R}$ differentiable. If $f(\alpha^+)$ and $f(\beta^-)$ exist and $f(\alpha^+) = f(\beta^-)$, then f has a critical point.*

Proof. According to Remark 2.5.12, f becomes maximal or minimal and thus, according to Theorem 3.1.17, a critical point exists.   □

As a result, we obtain

Theorem 3.1.20 *Let $I = (\alpha, \beta)$ be an open interval and $f, g : I \to \mathbb{R}$ differentiable. If $f(\alpha^+)$ and $f(\beta^-)$ as well as $g(\alpha^+)$ and $g(\beta^-)$ exist, then there exists a $\xi \in (\alpha, \beta)$ with*

$$f'(\xi)(g(\beta^-) - g(\alpha^+)) = (f(\beta^-) - f(\alpha^+))g'(\xi) .$$

Proof. We consider the function $\varphi : I \to \mathbb{R}$ with

$$\varphi(x) := f(x)\big(g(\beta^-) - g(\alpha^+)\big) - \big(f(\beta^-) - f(\alpha^+)\big)g(x) \qquad (x \in I).$$

Since $\varphi(\beta^-) = \varphi(\alpha^+) = f(\alpha^+)g(\beta^-) - f(\beta^-)g(\alpha^+)$, the statement follows by applying Rolle's theorem with φ instead of f.   □

Definition 3.1.21 Let V be a vector space and $u, v \in V$. We define $s_u^v : [0, 1] \to V$ by

$$s_u^v(t) := u + t(v - u) \quad (t \in [0, 1])$$

and call s_u^v **oriented segment** from u to v. Furthermore, we set

$$[u, v] := s_u^v([0, 1]) = \{u + t(v - u) : t \in [0, 1]\}$$

and

$$(u, v) := s_u^v((0, 1)) = \{u + t(v - u) : t \in (0, 1)\}.$$

With this, we come to a theorem that contains two further central results of analysis.

Theorem 3.1.22 *Let $a \in \mathbb{K}$ and $h \in \mathbb{K}^*$. Furthermore, let $f : [a, a + h] \to \mathbb{C}$ be continuous on $[a, a + h]$ and differentiable on $(a, a + h)$. Then the following holds*

1. **(Mean Value Theorem)** *If f is real-valued, there exists a $\xi \in (a, a + h)$ with*

$$f(a + h) - f(a) = f'(\xi) \cdot h .$$

2. **(Mean Value Inequality)** *There exists a $\xi \in (a, a + h)$ with*

$$|f(a + h) - f(a)| \leq |f'(\xi)| \cdot |h| .$$

Proof.
1. Results from the application of Theorem 3.1.20 with $I = (0, 1)$ and $f \circ s_a^{a+h}$ instead of f and $g(x) = x$.
2. Without restriction, we can assume $w := f(a + h) - f(a) \neq 0$. We set $\zeta := |w|/w$ and consider the function $\varphi := \mathrm{Re}(\zeta f \circ s_a^{a+h}) : [0, 1] \to \mathbb{R}$. According to 1., there exists a $\tau \in (0, 1)$ with

$$|w| = \mathrm{Re}(\zeta w) = \varphi(1) - \varphi(0) = \varphi'(\tau) = \mathrm{Re}\big(\zeta f'(s_a^{a+h}(\tau)) \cdot h\big) \leq |f'(s_a^{a+h}(\tau))| \cdot |h|.$$

With $\xi := s_a^{a+h}(\tau)$ the assertion follows. $\qquad\square$

Definition 3.1.23 A set $X \subset \mathbb{K}$ is called **star-shaped with respect to** $a \in X$, if

$$X = \bigcup_{x \in X} [a, x]$$

holds, and shortly **star-shaped,** if it is star-shaped with respect to some point a. The set X is called **convex,** if it is star-shaped with respect to all points $a \in X$, i.e., if $[a, x] \subset X$ for all $a, x \in X$. A non-empty set in $\mathbb{R}$ is star-shaped if and only if it is convex, and this holds if and only if it is an interval.

Theorem 3.1.24 *Let $X \subset \mathbb{K}$ be star-shaped and $f : X \to \mathbb{C}$. If f is differentiable with $f' = 0$, then f is constant.*

Proof. Let X be star-shaped with respect to a. If $x \in X$ with $x \neq a$, then from the mean value inequality with a $\xi \in (x, a)$

$$|f(x) - f(a)| \leq |f'(\xi)| \cdot |x - a| = 0$$

and thus $f(x) = f(a)$. Therefore, f is constant $(= f(a))$ on X. $\qquad\square$

Theorem 3.1.25 *Let I be an interval and $f : I \to \mathbb{R}$ be continuous on I and differentiable on I°. Then the following holds*

1. *If $f' \geq 0$ on $I°$, then f is increasing on I.*
2. *If $f' \leq 0$ on $I°$, then f is decreasing on I.*

If $f' > 0$ or $f' < 0$, then the monotonicity is strict.

Proof.
1. If $s, t \in I$ with $s < t$, then from the mean value theorem with a $\xi \in (s, t)$

$$f(t) - f(s) = f'(\xi)(t - s) \geq 0 \, ,$$

thus $f(s) \leq f(t)$. If $f'(\xi) > 0$, then $f(s) < f(t)$.
2. Results from applying 1. to $-f$.   □

Example 3.1.26 The function $\cosh|_{\mathbb{R}}$ is strictly increasing on $[0, \infty)$ and strictly decreasing on $(-\infty, 0]$, since $\cosh' = \sinh$ with $\sinh > 0$ on $(0, \infty)$ and $\sinh < 0$ on $(-\infty, 0)$. If $a < 0 < b$, the graph of $\cosh|_{[a,b]}$ is a so-called catenary. It describes the sag of a chain suspended at its ends $(a, \cosh(a))$ and $(b, \cosh(b))$ under the influence of gravity (see Fig. 3.1).

The following result provides a *sufficient* criterion for extreme points.

Theorem 3.1.27 (Sign Change Criterion) *Let I be an interval, $a \in I$ and $f : I \rightarrow \mathbb{R}$ continuous on I and differentiable on $I°$.*
  Then the following holds

Fig. 3.1 cosh on the interval $[-2, 2]$

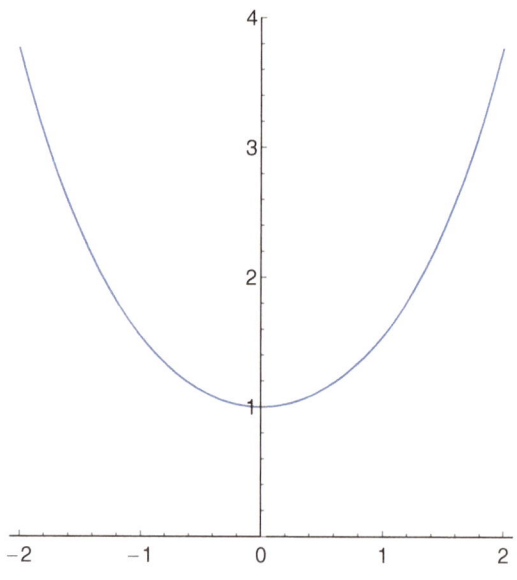

1. *If there exists a $\delta > 0$ with $f' \leq 0$ on $I \cap (a, a + \delta)$ and $f' \geq 0$ on $I \cap (a - \delta, a)$, then a is a maximum point of f.*
2. *If there exists a $\delta > 0$ with $f' \geq 0$ on $I \cap (a, a + \delta)$ and $f' \leq 0$ on $I \cap (a - \delta, a)$ then a is a minimum point of f.*

If $f' < 0$ or $f' > 0$, then the extremum is strict.

Proof.
1. According to Theorem 3.1.25, f is decreasing on $I \cap [a, a + \delta)$ and increasing on $I \cap (a - \delta, a]$. Therefore, a is a maximum point. If the strict inequality holds, then the maximum is strict.
2. Again by applying 1. to $-f$.   □

Example 3.1.28 Let $f : \mathbb{R} \to \mathbb{R}$ with

$$f(x) = 2x^3 + 3x^2 - 1 \qquad (x \in \mathbb{R}).$$

Then

$$f'(x) = 6x(x+1) \begin{cases} > 0, & \text{for } x \in (-\infty, -1) \\ < 0, & \text{for } x \in (-1, 0) \\ > 0, & \text{for } x \in (0, \infty) \end{cases}.$$

According to Theorem 3.1.25, f is strictly increasing on $(-\infty, -1]$, strictly decreasing on $[-1, 0]$ and strictly increasing on $[0, \infty)$. According to Theorem 3.1.27, 0 is also a minimum point with a strict local minimum $f(0)$ and -1 is a maximum point with a strict local maximum $f(-1)$ (see Fig. 3.2).

An elegant method for the possible calculation of certain limits results from

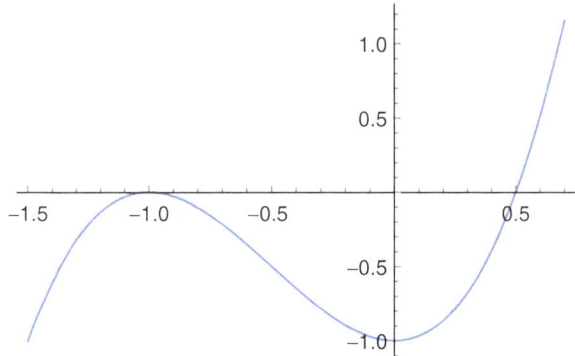

Fig. 3.2 Polynomial $x \mapsto 2x^3 + 3x^2 - 1$

Theorem 3.1.29 (Rules of the Hospital) *Let $I = (a, b) \subset \mathbb{R}$ an open interval and $f, g : I \to \mathbb{R}$ differentiable with*

$$f(a^+) = g(a^+) = 0 \quad or \quad g(x) \to \pm\infty \quad (x \to a).$$

If g' is zero-free and

$$f'(t)/g'(t) \to c \in \mathbb{R} \cup \{\pm\infty\} \qquad (t \to a),$$

it follows that

$$f(x)/g(x) \to c \qquad (x \to a) .$$

A corresponding statement applies for limit values $x \to b$.

Proof.
1. Let $f(a^+) = g(a^+) = 0$. If $x \in I$, then according to Theorem 3.1.20 there exists a $\xi(x) \in (a, x)$ with

$$\frac{f'(\xi(x))}{g'(\xi(x))} = \frac{f(x) - f(a^+)}{g(x) - g(a^+)} = \frac{f(x)}{g(x)} .$$

   Here, $a < \xi(x) \to a \ (x \to a)$. Therefore, with Remark 2.1.15 it follows that

$$\frac{f'(\xi(x))}{g'(\xi(x))} \to c \quad (x \to a).$$

2. Let $g(x) \to \infty$. If $\varepsilon > 0$ is given, there exists a $\delta > 0$ with $g(t) > 0$ and $f'(t)/g'(t) \in U_\varepsilon(c)$ for $t \in U_\delta(a)$. We choose an $s \in U_\delta(a)$. If $a < x < s$, then according to Theorem 3.1.20 there exists a $\xi(x) \in (x, s)$ with

$$f'(\xi(x))(g(s) - g(x)) = (f(s) - f(x))g'(\xi(x)) ,$$

   thus also

$$\frac{f'(\xi(x))}{g'(\xi(x))} \left(\frac{g(s)}{g(x)} - 1 \right) = \frac{f(s) - f(x)}{g(x)}$$

   and consequently

$$\frac{f(x)}{g(x)} = \frac{f(s)}{g(x)} + \frac{f'(\xi(x))}{g'(\xi(x))} \left(1 - \frac{g(s)}{g(x)} \right) .$$

   By assumption, $f(s)/g(x) \to 0$ and $g(s)/g(x) \to 0 \ (x \to a^+)$. Since

$$f'(\xi(x))/g'(\xi(x)) \in U_\varepsilon(c)$$

   holds, there exists an $\eta > 0$ with $f(x)/g(x) \in U_{2\varepsilon}(c)$ for $x \in U_\eta(a)$. Since $\varepsilon > 0$ was arbitrary, the assertion follows.
3. If $g(x) \to -\infty$, then $-g(x) \to \infty$ and thus the assertion follows from 2. □

Example 3.1.30

1. For all $\alpha > 0$, then

$$\lim_{x \to \infty} \frac{\ln x}{x^\alpha} = 0 ,$$

that is, the logarithm function grows slower than any power function $x \mapsto x^\alpha$ with positive α. This is a variant of the statement that exponential growth is stronger than polynomial (see Theorem 2.3.13).

Because: One has $\lim_{x \to \infty} x^\alpha = \infty$ and

$$\lim_{x \to \infty} \frac{x^{-1}}{\alpha x^{\alpha - 1}} = \lim_{x \to \infty} \frac{1}{\alpha x^\alpha} = 0 ,$$

thus the assertion follows with Theorem 3.1.29.

2. For all $c \in \mathbb{R}$ applies (Exercise 3.1.34.10)

$$x \ln(1 + c/x) \to c \qquad (x \to \infty)$$

and thus also

$$\left(1 + \frac{c}{x}\right)^x \to e^c \qquad (x \to \infty).$$

Remark 3.1.31 If $\zeta \in \mathbb{C}^*$ and $a \in \mathbb{C}$, then $a + \mathbb{R}\zeta$ is a straight line through the point a. If $\zeta = e^{i\theta}$ for a $\theta \in (-\pi/2, \pi/2)$, then $\tan(\theta)$ is the angle of inclination. If now $X \subset \mathbb{C}$ and $a \in X$ is an accumulation point of $M := (X - a) \cap \mathbb{R}\zeta$, then $f : X \to \mathbb{C}$ is called differentiable in **direction** ζ at a, if the limit

$$\lim_{M \ni h \to 0} (\tau_a f)(h)/h$$

exists. In this case,

$$(\partial_\zeta f)(a) := \zeta \cdot \lim_{M \ni h \to 0} (\tau_a f)(h)/h$$

is called the **directional derivative** of f in direction ζ at a. The term, which seems somewhat artificial from our point of view, proves to be very natural in real multi-dimensional analysis (which is not the subject of this book). It holds that

$$(\partial_\zeta f)(a) = \lim_{t \to 0} \frac{(\tau_a f)(t\zeta)}{t} ,$$

so that here only the real variable t appears in the denominator. This implies that for real-valued f also $(\partial_\zeta f)(a)$ is real-valued in case of existence.

The directional derivatives $(\partial_1 f)(a) = \partial_{(1,0)} f(a)$ and $(\partial_i f)(a) = \partial_{(0,1)} f\ (a)$ are called **partial derivatives** of f at the point a. If $\Omega \subset \mathbb{C}$ is open, then we write $C^1\ (\Omega)$ for the set of all $f : \Omega \to \mathbb{C}$ with the property that $\partial_{(1,0)} f$ and $\partial_{(0,1)} f$ exist on Ω and are continuous there.

In clear distinction to this, in the sense of Definition 3.1.1, continuously differentiable functions $f : \Omega \to \mathbb{C}$ are also called **holomorphic** (on or in Ω) and one sets

$$H(\Omega) := \{f : \Omega \to \mathbb{C} : f \text{ holomorphic}\}.$$

From these definitions it follows directly: If $f \in H(\Omega)$, then $f \in C^1\ (\Omega)$ with

$$\partial_\zeta f = \zeta \cdot f' \qquad (\zeta \in \mathbb{C}^*).$$

In particular, $\partial_{(1,0)} f = f'$ and $\partial_{(0,1)} f = i \cdot f'$, so

$$\partial_{(0,1)} f = i \cdot \partial_{(1,0)} f.$$

We call this equation the **Cauchy-Riemann equation.** It can be shown[2]: If $f \in C^1(\Omega)$ and the Cauchy-Riemann equation is satisfied, then $f \in H(\Omega)$. Thus, the holomorphy of f is equivalent to continuous partial differentiability and the validity of the Cauchy-Riemann equation.

Example 3.1.32 If $\Omega = \mathbb{C}$ and $f(z) = \bar{z}$ for $z \in \mathbb{C}$, then $\partial_\zeta f$ is constant with value $\bar{\zeta}$ on $\mathbb{C}$ for all $\zeta \in \mathbb{C}^*$. Therefore, $f \in C^1(\Omega)$. However,

$$\partial_{(0,1)} f = -i \neq i = i \cdot \partial_{(1,0)} f,$$

that is, the Cauchy-Riemann equation is not fulfilled. Thus, f is not holomorphic in $\mathbb{C}$.

Remark 3.1.33 If $f \in H\ (\Omega)$ is real-valued, then $f' = \partial_1 f$ and $if' = \partial_i f$ are also real-valued. Thus, $f' = 0$, so f is locally constant according to Theorem 3.1.24.

Exercises 3.1.34

1. Calculate the derivatives of $x \mapsto x^n$, where $n \in \mathbb{N}$, and $(0, \infty) \ni x \mapsto \sqrt{x}$, preferably both without and with the product or inversion rule.
2. Prove:
   a) $\cosh' = \sinh$ and $\sinh' = \cosh$,
   b) $\cosh^2 - \sinh^2 = 1$,
   c) $\sinh' = \sqrt{1 + \sinh^2}$ on $\mathbb{R}$.
3. Prove:
   a) $\sin' = \sqrt{1 - \sin^2}$ on $(-\pi/2,\ \pi/2)$,
   b) $\arcsin'(t) = 1/\sqrt{1 - t^2}$ for $t \in (-1, 1)$.

---

[2] Proofs can be found in almost all textbooks that deal with both real analysis of several variables and complex analysis.

4. Let $\alpha > 0$ and $f_\alpha : [0, \infty) \to \mathbb{R}$ be defined by

$$f_\alpha(x) = \begin{cases} x^\alpha \sin\left(\pi/x\right) & , \; x \neq 0 \\ 0 & , \; x = 0 \end{cases}.$$

For which α is f_α differentiable, for which α continuously differentiable?
5. For $n \in \mathbb{N}_0$ let $f_n : \mathbb{R} \to \mathbb{R}$ be defined by

$$f_n(x) := \left(\frac{2}{3}\right)^n \sin(2^n x) \qquad (x \in \mathbb{R}).$$

Prove:
a) $\sum_{v=0}^{\infty} f_v$ is uniformly convergent on $\mathbb{R}$.
b) $\sum_{v=0}^{\infty} f_v'(0)$ is divergent.
6. Show: For all $t \in \mathbb{R}$ the following holds

$$\arcsin\left(\frac{t}{\sqrt{1+t^2}}\right) = \arctan(t).$$

7. a) Let $X \subset \mathbb{K}$ be star-shaped with respect to 0 and $f : X \to \mathbb{C}$ differentiable
   with $f' = f$ on X and $f(0) = 1$. Show: Then $f = \exp|_X$ is true.
   Hint: Show that $\varphi : X \to \mathbb{C}$ with $\varphi(x) := f(x)e^{-x}$ is constant on X.
   b) Using a), prove again the functional equation of the exponential function:
   For all $z, w \in \mathbb{C}$ the following is true

$$e^{z+w} = e^z e^w.$$

8. The function $f : [0, \infty) \to \mathbb{R}$ is defined by

$$f(x) := \begin{cases} x^x, & x > 0 \\ 1, & x = 0 \end{cases}.$$

Examine f for continuity, differentiability, monotonicity, and extreme points.
9. Determine the limit values
   a) $\lim\limits_{x \to 1^+} \frac{\sqrt{x}-1}{\sqrt{x-1}}$,
   b) $\lim\limits_{x \to \infty} \frac{\sinh x}{\cosh x}$.
10. Show: For all $c \in \mathbb{R}$
    a) $\lim\limits_{x \to \infty} x \cdot \ln\left(1 + c/x\right) = c$,
    b) $\lim\limits_{x \to \infty} \left(1 + c/x\right)^x = e^c$.
11. Let $f : \mathbb{C} \to \mathbb{C}$ be defined by $f(z) = |z|^2$. Show
    a) It holds $f \in C^1(\mathbb{C})$ with

$$(\partial_\zeta f)(z) = 2\mathrm{Re}(\bar\zeta z) \quad (z \in \mathbb{C}, \; \zeta \in \mathbb{C}^*).$$

    b) For all $z \notin \mathbb{R}$ is $\partial_{(0,1)} f(z) \neq i \cdot \partial_{(1,0)} f(z)$.

3.2   Higher Derivatives and Analytic Functions

Definition 3.2.1 Let $X \subset \mathbb{K}$ with $X \subset X'$ and $f : X \to \mathbb{C}$. With $f^{(0)} := f$ one can define higher derivatives recursively: If $n \in \mathbb{N}$ and $f^{(n-1)}$ is differentiable, then f is called n-**times differentiable** and the function

$$f^{(n)} := (f^{(n-1)})' : X \to \mathbb{C}$$

is the n-**th derivative** of f. One usually writes $f'' := f^{(2)}$ and $f''' := f^{(3)}$. If $f^{(n)}$ is continuous, then one says f is n-times continuously differentiable and if $f^{(n)}$ exists for all $n \in \mathbb{N}_0$, then one says f is arbitrarily often differentiable.

In the case of $X \subset \mathbb{R}$ we further set

$$C^n(X) := \{f : X \to \mathbb{C} : f^{(n)} \text{ exists and is continuous}\}$$

for $n \in \mathbb{N}_0$ and

$$C^\infty(X) := \bigcap_{n \in \mathbb{N}_0} C^n(X).$$

Thus, we have the inclusion chain

$$C^\infty(X) \subset C^n(X) \subset C^{n-1}(X) \subset C^0(X) = C(X) \quad (n \in \mathbb{N}),$$

where all inclusions for $n > 1$ are strict (see Exercise 3.2.31.2).

Theorem 3.2.2 *Let $I \subset \mathbb{R}$ be an open interval, $f : I \to \mathbb{R}$ twice continuously differentiable and $a \in I$ a critical point. Then the following holds:*

1. *If $f''(a) > 0$, then f has a strict local minimum at a.*
2. *If $f''(a) < 0$, then f has a strict local maximum at a.*

Proof.
1. Since f'' is continuous at a, there exists a $\delta > 0$ with $f''(x) > 0$ for all $x \in U_\delta(a)$. Therefore, f' is strictly increasing according to Theorem 3.1.25 on $U_\delta(a)$. From $f'(a) = 0$ it follows

$$f'(x) \begin{cases} > 0, & a < x < a + \delta \\ < 0, & a - \delta < x < a \end{cases}.$$

According to the sign change criterion (3.1.27), f has a strict local minimum at a.
2. This follows again from 1. by applying to $-f$.   □

Example 3.2.3 We consider once again the polynomial f from Example 3.1.28, so

$$f(x) = 2x^3 + 3x^2 - 1 .$$

Here is

$$f''(x) = 12x + 6$$

and thus $f''(0) = 6 > 0$ at the critical point 0 and $f''(-1) = -6 < 0$ at the critical point -1. Therefore, f has a strict local minimum at 0 and a strict local maximum at -1.

We have already seen that functions that can be represented by power series are continuous on the circle of convergence. We now show, much more sharply, that such functions are actually arbitrarily often differentiable. We will limit ourselves to power series with expansion center $a = 0$. The general case can be easily reduced to this.

Theorem 3.2.4 *Let $\sum_{\nu=0}^{\infty} c_\nu x^\nu$ be a power series with convergence radius $R > 0$ and let $f(x) := \sum_{\nu=0}^{\infty} c_\nu x^\nu$ for $x \in U_R(0) = U_{R,\mathbb{K}}(0)$. Then f is arbitrarily often differentiable on $U_R(0)$ and it holds*

$$f^{(k)}(x) = \sum_{\nu=0}^{\infty} (\nu + 1)(\nu + 2) \ldots (\nu + k) c_{\nu+k} \, x^\nu \qquad (x \in U_R(0), \; k \in \mathbb{N}_0) .$$

In particular

$$f^{(k)}(0) = k! c_k \qquad (k \in \mathbb{N}_0) ,$$

thus

$$f(x) = \sum_{\nu=0}^{\infty} \frac{f^{(\nu)}(0)}{\nu!} x^\nu \qquad (x \in U_R(0)) .$$

Proof.
1. We show: f is differentiable on $U_R(0)$ and

$$f'(x) = \sum_{\nu=1}^{\infty} \nu c_\nu x^{\nu-1} = \sum_{\nu=0}^{\infty} (\nu + 1) c_{\nu+1} x^\nu \qquad (x \in U_R(0)) .$$

Let $a \in U_R(0)$. We choose an r with $|a| < r < R$. Then for $|h| \le \delta := r - |a|$ with

$$\phi_n(h) := c_n \sum_{k=0}^{n-1} (a + h)^k a^{n-1-k}$$

according to Theorem 1.3.1

$$f(a + h) - f(a) = \sum_{\nu=1}^{\infty} c_\nu ((a + h)^\nu - a^\nu) = h \sum_{\nu=1}^{\infty} \phi_\nu(h).$$

From $|(a + h)^k a^{n-1-k}| \le r^{n-1}$ for $k = 0, \ldots, n$ follows $|\phi_n(h)| \le n |c_n| r^{n-1}$. In Remark 2.6.12 we have seen that a $q < 1$ exists with $|c_n| r^n \le q^n$ for n sufficiently large. According to the root criterion, the series $\sum_{\nu=1}^{\infty} \nu |c_\nu| r^{\nu-1}$ converges and according to the Weierstrass criterion (Theorem 2.6.11) consequently the function series $\sum_{\nu=1}^{\infty} \phi_\nu$ converges uniformly on $B_\delta(0)$. With Theorem 2.6.6 it follows that the limit function $\phi := \sum_{\nu=1}^{\infty} \phi_\nu$ is continuous at 0. Therefore,

$$\lim_{h \to 0} \frac{f(a+h) - f(a)}{h} = \lim_{h \to 0} \phi(h) = \phi(0) = \sum_{\nu=1}^{\infty} \phi_\nu(0) = \sum_{\nu=1}^{\infty} \nu c_\nu a^{\nu-1}.$$

2. Inductively, using proof step 1, it is obtained that $f^{(k-1)}$ is differentiable for all $k \in \mathbb{N}$ on $U_R(0)$ with

$$f^{(k)}(x) = \sum_{\nu=0}^{\infty} (\nu+1)(\nu+2) \dots (\nu+k) c_{\nu+k} x^\nu.$$

The additional assertion $k! c_k = f^{(k)}(0)$ results for $x = 0$. $\square$

Example 3.2.5 For the function $f : \mathbb{C} \to \mathbb{C}$ with

$$f(z) = \begin{cases} \sin(z)/z, & z \neq 0 \\ 1, & z = 0 \end{cases}$$

one has

$$f(z) = \sum_{\mu=0}^{\infty} \frac{(-1)^\mu z^{2\mu}}{(2\mu+1)!} = \sum_{\nu=0}^{\infty} c_\nu z^\nu$$

with $c_k = (-1)^{k/2}/(k+1)!$ for even k and $c_k = 0$ for odd k. In particular, f is arbitrarily often differentiable on $\mathbb{C}$ with

$$f^{(k)}(0) = \begin{cases} \frac{(-1)^{k/2}}{k+1}, & \text{if } k \text{ is even} \\ 0, & \text{if } k \text{ is odd} \end{cases}.$$

Remark 3.2.6 As a simple but important consequence of Theorem 3.2.4, it is obtained: If $\sum_{\nu=0}^{\infty} c_\nu x^\nu$ is a power series with a positive radius of convergence and if $\sum_{\nu=0}^{\infty} c_\nu x^\nu = 0$ for all x from a neighborhood of 0, then $c_n = 0$ for all $n \in \mathbb{N}_0$. So: The coefficients c_n of the power series are already determined by the series values in an arbitrarily small neighborhood of 0. In fact, the statement can be significantly strenghtened (Exercise 3.2.31.7).

Example 3.2.7 Let $a := (1 + \sqrt{5})/2$ and $b := (1 - \sqrt{5})/2$. Then (geometric series)

$$\sum_{\nu=0}^{\infty} (a^{\nu+1} - b^{\nu+1}) x^\nu = \frac{a}{1-ax} - \frac{b}{1-bx} = \frac{a-b}{(1-ax)(1-bx)}$$

holds for $|x| < 1/a = (\sqrt{5}-1)/2$. With $a - b = \sqrt{5}$, $a + b = 1$ and $ab = -1$ follows

$$\frac{1}{\sqrt{5}} \sum_{\nu=0}^{\infty} (a^{\nu+1} - b^{\nu+1}) x^{\nu} = \frac{1}{1 - x - x^2} \ .$$

So if one sets $c_{-2} := c_{-1} := 0$ and

$$c_n := (a^{n+1} - b^{n+1})/\sqrt{5}$$

for $n \in \mathbb{N}$, then

$$1 = (1 - x - x^2) \sum_{\nu=0}^{\infty} c_{\nu} x^{\nu} = \sum_{\nu=0}^{\infty} (c_{\nu} - c_{\nu-1} - c_{\nu-2}) x^{\nu}$$

and thus by Remark 3.2.6

$$c_0 = 1 \quad \text{and} \quad c_n = c_{n-1} + c_{n-2} \quad (n \in \mathbb{N}).$$

The sequence (c_n) is called **Fibonacci sequence** and the number $1/a = (\sqrt{5} - 1)/2$ (alternatively also $a = (1 + \sqrt{5})/2 = 1 + 1/a$) **Golden Ratio**.[3]

Remark and Definition 3.2.8 Let $X \subset \mathbb{K}$ and $f : X \to \mathbb{C}$. A function $F : X \to \mathbb{C}$ is called an **antiderivative** of f (on X), if F is differentiable with $F' = f$ on X. If $X \subset \mathbb{K}$ is star-shaped and F and G are antiderivatives of f on X, then $(F - G)' = 0$ on X, and thus $F - G$ is constant on X according to Theorem 3.1.24, in other words, F and G differ only by an additive constant.

Remark 3.2.9 If $\sum_{\nu=0}^{\infty} c_{\nu} x^{\nu}$ is a power series with radius of convergence $R > 0$ and if

$$f(x) := \sum_{\nu=0}^{\infty} c_{\nu} x^{\nu} \qquad (x \in U_R(0)),$$

then according to Theorem 3.2.4

$$F(x) := \sum_{\nu=0}^{\infty} \frac{c_{\nu}}{\nu + 1} x^{\nu+1} = \sum_{\nu=1}^{\infty} \frac{c_{\nu-1}}{\nu} x^{\nu} \qquad (x \in U_R(0))$$

an antiderivative of f on $U_R(0)$ is defined by (it is easy to see that the radius of convergence is also R).

Example 3.2.10
1. If $f(z) := 1/(1 - z)$ for $z \in \mathbb{C} \setminus \{1\}$, then according to Remark 3.2.9 by

$$F(z) = \sum_{\nu=1}^{\infty} \frac{z^{\nu}}{\nu} \qquad (z \in \mathbb{D})$$

---

[3] A multitude of illustrations on the fascinating topic of the Golden Ratio can be found in [1].

an antiderivative to f on $\mathbb{D}$ is given. With the chain rule, it is immediately apparent that on $(-\infty, 1)$ also by

$$G(x) = \ln(1/(1-x)) \qquad (x < 1)$$

an antiderivative to f is given. According to Remark 3.2.8, F and G on $I = (-1, 1)$ are equal except for an additive constant. Since $F(0) = 0 = G(0)$, the additive constant $= 0$, and thus F and G agree on I. Therefore,

$$\ln\left(\frac{1}{1-x}\right) = \sum_{v=1}^{\infty} \frac{x^v}{v} \qquad (-1 < x < 1).$$

This is also referred to as the **logarithmic series.**

2. If

$$f(z) := e^{-z^2} = \sum_{k=0}^{\infty} \frac{(-1)^k z^{2k}}{k!} \qquad (z \in \mathbb{C}),$$

then by

$$F(z) = \sum_{k=0}^{\infty} \frac{(-1)^k}{k!(2k+1)} z^{2k+1} \qquad (z \in \mathbb{C})$$

an antiderivative to f on $\mathbb{C}$ is given. The function

$$\mathrm{erf} := \frac{2}{\sqrt{\pi}} F$$

is called the **error function** (error function). The error function is of central importance in statistics due to its direct relationship to the distribution functions of normal distributions.

At the boundary of their circle of convergence, functions defined by power series can exhibit very complex behavior. If the power series converges at a boundary point ζ of the circle of convergence, the so-called radial boundary value of the limit function at the point ζ exists. More precisely:

Theorem 3.2.11 (Abel's Limit Theorem) *Let* $\sum_{v=0}^{\infty} c_v x^v$ *a power series with radius of convergence* $0 < R < \infty$ *and*

$$f(x) := \sum_{v=0}^{\infty} c_v x^v \qquad (x \in U_R(0)).$$

If for a $\zeta \in K_R(0)$ *the series* $\sum_{v=0}^{\infty} c_v \zeta^v$ *is convergent, then*

$$\lim_{r \to 1^-} f(r\zeta) = \sum_{v=0}^{\infty} c_v \zeta^v .$$

Proof. Without restriction, we can assume $R = 1$ and $\zeta = 1$ (otherwise consider $g(x) := f(\zeta x)$).

We set $s_n := \sum_{\nu=0}^{n} c_\nu$ for $n \in \mathbb{N}_0$ and $s := \sum_{\nu=0}^{\infty} c_\nu = \lim s_n$. Since (s_n) is bounded, the power series $\sum_{\nu=0}^{\infty} s_\nu x^\nu$ has a convergence radius ≥ 1. Thus, with $s_{-1} := 0$ for $|x| < 1$

$$(1 - x) \sum_{\nu=0}^{\infty} s_\nu x^\nu = \sum_{\nu=0}^{\infty} s_\nu x^\nu - \sum_{\nu=0}^{\infty} s_\nu x^{\nu+1} = \sum_{\nu=0}^{\infty} (s_\nu - s_{\nu-1}) x^\nu = \sum_{\nu=0}^{\infty} c_\nu x^\nu = f(x) .$$

Let $\varepsilon > 0$ be given. Then there exists an $n \in \mathbb{N}$ with $|s_\nu - s| < \varepsilon$ for all $\nu > n$. For $0 < r < 1$, with $1 = (1 - r) \sum_{\nu=0}^{\infty} r^\nu$

$$|f(r) - s| = |(1 - r) \sum_{\nu=0}^{\infty} (s_\nu - s) r^\nu|$$

$$\leq (1 - r) \sum_{\nu=0}^{n} |s_\nu - s| + \varepsilon (1 - r) \sum_{\nu=n+1}^{\infty} r^\nu \leq (1 - r) \sum_{\nu=0}^{n} |s_\nu - s| + \varepsilon.$$

From $(1 - r) \sum_{\nu=0}^{n} |s_\nu - s| \to 0$ for $r \to 1^-$ follows the existence of a $\delta > 0$ with $|f(r) - s| < 2\varepsilon$ for $1 - \delta < r < 1$. □

Example 3.2.12 According to Example 3.2.10,

$$\ln \left(\frac{1}{1 - x} \right) = \sum_{\nu=1}^{\infty} \frac{x^\nu}{\nu} \qquad \left(x \in (-1, 1) \right).$$

Since the alternating series $\sum_{\nu=1}^{\infty} (-1)^\nu / \nu$ converges according to the Leibniz criterion, it follows with the Abel's limit theorem for $\zeta = -1$

$$\ln(1/2) = \lim_{r \to 1^-} \ln \left(\frac{1}{1 + r} \right) = \sum_{\nu=1}^{\infty} \frac{(-1)^\nu}{\nu}$$

and thus also

$$\ln(2) = -\ln(1/2) = \sum_{\nu=1}^{\infty} \frac{(-1)^{\nu-1}}{\nu}.$$

Remark and Definition 3.2.13 Let $X \subset \mathbb{K}$ be open and $f : X \to \mathbb{C}$. Then f is called **analytic at the point** $x \in X$, if an $R > 0$ and a sequence (c_k) in $\mathbb{C}$ exist such that

$$f(x + h) = \sum_{\nu=0}^{\infty} c_\nu h^\nu \qquad (|h| < R)$$

holds. In this case, f is, according to Theorem 3.2.4, in particular arbitrarily often differentiable on $U_R(x)$ and

$$c_k = c_k(f, x) := f^{(k)}(x)/k! \qquad (k \in \mathbb{N}_0).$$

holds. Furthermore, f is called **analytic in** X, if f is analytic at every point $x \in X$. We set

$$C^\omega(X) := \{f \in C(X) : f \text{ analytic in } X\}.$$

Example 3.2.14

1. We consider for fixed $a \in \mathbb{C}$ the function $f : \mathbb{C} \setminus \{a\} \to \mathbb{C}$ with

$$f(z) = \frac{1}{a - z} .$$

Here, for all $z \neq a$ and all h with $|h| < |a - z|$

$$f(z + h) = \frac{1}{a - z} \cdot \frac{1}{1 - \frac{h}{a-z}} = \sum_{\nu=0}^{\infty} \frac{1}{(a - z)^{\nu+1}} h^\nu .$$

Thus, f is analytic in $\mathbb{C} \setminus \{a\}$.

2. The function exp is analytic in $\mathbb{C}$ (Exercise 3.2.31.9).

Remark and Definition 3.2.15 According to Remark 3.2.13, every function in X that is analytic can be differentiated arbitrarily often on X. This implies that for open sets $X \subset \mathbb{R}$

$$C^\omega(X) \subset C^\infty(X).$$

The function $f : \mathbb{R} \to \mathbb{R}$, defined by

$$f(x) := \begin{cases} e^{-1/x}, & \text{if } x > 0 \\ 0, & \text{if } x \leq 0 \end{cases},$$

can be differentiated arbitrarily often on $\mathbb{R}$ with $f^{(k)}(0) = 0$ for all $k \in \mathbb{N}_0$ (Exercise 3.2.31.10). However, f is not analytic at the point 0, otherwise an $R > 0$ would exist with

$$f(h) = \sum_{\nu=0}^{\infty} \frac{f^{(\nu)}(0)}{\nu!} h^\nu = 0 \qquad (|h| < R),$$

which is not the case. Therefore, $f \in C^\infty(\mathbb{R}) \setminus C^\omega(\mathbb{R})$.

Remark and Definition 3.2.16 Let $X \subset \mathbb{K}$ be open and $f \in C^\omega(X)$. We set with $\min \emptyset := \infty$

$$n_f(x) = \min\{k \in \mathbb{N}_0 : f^{(k)}(x) \neq 0\} \in \mathbb{N}_0 \cup \{\infty\} \quad (x \in X)$$

and call $n_f(x)$ the **order** at f at x. Here, $n_f(x) > 0$ exactly when x is a root of f. In this case, $n_f(x)$ is also called the **multiplicity** of the root.

Theorem 3.2.17 *Let $X \subset \mathbb{K}$ be open and $f : X \to \mathbb{C}$ analytic at the point $x \in X$. Then the following holds:*

1. *$n_f(x) < \infty$ if and only if there exist an open neighborhood U of 0 and a function analytic at 0 named $g : U \to \mathbb{C}$ such that $g(0) \neq 0$ and*

$$f(x + h) = h^n g(h) \qquad (h \in U).$$

2. *If $0 < n_f(x) < \infty$, then x is an isolated point of $Z(f)$.*
3. *If $n_f(x) = \infty$, then f is locally constant (with value 0) at x, thus x is an interior point of $Z(f)$.*

Proof.
1. Let $n := n_f(x)$. Initially, $n < \infty$ holds true if and only if an open neighborhood U of 0 exists with

$$f(x + h) = \sum_{\nu=n}^{\infty} c_\nu h^\nu = h^n \sum_{\mu=0}^{\infty} c_{\mu+n} h^\mu \quad (h \in U)$$

and $c_n \neq 0$ (note: $k! c_k = f^{(k)}(x)$ for all k).
So if on the one hand $n < \infty$, we set

$$g(h) := \begin{cases} h^{-n} f(x + h) & \text{for } h \in U \setminus \{0\} \\ c_n & \text{for } h = 0 \end{cases}.$$

Then g is as required. If on the other hand g is as required with $g(h) = \sum_{\mu=0}^{\infty} b_\mu h^\mu$ for h from a neighborhood $U_0 \subset U$ of 0, then

$$f(x + h) = h^n g(h) = \sum_{\nu=n}^{\infty} b_{\nu-n} h^\nu$$

for $h \in U_0$ with $f^{(n)}(x)/n! = b_0 = g(0) \neq 0$.
2. Results from 1. and the continuity of g at 0.
3. Results from the analyticity and $k! c_k = f^{(k)}(x)$ for all k.   □

Remark 3.2.18 From Theorem 3.2.17 it follows in particular: If $X \subset \mathbb{K}$ is open and if $f \in C^\omega(X)$ is not locally constant at any point, then $Z(f)$ is discrete in X.

Example 3.2.19 If one defines $f : \mathbb{R} \to \mathbb{R}$ by

$$f(x) := \begin{cases} e^{-1/|x|} \sin(\pi/x), & \text{if } x \neq 0 \\ 0, & \text{if } x = 0 \end{cases},$$

then $f \in C^\infty(\mathbb{R})$ with Exercise 3.2.31.1 and 3.2.31.10. The function f is not locally constant at any point, but 0 is an accumulation point of $Z(f)$. Thus, it can be seen that a statement like in Remark 3.2.18 does not generally apply to functions in $C^\infty(\mathbb{R})$!

Remark and Definition 3.2.20 Let (X, d) be a metric space.

1. (X, d) is called **connected,** if the following holds: If $U, V \subset X$ are open with $X = U \cup V$ and $U \cap V = \emptyset$, then $U = \emptyset$ or $V = \emptyset$. Otherwise, X is called **disconnected.**
2. $M \subset X$ is called **connected,** if (M, d_M) is connected (or $M = \emptyset$).

From the definition, it immediately follows that X is connected if and only if X and $\emptyset$ are the only sets in (X, d) that are both open and closed. Furthermore, it is easy to see (Exercise 3.2.31.11) that a subset of M is open in (M, d_M) if and only if it is of the form $U \cap M$ for an open set U in (X, d). Thus, from the above definition, it follows that $M \subset X$ is disconnected if and only if open sets $U, V \subset X$ exist with $M \subset U \cup V$, $U \cap M \neq \emptyset$, $V \cap M \neq \emptyset$ and $U \cap V \cap M = \emptyset$.

Remark 3.2.21 If X, Y are metric spaces, if X is connected and if $f\colon X \to Y$ is locally constant, then f is constant.

> Because: Let c be an element of $f(X)$. Then $A := \{x\colon f(x) = c\}$ is non-empty, closed (since f is continuous) and open (since f is locally constant). Therefore, $A = X$.

Theorem 3.2.22 *A non-empty subset M of $\mathbb{R}$ is connected if and only if it is an interval.*

Proof. $\Rightarrow$: Let $M \subset \mathbb{R}$ not an interval. Then there exist points a, b, c with $a < c < b$ and $a, b \in M$, $c \notin M$. Consequently, for $U := (-\infty, c)$ and $V := (c, \infty)$

$$M \subset U \cup V, \quad U \cap M \neq \emptyset, \quad V \cap M \neq \emptyset \quad \text{and} \quad U \cap V \cap M = \emptyset,$$

thus M is disconnected.

$\Leftarrow$: Assume, M is disconnected. Then there exist open sets U, V in $\mathbb{R}$ with $M \subset U \cup V$, $U \cap M \neq \emptyset$, $V \cap M \neq \emptyset$ and $U \cap V \cap M = \emptyset$.

Let $a \in U \cap M$, $b \in V \cap M$. Then $a \neq b$ (and then without restriction $a < b$). Since M is an interval, $[a, b] \subset M$. We set $\xi := \sup(U \cap [a, b])$. Since U is open, $\xi > a$. Assume, $\xi \in V$. Since V is open, then there exists a $a < c < \xi$ with $(c, \xi] \subset V$. According to the definition of the supremum, $(c, \xi] \cap U \neq \emptyset$ and thus also $U \cap V \cap M \neq \emptyset$. Contradiction. Thus $\xi \notin V$. Since $M \subset U \cup V$, it follows $\xi \in U$. Since U is open and $\xi < b$, this contradicts the definition of ξ.   □

The following theorem shows that the connectedness of a set is transferred to the image set under continuous mappings.

Theorem 3.2.23 *Let (X, d), (Y, d_Y) be metric spaces and $f : X \to Y$ continuous. If X is connected, then $f(X)$ is also connected.*

Proof. Let $B \subset f(X)$ be open and closed. Then there exist an open set U and a closed set A in (Y, d_Y) with $B = U \cap f(X) = A \cap f(X)$. From the continuity of f, it follows that $f^{-1}(B) = f^{-1}(U) = f^{-1}(A)$ is open and closed in (X, d). Since X is connected, $f^{-1}(B) = \emptyset$ or $f^{-1}(B) = X$. In the first case, $B = \emptyset$ and in the second, $f(X) = f(f^{-1}(B)) \subset B$, thus $B = f(X)$. Therefore, $f(X)$ is connected. □

As a consequence of Theorem 3.2.22 and 3.2.23, we obtain a generalization of the intermediate value theorem:

Theorem 3.2.24 *Let (X, d) be a connected metric space. If $f : X \to \mathbb{R}$ is continuous, then $f(X)$ is an interval.*

Proof. According to Theorem 3.2.23, $f(X) \subset \mathbb{R}$ is connected, thus an interval according to Theorem 3.2.22. □

Remark 3.2.25 Let (X, d) be a metric space. If A_α ($\alpha \in I$) are connected sets in X with $\bigcap_{\alpha \in I} A_\alpha \neq \emptyset$, then $\bigcup_{\alpha \in I} A_\alpha$ is also connected.

Because: We set $A := \bigcup_{\alpha \in I} A_\alpha$. Let U and V be open sets in X with $A \subset U \cup V$ and $A \cap U \neq \emptyset$ as well as $A \cap V \neq \emptyset$. If $x \in \bigcap_{\alpha \in I} A_\alpha$, then $x \in U \cup V$. Without loss of generality, let $x \in U$. Furthermore, there exists an $\alpha \in I$ with $A_\alpha \cap V \neq \emptyset$. From $x \in A_\alpha \cap U$ it also follows that $A_\alpha \cap U \neq \emptyset$. Since A_α is connected, it follows that $A_\alpha \cap U \cap V \neq \emptyset$. Therefore, $A \cap U \cap V \neq \emptyset$ as well.

Remark and Definition 3.2.26   A set $G \subset \mathbb{K}$ is called a **region**, if G is open, non-empty, and connected. According to Theorem 3.2.22, $G \subset \mathbb{R}$ is a region if and only if G is an open interval. Since lines $[u, v] \subset \mathbb{C}$ are connected according to Theorems 3.2.22 and 3.2.23, star-shaped open sets $X \subset \mathbb{C}$ are connected according to Remark 3.2.25, thus they are regions (if not empty).

When we subsequently speak of a subset M of $\mathbb{K}$ as a metric space, unless otherwise stated, always (M, d_M) with $d_M = d_{|\cdot|}|_{M \times M}$ should be meant.

Theorem 3.2.27 *Let $G \subset \mathbb{K}$ be a domain and $f : G \to \mathbb{C}$ analytic. Then the following holds: Either $f = 0$ or $Z(f)$ is discrete in G.*

Proof. Let A be the set of accumulation points of $Z(f)$ in G. Then A is closed in G. Since f is continuous, $A \subset Z(f)$. If $A \neq \emptyset$ and $a \in A$, then a is an inner point of A according to Theorem 3.2.17. Consequently, A is also open in G. Since G is connected, $A = G$ already holds. Thus, $Z(f) = G$, so $f = 0$. □

Example 3.2.28 The function $f \in C^\infty(\mathbb{R})$ from Remark 3.2.15 shows that the statement of Theorem 3.2.27 is in general false for non-analytic functions.

If p is a polynomial of degree $\leq d$, then p is already completely determined by the function values on a $(d + 1)$-element set M in the sense that for every polynomial q of degree $\leq d$ with $p|_M = q|_M$ already $p = q$ follows. As a consequence of Theorem 3.2.27, we immediately obtain the following important result, which shows that analytic functions on domains are already completely determined by their values on a set with an accumulation point in G. The result clearly shows how much the function values of analytic functions depend on each other.

Theorem 3.2.29 (Identity Theorem) *Let $G \subset \mathbb{K}$ be a domain and let $f, g : G \to \mathbb{C}$ be analytic. If there exists a set $M \subset G$ with an accumulation point in G and $f|_M = g|_M$, then $f = g$. In particular, f is constant if f is constant at a point $a \in G$.*

Proof. With f and g, it is clear that $f - g$ is analytic in G. From $M \subset Z(f - g)$ the first assertion follows immediately with Theorem 3.2.27. The second follows from the first by choosing g as a constant function with value $f(a)$.   □

Example 3.2.30 If f is analytic in $\mathbb{K}$ with $f(1/n) = 0$ for all $n \in \mathbb{N}$, then f is the zero function.

Exercise 3.2.31

1. **(Leibniz's Formula)** Let $X \subset \mathbb{K}$ with $X \subset X'$ and f, g n-times differentiable on X. Show: $f \cdot g$ is n-times differentiable on X with

$$(f \cdot g)^{(n)} = \sum_{\nu=0}^{n} \binom{n}{\nu} f^{(\nu)} g^{(n-\nu)}.$$

2. For $n \in \mathbb{N}$ let $f_n : \mathbb{R} \to \mathbb{R}$ be defined by

$$f_n(x) := \begin{cases} x^n, & \text{if } x > 0 \\ 0, & \text{if } x \leq 0 \end{cases}.$$

   Prove that $f_n \in C^{n-1}(\mathbb{R}) \setminus C^n(\mathbb{R})$ holds.
3. Show using Theorem 3.2.4 (once again):
   a) $\exp' = \exp$,
   b) $\sin' = \cos$.
4. Consider that the function $f : \mathbb{C} \to \mathbb{C}$ with

$$f(z) := \begin{cases} (e^z - 1)/z, & z \neq 0 \\ 1, & z = 0 \end{cases}$$

   is analytic at the point 0 and determine the derivatives $f^{(k)}(0)$ for all $k \in \mathbb{N}$.

5. Determine an antiderivative for $f : \mathbb{C} \to \mathbb{C}$ with

$$f(z) := \begin{cases} \sin(z)/z, & z \neq 0 \\ 1, & z = 0 \end{cases}$$

in the form of a power series.

6. Show: If (c_n) is the sequence of Fibonacci numbers, then the sequence of quotients c_n/c_{n+1} converges to the golden ratio, in other words

$$\lim_{n \to \infty} c_n/c_{n+1} = (\sqrt{5} - 1)/2$$

7. Show: If $\sum_{\nu=0}^{\infty} c_\nu x^\nu$ is a power series with radius of convergence $R > 0$ and there exists a set $M \subset U_R(0)$ with $0 \in M'$ and $\sum_{\nu=0}^{\infty} c_\nu x^\nu = 0$ for all $x \in M$, then $c_n = 0$ for all $n \in \mathbb{N}_0$.

8. Show:
   a) For $x \in (-1, 1)$,

$$\arctan x = \sum_{\mu=0}^{\infty} \frac{(-1)^\mu x^{2\mu+1}}{2\mu + 1}$$

   b) It holds that

$$\sum_{\mu=0}^{\infty} \frac{(-1)^\mu}{2\mu + 1} = \frac{\pi}{4}.$$

9. Prove that exp, cos and sin are analytic in $\mathbb{C}$.

10. Let $c \in \mathbb{C}$ with $\mathrm{Re}(c) > 0$ and $f : \mathbb{R} \to \mathbb{C}$ defined by

$$f(x) := \begin{cases} e^{-c/x}, & \text{if } x > 0 \\ 0, & \text{if } x \leq 0 \end{cases}.$$

Prove:
   a) For every $n \in \mathbb{N}_0$ there exists a polynomial p_n with $f^{(n)}(x) = p_n(1/x)e^{-c/x}$ for $x > 0$.
   b) f is arbitrarily often differentiable on $\mathbb{R}$ with $f^{(n)}(0) = 0$ for all $n \in \mathbb{N}_0$.

11. Let (X, d) be a metric space, $M \subset X$ non-empty and $A \subset M$. Prove: A is open in (M, d_M) if and only if there exists an open set U in (X, d) with $A = M \cap U$.

3.3 Integral Calculus

Integral calculus originally arose from the question of defining and calculating areas. Similar to differential calculus, we will introduce integrals through a certain limit process. For this, we first consider particularly simple functions, for which we can define the "oriented area under the graph" in a very natural way via the areas of rectangles.

If $I \subset \mathbb{R}$ is an interval, we write $|I| := \mathrm{diam}(I)$ and call $|I|$ the **length** of I.

Remark and Definition 3.3.1 Let $[\alpha, \beta] \subset \mathbb{R}$ be a compact interval.

1. A finite decomposition $(I_j)_{j \in J}$ of $[\alpha, \beta]$, consisting of *intervals* I_j for $j \in J$, we call an **interval decomposition** or briefly **decomposition** of $[\alpha, \beta]$. If $(I_j)_{j \in J}$ is a decomposition of $[\alpha, \beta]$ and $I \subset [\alpha, \beta]$ is another interval, then (with $|\emptyset| := 0$)

$$|I| = \sum_{j \in J} |I \cap I_j|.$$

2. A function $\varphi \in B[\alpha, \beta] := B([\alpha, \beta], \mathbb{C})$ is called a **step function** (on $[\alpha, \beta]$), if a decomposition $(I)_{I \in E}$ of $[\alpha, \beta]$ and constants $c(I) = c_\varphi(I) \in \mathbb{C}$ exist with

$$\varphi = \sum_{I \in E} c(I) \cdot \mathbb{1}_I = \sum_{I \in E} c(I) \cdot \mathbb{1}_{I, [\alpha, \beta]},$$

so that φ is constant with value $c(I)$ on I. A decomposition for which corresponding constants $c(I)$ exist, we call **admissible** for the function φ. We write $T[\alpha, \beta]$ for the set of step functions on $[\alpha, \beta]$.

Example 3.3.2 We consider $[\alpha, \beta] = [0, 1]$ and the step function $\varphi = \mathbb{1}_{(1/2, 1]}$. Then $(I)_{I \in E}$ with

$$E := \{[0, 1/2], (1/2, 1]\}$$

is a admissible decomposition for φ, where here $c([0, 1/2]) = 0$ and $c((1/2, 1]) = 1$. Another one is $(J)_{J \in F}$ with

$$F = \{[0, 1/2], (1/2, 3/4], (3/4, 1]\},$$

where then $c([0, 1/2]) = 0$ and $c((1/2, 3/4]) = c((3/4, 1]) = 1$. By the way, $\{[0, 1/2), \{1/2\}, (1/2, 1]\}$ is also admissible, as single-point intervals are not excluded.

Remark and Definition 3.3.3 Let $(I)_{I \in E}$ and $(J)_{J \in F}$ be decompositions of $[\alpha, \beta]$. If

$$G := \{(I, J) \in E \times F : I \cap J \neq \emptyset\},$$

then $(I \cap J)_{(I,J) \in G}$ is called the **common refinement** of $(I)_{I \in E}$ and $(J)_{J \in F}$. From Theorem 1.1.16 it follows that the common refinement is also a decomposition of $[\alpha, \beta]$. If $(I)_{I \in E}$ and $(J)_{J \in F}$ are permissible for φ, then the common refinement is also admissible for φ, and $\varphi|_{I \cap J} = c(I) = c(J)$ for $(I, J) \in G$. With Remark 3.3.1 it follows

$$\sum_{I \in E} c(I)|I| = \sum_{(I,J) \in G} c(I)|I \cap J| = \sum_{(I,J) \in G} c(J)|I \cap J| = \sum_{J \in F} c(J)|J|.$$

Remark and Definition 3.3.4

1. If $\varphi : [\alpha, \beta] \to \mathbb{C}$ is a step function as in Remark 3.3.1, then

$$\int_\alpha^\beta \varphi := \int_\alpha^\beta \varphi(t)\, dt := \sum_{I \in E} c(I) \cdot |I| \,,$$

is called the **integral** of φ (on $[\alpha, \beta]$). It is important to note: The sum on the right side is independent of the choice of the partition according to Remark 3.3.3!

2. Let $X \subset \mathbb{K}$ and $U \subset \mathbb{C}^X$ be a subspace. If $\ell : U \to \mathbb{C}$ is a linear mapping, we say ℓ is **nonnegative,** if $\ell(f) \geq 0$ for all f with $f \geq 0$. Due to linearity, in this case ℓ is also **monotonic** in the sense that $\ell(f) \leq \ell(g)$ for all real-valued f, g with $f \leq g$.

Example 3.3.5 In the situation of Example 3.3.2,

$$\int_0^1 \varphi = \sum_{I \in E} c(I) \cdot |I| = \frac{1}{2} = \sum_{J \in F} c(J) \cdot |J|$$

Theorem 3.3.6 *The mapping* $\int_\alpha^\beta : T[\alpha, \beta] \to \mathbb{C}$ *is linear and nonnegative. Moreover, for every step function* φ

1. $|\varphi|$ *is a step function with*

$$\left| \int_\alpha^\beta \varphi \right| \leq \int_\alpha^\beta |\varphi| \leq (\beta - \alpha) \max_{[\alpha,\beta]} |\varphi| \,.$$

2. *For* $\tau \in [\alpha, \beta]$ *is*

$$\int_\alpha^\beta \varphi = \int_\alpha^\tau \varphi + \int_\tau^\beta \varphi \,.$$

Proof. Let φ, ψ be step functions and $\lambda \in \mathbb{C}$. If $(I)_{I \in E}$ and $(J)_{J \in F}$ are admissible decompositions for φ and ψ respectively, then the common refinement $(I \cap J)_{(I,J) \in G}$ is admissible for both φ and ψ. If $c_\varphi(I) \in \mathbb{C}$ and $c_\psi(J) \in \mathbb{C}$ are as in Remark 3.3.1 for φ and ψ respectively, then $\lambda\varphi + \psi$ is constant $= \lambda c_\varphi(I) + c_\psi(J)$ on $I \cap J$ for $(I, J) \in G$. Thus, $\lambda\varphi + \psi$ is a step function (and thus $T[\alpha, \beta]$ is a subspace of $B[\alpha, \beta]$) and it holds that

$$\int_\alpha^\beta (\lambda\varphi + \psi) = \sum_{(I,J) \in G} (\lambda c_\varphi(I) + c_\psi(J)) |I \cap J|$$

$$= \lambda \sum_{(I,J) \in G} c_\varphi(I)|I \cap J| + \sum_{(I,J) \in G} c_\psi(J)|I \cap J| = \lambda \int_\alpha^\beta \varphi + \int_\alpha^\beta \psi.$$

The non-negativity and 1. result directly from the definition and corresponding properties of sums. Statement 2. follows with

$$\varphi = \varphi \cdot \mathbb{1}_{[\alpha,\tau]} + \varphi \cdot \mathbb{1}_{(\tau,\beta]}$$

and $\int_\alpha^\tau \varphi = \int_\alpha^\beta \varphi \cdot \mathbb{1}_{[\alpha,\tau]}$ as well as $\int_\tau^\beta \varphi = \int_\alpha^\beta \varphi \cdot \mathbb{1}_{(\tau,\beta]}$ from linearity.   □

We will now consider more general functions that can be approximated in a suitable way by step functions. For these functions, we can then define the integral over the integrals of the corresponding step functions.

Remark and Definition 3.3.7 A function $f \in B[\alpha, \beta]$ is called a **regulated function** (on $[\alpha, \beta]$), if there exists a sequence (φ_n) of step functions with

$$||f - \varphi_n||_\infty \to 0$$

for $n \to \infty$, thus $\varphi_n \to f$ uniformly on $[\alpha, \beta]$. We write $R[\alpha, \beta]$ for the set of regulated functions. Thus, $R[\alpha, \beta]$ is the closure of $T[\alpha, \beta]$ in the Banach space $(B\,[\alpha, \beta], ||\cdot||_\infty)$, thus also a Banach space (see Exercise 2.4.33.6).

Theorem 3.3.8 *Let* $f : [0, 1] \to \mathbb{C}$ *be continuous. Then with* $I_0^{(n)} := \{0\}$ *and*

$$I_j^{(n)} := ((j - 1)/n, j/n] \quad (j = 1, \dots, n)$$

for $n \in \mathbb{N}$ *through*

$$\varphi_n(t) := f(j/n)   \quad (t \in I_j^{(n)}, j = 0, \dots, n)$$

a sequence (φ_n) *of step functions is given with* $||\varphi_n - f||_\infty \to 0$. *In particular, f is a regulated function.*

Proof. Let $\varepsilon > 0$ be given. Since f is uniformly continuous on $[0, 1]$, there exists a $\delta > 0$ such that $|f(t) - f(s)| < \varepsilon$ for all t, s with $|t - s| < \delta$. If $n > 1/\delta$ and $t \in [0, 1]$, then there exists a $j = j_{n,t}$ such that $t \in I_j^{(n)}$ and thus (since $0 \le j/n - t < 1/n < \delta$)

$$|f(t) - \varphi_n(t)| = |f(t) - f(j/n)| < \varepsilon.$$

Since t was arbitrary, it follows that $||f - \varphi_n||_\infty \le \varepsilon$ for $n > 1/\delta$.   □

Remark 3.3.9 From Theorem 3.3.8 it easily follows that $C[\alpha, \beta] \subset R[\alpha, \beta]$ for any compact interval $[\alpha, \beta]$. It can be shown (Exercise 3.3.24.4) that also monotone functions $f : [\alpha, \beta] \to \mathbb{R}$ are regular functions.

Remark and Definition 3.3.10 Let f be a regular function and (φ_n) be a sequence of step functions with $\varphi_n \to f$ uniformly on $[\alpha, \beta]$. Then the following holds:

1. The sequence $(\int_\alpha^\beta \varphi_n)_n$ converges in $\mathbb{C}$, because for $n, n' \in \mathbb{N}$ according to Theorem 3.3.6

$$\left| \int_\alpha^\beta \varphi_n - \int_\alpha^\beta \varphi_{n'} \right| = \left| \int_\alpha^\beta (\varphi_n - \varphi_{n'}) \right| \le ||\varphi_n - \varphi_{n'}||_\infty (\beta - \alpha),$$

and since (φ_n) is a Cauchy sequence in $B[\alpha, \beta]$, $(\int_\alpha^\beta \varphi_n)$ is a Cauchy sequence in $\mathbb{C}$, thus convergent.

2. If (ψ_n) is another sequence of step functions with $\psi_n \to f$ uniformly on $[\alpha, \beta]$, then

$$\left| \int_\alpha^\beta \varphi_n - \int_\alpha^\beta \psi_n \right| \leq ||\varphi_n - \psi_n||_\infty (\beta - \alpha) \leq \left(||\varphi_n - f||_\infty + ||f - \psi_n||_\infty \right)(\beta - \alpha) \to 0 \,,$$

thus $\lim\limits_{n\to\infty} \int_\alpha^\beta \varphi_n = \lim\limits_{n\to\infty} \int_\alpha^\beta \psi_n$.

With this, we set

$$\int_\alpha^\beta f := \int_\alpha^\beta f(t)\, dt := \lim_{n\to\infty} \int_\alpha^\beta \varphi_n$$

and call $\int_\alpha^\beta f$ the **Riemann integral** or briefly **integral** of f on $[\alpha, \beta]$. According to 2., the value is independent of the specific choice of the sequence of step functions.

Example 3.3.11 We consider $f(t) = t$ on $[0, 1]$. Then, according to Theorem 3.3.8, a sequence of step functions on $[0, 1]$ given by

$$\varphi_n(t) := \begin{cases} j/n \,, & t \in ((j-1)/n, j/n], j = 1, \ldots, n \\ 0 \,, & t = 0 \end{cases}$$

satisfies $\varphi_n \to f$ uniformly on $[0, 1]$. It holds that

$$\int_0^1 \varphi_n = \sum_{j=1}^n \frac{j}{n} \cdot |I_j^{(n)}| = \frac{1}{n} \sum_{j=1}^n \frac{j}{n} = \frac{n(n+1)}{2n^2} = \frac{1}{2} + \frac{1}{2n} \to \frac{1}{2} \quad (n \to \infty) \,,$$

so

$$\int_0^1 f = \int_0^1 t\, dt = 1/2.$$

Fig. 3.3 shows the integral $\int_0^1 \varphi_{10}$ as an approximation for $\int_0^1 f$.

We compile some properties for integrals, which result from the approximation by step functions.

Theorem 3.3.12 *The mapping $\int_\alpha^\beta : R[\alpha, \beta] \to \mathbb{C}$ is linear and non-negative. Moreover, for every regulated function f*

1. $|f|$ *is a regulated function and*

$$\left| \int_\alpha^\beta f \right| \leq \int_\alpha^\beta |f| \leq (\beta - \alpha) \sup_{[\alpha,\beta]} |f|.$$

Fig. 3.3 Step function
φ_{10} and $\int_0^1 \varphi_{10}$ as an
approximation of $\int_0^1 t\,dt$.

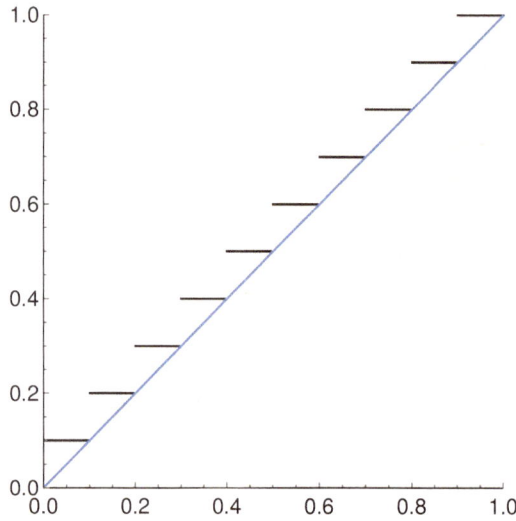

2. *For $\tau \in [\alpha, \beta]\, f|_{[\alpha, \tau]} \in R[\alpha, \tau]$ and $f|_{[\tau, \beta]} \in R[\tau, \beta]$ and it holds*

$$\int_\alpha^\beta f = \int_\alpha^\tau f + \int_\tau^\beta f \ .$$

Proof. According to Remark 3.3.7, $R[\alpha, \beta]$ is a linear space. If $f, g \in R[\alpha, \beta]$ and $\lambda \in \mathbb{C}$, then there exist sequences of step functions (φ_n) and (ψ_n) with $\varphi_n \to f$ and $\psi_n \to g$ uniformly on $[\alpha, \beta]$ and thus

$$||\lambda f + g - (\lambda \varphi_n + \psi_n)||_\infty \leq |\lambda| \cdot ||f - \varphi_n||_\infty + ||g - \psi_n||_\infty \to 0$$

for $n \to \infty$. Therefore, with Theorem 3.3.6

$$\int_\alpha^\beta \lambda f + g = \lim_{n\to\infty} \int_\alpha^\beta (\lambda \varphi_n + \psi_n) = \lambda \int_\alpha^\beta f + \int_\alpha^\beta g \ .$$

Now let $f \geq 0$. We set

$$\varphi_n^+ := \varphi_n + ||f - \varphi_n||_\infty \ .$$

Then φ_n^+ are step functions with $\varphi_n^+ \geq f \geq 0$ and $||f - \varphi_n^+||_\infty \to 0$. Therefore, with Theorem 3.3.6

$$\int_\alpha^\beta f = \lim_{n\to\infty} \int_\alpha^\beta \varphi_n^+ \geq 0 \ .$$

1. From $||f| - |\varphi_n|| \leq |f - \varphi_n|$ it follows that $|f|$ is also a regulated function and that $|\varphi_n| \to |f|$ uniformly on $[\alpha, \beta]$. This results with $|f| \leq \sup_{[\alpha,\beta]} |f|$ and Theorem 3.3.6.1 in

$$\left| \int_\alpha^\beta f \right| = \lim_{n\to\infty} \left| \int_\alpha^\beta \varphi_n \right| \le \lim_{n\to\infty} \int_\alpha^\beta |\varphi_n| = \int_\alpha^\beta |f| \le (\beta - \alpha) \sup_{[\alpha,\beta]} |f| \ .$$

2. The statement results as above from Theorem 3.3.6.2 by limit transition $\varphi_n \to f$. □

We now come to central theorems of one-dimensional analysis, which establish the relationship between differential and integral calculus.

Remark and Definition 3.3.13 We set for $f \in R[\alpha, \beta]$

$$\int_\beta^\alpha f := - \int_\alpha^\beta f$$

and for general intervals I

$$R(I) := \{f : I \to \mathbb{C} : f|_{[\alpha,\beta]} \in R[\alpha, \beta] \text{ for all } [\alpha, \beta] \subset I\}.$$

If $f \in R(I)$, then

$$\int_u^w f = \int_u^v f + \int_v^w f$$

applies for any $u, v, w \in I$. If $u \in I$ is fixed, we call the function $Vf = V_u f : I \to \mathbb{C}$, defined by

$$(Vf)(x) := \int_u^x f \qquad (x \in I),$$

the **integral function** of f (with respect to u). If $w \in I$, the functions $V_u f$ and $V_w f$ differ only by an additive constant (more precisely, $V_u f = V_w f + \int_u^w f$).

Theorem 3.3.14 (Main theorem on integral functions) *Let I be an interval, $f \in R(I)$ and $u \in I$. Then the integral function $V f = V_u f$ is continuous on I.*[4] *Moreover, if f is continuous at the point $x \in I$, then $V f$ is differentiable at x with*

$$(Vf)'(x) = f(x) \ .$$

Proof. Let $x \in I$ be arbitrary. Then there exists a $\delta > 0$ such that $J := I \cap [x - \delta, x + \delta]$ is a compact interval. Thus, f is bounded on J. For $h \in J - x$, the following holds

$$|(Vf)(x + h) - (Vf)(x)| = \left| \int_u^{x+h} f - \int_u^x f \right|$$

$$= \left| \int_x^{x+h} f \right| \le \sup_J |f| \cdot |h| \to 0 \quad (h \to 0).$$

---

[4] The linear mapping $V : R(I) \to C(I)$ is called the Volterra operator on $R(I)$; hence the V.

Now let f be continuous at the point x. Then from $\int_x^{x+h} f(x)\,dt = f(x)h$ follows

$$\left| \frac{(Vf)(x+h) - (Vf)(x)}{h} - f(x) \right| = \left| \frac{1}{h} \int_x^{x+h} (f - f(x)) \right| \leq \sup_{[x,x+h]} |f - f(x)| \to 0$$

for $h \to 0$.   □

Remark and Definition 3.3.15 According to Theorem 3.3.14, the integral functions $V_u f$ in the case of a *continuous* function f are also antiderivatives of f on I. For non-continuous f, integral functions are not always antiderivatives: If $f = \mathbb{1}_{[0,\infty)}$, then $V_0 f = \mathrm{id}_{\mathbb{R}} f$ is not differentiable at the jump point 0 of f.

The following theorem contains *the* central result for the calculation of integrals.

Theorem 3.3.16 (Fundamental Theorem of Calculus) *Let I be an interval and $f : I \to \mathbb{C}$ continuous. If F is an antiderivative of f on I, then*

$$\int_u^v f = F(v) - F(u) =: F(t)\big|_u^v =: F\big|_u^v$$

for all $u, v \in I$.

Proof. Since f is continuous, according to Remark 3.3.15, $V_u f$ is also an antiderivative of f on I. According to Remark 3.2.8, the difference $F - V_u f$ is constant on I. This results in

$$\int_u^v f = (V_u f)(v) = (V_u f)(v) - (V_u f)(u) = F(v) - F(u)$$

for all $u, v \in I$.   □

Example 3.3.17
1. Let $f(t) = 1/t$ for $t > 0$. Then $t \mapsto \ln(t)$ is an antiderivative of f on $(0, \infty)$. According to the fundamental theorem of calculus, for $0 < u, v < \infty$

$$\int_u^v f = \int_u^v \frac{1}{t}\,dt = \ln t\big|_u^v = \ln(v) - \ln(u) \ .$$

2. Let $\alpha \in \mathbb{C} \setminus \{-1\}$ and $f(t) = t^\alpha$ for $t > 0$. Then $t \mapsto \frac{1}{\alpha+1} t^{\alpha+1}$ is an antiderivative of f on $(0, \infty)$ and consequently for $0 < u, v < \infty$

$$\int_u^v t^\alpha\,dt = \frac{1}{\alpha+1} t^{\alpha+1}\big|_u^v = \frac{1}{\alpha+1}\left(v^{\alpha+1} - u^{\alpha+1}\right) \ .$$

In the case $\mathrm{Re}(\alpha) \geq 0$, this also holds for $u = 0$.

Remark 3.3.18 Let I be an interval.

1. **(Substitution rule)** If $\gamma : I \to \mathbb{R}$ is continuously differentiable and $f : \gamma(I) \to \mathbb{C}$ is continuous, then for $u, v \in I$

$$\int_u^v f(\gamma(t))\gamma'(t)\, dt = \int_u^v (f \circ \gamma)\gamma' = \int_{\gamma(u)}^{\gamma(v)} f \, .$$

Because: According to the main theorem on integral functions, there exists an antiderivative F for f on (the interval) $\gamma(I)$. Moreover, $F \circ \gamma$ is an antiderivative for the continuous function $(f \circ \gamma)\gamma'$ on I according to the chain rule. Thus, both integrals have the value $F(\gamma(v)) - F(\gamma(u))$ according to the main theorem of calculus.

2. **(partial integration)** If $f, g : I \to \mathbb{C}$ and F and G are antiderivatives for f and g on I, then the product rule implies that FG is an antiderivative for $fG + Fg$ on I. If f, g are continuous, then the fundamental theorem of calculus yields for $u, v \in I$

$$\int_u^v fG = FG\big|_u^v - \int_u^v Fg \, .$$

Thus, under certain circumstances, the calculation of the integral $\int_u^v fG$ can be reduced to that of $\int_u^v fG$.

Example 3.3.19

1. With $\gamma(t) := t^2$ on $(-1, 1)$ and $f(s) := 1/\sqrt{1-s}$ on $[0, 1)$ applies according to the substitution rule for $u, v \in (-1, 1)$

$$\int_u^v \frac{2t}{\sqrt{1-t^2}}\, dt = \int_u^v \frac{\gamma'(t)}{\sqrt{1-\gamma(t)}}\, dt = \int_{u^2}^{v^2} \frac{1}{\sqrt{1-s}}\, ds = -2\sqrt{1-s}\big|_{u^2}^{v^2} \, .$$

2. For $\alpha \neq -1$ and $u, v > 0$ applies with partial integration

$$\int_u^v t^\alpha \ln t \, dt = \frac{t^{\alpha+1}}{\alpha+1} \ln t\big|_u^v - \frac{1}{\alpha+1} \int_u^v t^\alpha \, dt = \frac{t^{\alpha+1}}{\alpha+1}\Big(\ln t - \frac{1}{\alpha+1}\Big)\big|_u^v \, .$$

Remark 3.3.20 If $a, b \in \mathbb{R}$ and $f \in C[a, b]$, then with the substitution rule and $\gamma := s_a^b$

$$\int_a^b f = \int_0^1 f(s_a^b(t))(b-a)\, dt = (b-a) \int_0^1 f(a + t(b-a))\, dt.$$

We set for any $a, b \in \mathbb{K}$ and $f \in C[a, b]$

$$\int_a^b f := (b-a) \int_0^1 f(a + t(b-a))\, dt.$$

If F is a antiderivative to f on $[a, b]$, then $F \circ s_a^b$ is an antiderivative to $(b-a)(f \circ s_a^b)$ on $[0, 1]$ and thus according to the fundamental theorem calculus

$$\int_a^b f = F(b) - F(a).$$

Theorem 3.3.21 (Taylor) *Let $X \subset \mathbb{K}$ be star-shaped with respect to a and $n \in \mathbb{N}_0$. Then for all $(n+1)$-times continuously differentiable functions f on X*

$$f(a + h) = \sum_{\nu=0}^{n} \frac{f^{(\nu)}(a)}{\nu!} h^{\nu} + \frac{h^{n+1}}{n!} \int_0^1 (1 - t)^n f^{(n+1)}(a + th)\, dt \qquad (h \in X - a).$$

Proof. We prove the assertion by induction on n. For $n = 0$ and f continuously differentiable on X, we obtain with Remark 3.3.20

$$f(a + h) - f(a) = \int_a^{a+h} f' = h \int_0^1 f'(a + th)\, dt\ .$$

If the assertion holds for $n - 1$ and f is $(n + 1)$-times continuously differentiable, then it follows with partial integration

$$\begin{aligned}
f(a + h) - \sum_{\nu=0}^{n-1} \frac{f^{(\nu)}(a)}{\nu!} h^{\nu} &= \frac{h^n}{(n-1)!} \int_0^1 (1 - t)^{n-1} f^{(n)}(a + th)\, dt \\
&= \frac{h^n}{(n-1)!} \left(-\frac{(1-t)^n}{n} f^{(n)}(a + th)\big|_0^1 + h \int_0^1 \frac{(1-t)^n}{n} f^{(n+1)}(a + th)\, dt \right) \\
&= \frac{h^n}{n!} f^{(n)}(a) + \frac{h^{n+1}}{n!} \int_0^1 (1 - t)^n f^{(n+1)}(a + th)\, dt\ .
\end{aligned}$$

$\square$

Remark and Definition 3.3.22 In the situation of Theorem 3.3.21,

$$c_k(f, a) := f^{(k)}(a)/k!$$

for $k = 0, \ldots, n$ is called the k-th **Taylor coefficient** of f with respect to a. The polynomial $T_n(f, a)$, defined by

$$T_n(f, a)(h) := \sum_{\nu=0}^{n} c_\nu(f, a) h^{\nu} \qquad (h \in \mathbb{C}),$$

is called the n-th **Taylor polynomial** of f with respect to a and

$$R_n(f, a)(h) := f(a + h) - T_n(f, a)(h) \qquad (h \in X - a)$$

the n-th **remainder term**. If f is differentiable arbitrarily often, the function series $\sum_{\nu=0}^{\infty} c_\nu(f, a) h^{\nu}$ is called the **Taylor series** of f with respect to a. If $p : \mathbb{C} \to \mathbb{C}$ is a polynomial of degree d, then $p^{(d+1)} = 0$, thus $R_d(p, a)(h) = 0$ for any $a, h \in \mathbb{C}$, and thus

$$p(a + h) = \sum_{\nu=0}^{d} c_\nu(p, a) h^{\nu} \qquad (a, h \in \mathbb{C}).$$

From Theorem 3.2.4 it follows: If $\sum_{\nu=0}^{\infty} c_\nu(x - a)^{\nu}$ is a power series with a positive radius of convergence and f is the limit function on the circle of convergence, then $c_k = c_k(f, a)$ for all $k \in \mathbb{N}_0$ and thus the power series $\sum_{\nu=0}^{\infty} c_\nu h^{\nu}$ is the Taylor series of f with respect to a.

Remark 3.3.23 (Lagrange form of the remainder) It can be shown (Exercise 3.3.24.12) that the following variant of a mean value theorem for integrals holds: If $[\alpha, \beta] \subset \mathbb{R}$ and $\varphi : [\alpha, \beta] \to \mathbb{R}$ are continuous and $\psi \in R\,[\alpha, \beta]$ with $\psi \geq 0$ on $[\alpha, \beta]$, then there exists a $\tau \in [\alpha, \beta]$ with

$$\int_\alpha^\beta \varphi\psi = \varphi(\tau) \int_\alpha^\beta \psi \ .$$

If this is applied to $\varphi(t) := f^{(n+1)}(a + th)$ and $\psi(t) := (1 - t)^n$ with $[\alpha, \beta] = [0, 1]$ and it is noted that

$$\int_0^1 \psi = \int_0^1 (1 - t)^n \, dt = \frac{1}{n + 1}$$

then under the conditions of Taylor's theorem for real-valued $f^{(n+1)}$ and $h \in X - a$ there exists a $\tau \in [0, 1]$ (dependent on f, a, h and n) with

$$R_n(f, a)(h) = \frac{f^{(n+1)}(a + \tau h)}{(n + 1)!} h^{n+1} \ .$$

Exercises 3.3.24

1. Let $f : [0, 1] \to \mathbb{R}$ be defined by

$$f(t) := \begin{cases} \cos(\pi/t), & t > 0 \\ 0, & t = 0 \end{cases} \ .$$

   Investigate whether the functions f or $\mathrm{id}_{[0,1]}f$ are regulated functions on $[0, 1]$.

2. Show that with f, $g \in R\,[\alpha, \beta]$ also $f \cdot g \in R[\alpha, \beta]$ and that $R[\alpha, \beta]$ is a unitary $\mathbb{C}$-algebra.

3. Show: If $f \in R[\alpha, \beta]$, then one-sided limits exist at all points, i.e., $f(a^+)$ for $a \in [\alpha, \beta)$ and $f(b^-)$ for $b \in (\alpha, \beta]$.
   Hint: Use Theorem 2.3.4.

4. a) Prove: If $f : [\alpha, \beta] \to \mathbb{R}$ is monotonic, then f is a regulated function.
   b) Find a monotonic function $f \in R[0, 1]$ with infinitely many discontinuity points.

5. Prove: If $f \in R[\alpha, \beta]$, then f has at most countably many discontinuity points.

6. (Thomae function) Let $f : [0, 1] \to [0, 1]$ be defined by

$$f(t) := \begin{cases} 1/q, & \text{if } 0 < t = p/q \in \mathbb{Q} \text{ with } p, q \text{ coprime} \\ 0, & \text{if } t \in (0, 1] \setminus \mathbb{Q} \\ 1, & \text{if } t = 0 \end{cases} \ .$$

   Show:
   a) For all $\varepsilon > 0$, the set $\{t \in [0, 1] : f(t) \geq \varepsilon\}$ is finite.
   b) f has the limit 0 at all points $a \in [0, 1]$.
   c) f is continuous at a if and only if a is irrational.
   d) f is a regulated function and $\int_0^1 f = 0$ holds.

7. Let $\alpha, \beta \in \mathbb{R}$ with $\alpha < \beta$. Show:

   a) If $f \in C[\alpha, \beta]$ with $f \geq 0$ and $f(t) > 0$ for a $t \in [\alpha, \beta]$, then

   $$\int_\alpha^\beta f > 0.$$

   b) By $\|f\| := \int_\alpha^\beta |f|$ for $f \in C[\alpha, \beta]$ a norm on $C[\alpha, \beta]$ is defined.

8. Calculate:

   a) $\int_u^v \frac{t^2}{1+t^3}\, dt$ for $u, v > -1$,

   b) $\int_u^v t e^t\, dt$ for $u, v \in \mathbb{R}$,

   c) $\int_u^v t^{-1} \ln t\, dt$ for $u, v > 0$,

   d) $\int_0^1 \arctan(t)\, dt$.

9. (Fundamental theorem of calculus for regulated functions) Let $[\alpha, \beta] \subset \mathbb{R}$ and $f \in R[\alpha, \beta]$. Show: If F is an antiderivative of f on I, then

   $$\int_\alpha^\beta f = F(\beta) - F(\alpha).$$

   Hint: Assume that F is real-valued, and then apply the mean value theorem to $F|_{\bar{I}}$ for I from suitable decompositions of $[\alpha, \beta]$.

10. (**Euler's summation formula**) Let f be continuously differentiable on $[1, \infty)$. Show: With $b(t) := t - \lfloor t \rfloor - 1/2$, for $n \in \mathbb{N}$

    $$\sum_{\nu=1}^n f(\nu) = \int_1^n f + \frac{1}{2}\left(f(1) + f(n)\right) + \int_1^n bf'.$$

11. (**Binomial series**) Let $\alpha \in \mathbb{C}$ and $f : (-1, \infty) \to \mathbb{R}$ be defined by

    $$f(x) = (1+x)^\alpha$$

    for $x > -1$. Show:

    a) For the Taylor coefficients $c_k\,(f, 0) = f^{(k)}(0)/k!$ one has $c_k(f, 0) = \binom{\alpha}{k}$.

    b) For $|h| < 1$ one has $n\binom{\alpha}{n} h^n \to 0\ (n \to \infty)$.

    c) For $-1 < h < 1$

    $$(1+h)^\alpha = \sum_{\nu=0}^\infty \binom{\alpha}{\nu} h^\nu.$$

    Note: For $-1 < h < 1$ and $0 \leq t \leq 1$ is $1 - t \leq 1 + th$.

12. (Mean value theorem for integrals) Let $\varphi, \psi \in R[\alpha, \beta]$ with $\psi \geq 0$ on $[\alpha, \beta]$ and φ real-valued. Show:

    a) It holds

    $$\inf_{[\alpha,\beta]} \varphi \cdot \int_\alpha^\beta \psi \leq \int_\alpha^\beta \varphi\psi \leq \sup_{[\alpha,\beta]} \varphi \cdot \int_\alpha^\beta \psi.$$

b) If φ is continuous, then there exists a $\tau \in [\alpha, \beta]$ with

$$\int_\alpha^\beta \varphi\psi = \varphi(\tau) \int_\alpha^\beta \psi .$$

3.4   Improper Integrals

So far, we have only defined integrals on compact intervals. We now want to consider non-compact intervals as well.

Remark and Definition 3.4.1 Let I be an interval and $a := \inf I$, $b := \sup I$. A function $f \in R\,(I)$ is called **integrable** on I, if $(V_u f)(a^+)$ and $(V_u f)(b^-)$ exist for a $u \in I$. In this case, the two limits exist for every $u \in I$, and the difference $(V_u f)(b^-) - (V_u f)(a^+)$ is independent of u according to Remark 3.3.13. It is then also said that the improper integral $\int_{a^+}^{b^-} f$ exists (or converges), and the number

$$\int_{a^+}^{b^-} f := \int_{a^+}^{b^-} f(t)\, dt := (V_u f)(b^-) - (V_u f)(a^+)$$

is called **improper integral** of f on I. If $b = \infty$, one usually writes ∞ instead of ∞^- and correspondingly for $a = -\infty$. From Theorem 2.1.9, Remark 2.1.10 and Theorem 3.3.12 it easily follows that by $f \mapsto \int_{a^+}^{b^-} f$ a linear and non-negative mapping is defined on the set of functions integrable on I.

Remark 3.4.2
1. If $b \in I$, then $(V_u f)(b) = (V_u f)(b^-)$ according to the main theorem on integral functions. Accordingly, $(V_u f)(a) = (V_u f)(a^+)$ in the case $a \in I$. So in the case $I = [a, b]$

$$\int_{a^+}^{b^-} f = (V_u f)(b^-) - (V_u f)(a^+) = \int_a^b f,$$

   that is, "proper" and improper integral coincide. Therefore, one sometimes also briefly speaks of the **integral** of f and also writes—in the case $b \in I$ mostly—briefly b instead of b^-. The same applies to a.
2. **(Extended fundamental theorem of calculus)** Let $f : I \to \mathbb{C}$ be continuous. If F is any antiderivative of f on I, then $F - V_u f$ is constant on I. Therefore, f is integrable exactly when $F(b^-)$ and $F(a^+)$ exist. Moreover, then

$$\int_{a^+}^{b^-} f = F(b^-) - F(a^+) =: F(t)\big|_a^b.$$

Example 3.4.3
1. For $\alpha \in \mathbb{R}$ let $f_\alpha(t) := t^{-\alpha}$ on $I = (0, \infty)$. Then by

$$F_\alpha(t) := \begin{cases} \frac{t^{1-\alpha}}{1-\alpha}, & \alpha \neq 1 \\ \ln(t), & \alpha = 1 \end{cases}$$

an antiderivative F_α to f_α on I is defined. Furthermore, for $t \to \infty$ we get

$$F_\alpha(t) \to \begin{cases} 0, & \alpha > 1 \\ \infty, & \alpha \le 1 \end{cases}$$

and for $t \to 0^+$

$$F_\alpha(t) \to \begin{cases} 0, & \alpha < 1 \\ -\infty, & \alpha \ge 1 \end{cases}.$$

Therefore, f_α is integrable on $[1, \infty)$ according to Remark 3.4.2.2 if and only if $\alpha > 1$, and in this case

$$\int_1^\infty t^{-\alpha}\, dt = F_\alpha(t)\big|_1^\infty = 0 - \frac{1}{1-\alpha} = \frac{1}{\alpha - 1}\,.$$

Accordingly, f_α is integrable on $(0, 1]$ if and only if $\alpha < 1$, and then

$$\int_{0^+}^1 t^{-\alpha}\, dt = F_\alpha(t)\big|_0^1 = \frac{1}{1-\alpha}\,.$$

This also implies that f_α is not integrable on $(0, \infty)$ for any $\alpha \in \mathbb{R}$.

2. We consider $f(t) = 1/\sqrt{1 - t^2}$ on $I = (-1, 1)$. Then $\arcsin' = f$ on $(-1, 1)$. Since $\arcsin$ is continuous on $[-1, 1]$, f is integrable according to Remark 3.4.2.2 and

$$\int_{-1^+}^{1^-} \frac{dt}{\sqrt{1-t^2}} = \arcsin(t)\big|_{-1}^1 = \pi\,.$$

Theorem 3.4.4 (Improper partial integration) *Let $I \subset \mathbb{R}$ be an interval, $a := \inf I$, $b := \sup I$ and $f, g : I \to \mathbb{C}$ continuous. If F, G are corresponding antiderivatives on I and if $(FG)(b^-)$ and $(FG)(a^+)$ exist, then: fG is integrable on I if and only if Fg is integrable on I, and in this case*

$$\int_{a^+}^{b^-} fG = FG\big|_a^b - \int_{a^+}^{b^-} Fg.$$

Proof. Since F, G are continuous on I, $fG + Fg$ is continuous. Also, FG is an antiderivative of $fG + Fg$. According to Remark 3.4.2.2, $fG + Fg$ is integrable with

$$\int_{a^+}^{b^-} (fG + Fg) = FG\big|_a^b.$$

If, for example, fG is integrable, then $Fg = (Fg + fG) - fG$ is also integrable and it holds

$$\int_{a^+}^{b^-} fG = FG\big|_a^b - \int_{a^+}^{b^-} Fg.$$

The same applies if Fg is integrable. $\square$

Example 3.4.5 (Area of the unit disk) We consider the continuous functions $f(t) := 1$ and $G(t) := \sqrt{1 - t^2}$ on $[-1, 1]$. Then the (proper) integral $\int_{-1}^{1} fG$ exists. With $F(t) = t$ on $[-1, 1]$, according to Theorem 3.4.4 (since $t^2 = 1 - (1 - t^2)$)

$$\int_{-1}^{1} \sqrt{1 - t^2}\, dt = t\sqrt{1 - t^2}\Big|_{-1}^{1} + \int_{-1^+}^{1^-} \frac{t^2}{\sqrt{1 - t^2}}\, dt$$

$$= \int_{-1^+}^{1^-} \frac{dt}{\sqrt{1 - t^2}} - \int_{-1}^{1} \sqrt{1 - t^2}\, dt$$

and thus

$$2\int_{-1}^{1} \sqrt{1 - t^2}\, dt = \int_{-1^+}^{1^-} \frac{dt}{\sqrt{1 - t^2}} = \pi.$$

Theorem 3.4.6 (Majorant criterion for integrals) *Let I be an interval and $a := \inf I$, $b := \sup I$. If $f \in R(I)$ and g is integrable on I with $|f| \leq g$, then f is also integrable on I with*

$$\left| \int_{a^+}^{b^-} f \right| \leq \int_{a^+}^{b^-} g.$$

Proof. Let $F := V_u f$ and $G := V_u g$ for a $u \in I$. If $s, t \in I$ with $s < t$, then

$$|F(t) - F(s)| = \left| \int_s^t f \right| \leq \int_s^t |f| \leq \int_s^t g = G(t) - G(s) \qquad (3.4.1)$$

and thus $|F(t) - F(s)| \leq |G(t) - G(s)|$ for any $s, t \in I$.

We show: $F(b^-)$ exists. For this, let (t_n) be a sequence in I with $b > t_n \to b$. Then $|F(t_n) - F(t_{n'})| \leq |G(t_n) - G(t_{n'})|$ for $n, n' \in \mathbb{N}$. Since $(G(t_n))$ is a Cauchy sequence according to Theorem 2.3.4, $(F(t_n))$ is also a Cauchy sequence and thus $F(b^-)$ exists, again according to Theorem 2.3.4. Similarly, it can be seen that $F(a^+)$ exists. Finally, from (3.4.1) it also follows that $|F(b^-) - F(a^+)| \leq G(b^-) - G(a^+)$. □

Remark and Definition 3.4.7 In particular, with $g := |f|$ we obtain from Theorem 3.4.6: If $f \in R(I)$ (and thus also $|f| \in R(I)$), the integrability of $|f|$ implies that of f. If $|f|$ is integrable, we also call f **absolutely integrable**. As with series, if f is absolutely integrable, then f is integrable. Moreover, then

$$\left| \int_{a^+}^{b^-} f \right| \leq \int_{a^+}^{b^-} |f|.$$

Example 3.4.8 For $\alpha > 1$, we consider the function $f : [1, \infty) \to \mathbb{R}$ with

$$f(t) := t^{-\alpha} \cos(t) \qquad (t \geq 1).$$

It holds $|\cos t| t^{-\alpha} \leq t^{-\alpha}$ for $t \geq 1$. Since $\int_1^{\infty} t^{-\alpha}\, dt$ exists according to Example 3.4.3.1, the absolute integrability of f follows from Theorem 3.4.6. The same applies to the function $t \mapsto t^{-\alpha} \sin(t)$ on $[1, \infty)$.

Example 3.4.9 Let $f : [1, \infty) \to \mathbb{C}$ be continuous. If F is an antiderivative to f and if F is bounded, then $t \mapsto t^{-1}f(t)$ is integrable on $[1, \infty)$.

Because: Since F is bounded (and continuous), the integral $\int_1^\infty F(t)t^{-2}\,dt$ exists according to the majorant criterion. With $G(t) = t^{-1}$ and $g(t) = -t^{-2}$, the assertion follows from Theorem 3.4.4 and $(FG)(t) = F(t)t^{-1} \to 0$ for $t \to \infty$. We consider $f(t) = \sin t$ for $t \geq 1$. Here, $F(t) = -\cos t$ is bounded on $[1, \infty)$. Therefore, $t \mapsto t^{-1}\sin(t)$ is integrable on $[1, \infty)$. It can be shown that the function $t \mapsto t^{-1}|\sin t|$ is not integrable on $[1, \infty)$ (Exercise 3.4.14.3). Thus: Absolute integrability is a strictly stronger property than integrability.

The following theorem establishes a connection between the convergence of series and the existence of improper integrals:

Theorem 3.4.10 (Integral criterion) *Let $f : [1, \infty) \to [0, \infty)$ be decreasing. Then there exists $c := \lim\limits_{n \to \infty} \left(\sum_{\nu=1}^n f(\nu) - \int_1^{n+1} f \right)$ and it holds $0 \leq c \leq f(1)$.*

Proof. We set $a_n := f(n)$ for $n \in \mathbb{N}$. From $a_n \geq f(t) \geq a_{n+1}$ for $t \in [n, n+1]$ follows $a_n \geq \int_n^{n+1} f \geq a_{n+1}$ and thus

$$0 \leq a_n - \int_n^{n+1} f \leq a_n - a_{n+1}.$$

Therefore, the sequence (s_n) with

$$s_n := \sum_{\nu=1}^n a_\nu - \int_1^{n+1} f = \sum_{\nu=1}^n \left(a_\nu - \int_\nu^{\nu+1} f \right)$$

is increasing with $0 \leq s_n \leq a_1 - a_{n+1} \leq a_1$. According to the monotone convergence theorem, the sequence (s_n) is convergent with $0 \leq \lim s_n \leq a_1$. □

Example 3.4.11 Let $\alpha > 0$ and $f(t) := t^{-\alpha}$ for $t \geq 1$. Then f is decreasing on $[1, \infty)$ and $f \geq 0$. Therefore, according to Theorem 3.4.10

$$c = \lim_{n \to \infty} \left(\sum_{\nu=1}^n \nu^{-\alpha} - \int_1^{n+1} t^{-\alpha}\,dt \right).$$

If $\alpha > 1$, then $\lim_{n\to\infty} \int_1^{n+1} t^{-\alpha}\,dt = \int_1^\infty t^{-\alpha}\,dt = (\alpha-1)^{-1}$ and $\zeta(\alpha) = \sum_{\nu=1}^\infty \nu^{-\alpha}$. According to Theorem 3.4.10,

$$0 \leq \zeta(\alpha) - \frac{1}{\alpha-1} \leq 1 \qquad (\alpha > 1).$$

If $\alpha = 1$, the convergence of

$$s_n := \sum_{\nu=1}^n \frac{1}{\nu} - \int_1^{n+1} \frac{dt}{t} = \sum_{\nu=1}^n \frac{1}{\nu} - \ln(n+1).$$

The limit $c = \lim s_n$ is called **Euler-Mascheroni Constant**. If $0 < \alpha < 1$, the convergence of

$$s_n := \sum_{v=1}^{n} v^{-\alpha} - \int_1^{n+1} t^{-\alpha}\, dt = \sum_{v=1}^{n} v^{-\alpha} - \frac{(n+1)^{1-\alpha} - 1}{1-\alpha}$$

is obtained.

Theorem 3.4.12 *The function $t \mapsto e^{-t}t^{z-1}$ is for all $z \in \mathbb{C}$ absolutely integrable on $[1, \infty)$ and for $\mathrm{Re}(z) > 0$ absolutely integrable on $(0, \infty)$.*

Proof. We set $f(t) := e^{-t}t^{z-1}$ for $t > 0$. From $t^{z+1}e^{-t} \to 0$ for $t \to \infty$ it follows that $t \mapsto |e^{-t}t^{z+1}|$ becomes maximal on $[1, \infty)$. Therefore, there exists a constant $M > 0$ such that $| f(t)| \le Mt^{-2}$ for all $t \in [1, \infty)$. With the existence of $\int_1^\infty t^{-2}\, dt$ the majorant criterion implies the absolute integrability of f on $[1, \infty)$. Furthermore, $| f(t)| \le t^{\mathrm{Re}(z)-1}$ for $t \in (0, 1]$. If $\mathrm{Re}(z) > 0$, then from the existence of $\int_0^1 t^{\mathrm{Re}(z)-1}\, dt$ (according to Example 3.4.3.1) again with the majorant criterion the absolute integrability of f on $(0, 1]$ and thus also on $(0, \infty)$ follows. □

Remark and Definition 3.4.13 Let $\Omega := \{z \in \mathbb{C} : \mathrm{Re}(z) > 0\}$ be the open right half-plane. The function $\Gamma : \Omega \to \mathbb{C}$ with

$$\Gamma(z) := \int_{0^+}^\infty e^{-t}t^{z-1}\, dt \qquad (\mathrm{Re}(z) > 0)$$

is called **(Euler's) Gamma function.** By improper partial integration, one directly obtains

$$\Gamma(z + 1) = z \cdot \Gamma(z) \qquad (\mathrm{Re}(z) > 0) . \tag{3.4.2}$$

Specifically, $\Gamma(1) = \int_0^\infty e^{-t}\, dt = -e^{-t}\big|_0^\infty = 1$ holds, from which, in turn, with (3.4.2) inductively

$$\Gamma(n + 1) = n!$$

for all $n \in \mathbb{N}$ results. The Gamma function thus "interpolates" the factorials; one can consider the values $\Gamma(z)$ as generalized factorials.

Exercises 3.4.14
1. Examine the following functions for integrability and, if applicable, calculate the improper integral
   a) $f(t) = 1/(1 + t^2)$ $(t \in \mathbb{R})$,
   b) $f(t) = \cos(t)/t^2$ $(t \in (0, 1])$,
   c) $f(t) = \ln(t)/t^2$ $(t \in [1, \infty))$.
2. Prove: For all $n \in \mathbb{N}$ is

$$\int_0^{\pi/2} \sin^{2n+1}(t)\, dt = \int_0^1 (1 - s^2)^n\, ds$$

and

$$\int_0^{\pi/2} \sin^{2n-2}(t)\, dt = \int_0^\infty \frac{ds}{(1 + s^2)^n} .$$

3. Show: The function $f : [1, \infty) \to \mathbb{R}$ with $f(t) := \sin(\pi t)/t$ for $t \geq 1$ is not absolutely integrable on $[1, \infty)$.

   Hint: For $k \in \mathbb{N}$ is $\int_k^{k+1} |\sin(\pi t)| \, dt = \int_0^1 \sin(\pi s) \, ds$.

4. Show that the sequence

$$\left(\sum_{v=1}^{n} \frac{1}{(v+1)\ln(v+1)} - \ln\left(\ln(n+2)\right) \right)_{n \in \mathbb{N}}$$

   converges.

5. Let b be as in Exercise 3.3.24.10. Show:

   a) For $n \in \mathbb{N}$ is

$$\sum_{v=1}^{n} \ln v = n \ln n - n + 1 + \frac{1}{2} \ln n + \int_1^n t^{-1} b(t) \, dt \, .$$

   b) For $B := V_1 b$

$$B(x) = \frac{1}{2}\left(x - \lfloor x \rfloor - \frac{1}{2} \right)^2 - \frac{1}{8} \qquad (x \geq 1)$$

   and

$$\int_1^n t^{-1} b(t) \, dt = \int_1^n t^{-2} B(t) \, dt \quad (n \in \mathbb{N}).$$

   c) The sequence (a_n) with $a_n := \frac{n!}{n^n e^{-n} \sqrt{n}}$ is convergent.[5]

3.5   Path Integrals

Definition 3.5.1

1. Let $[\alpha, \beta] \subset \mathbb{R}$ be a compact interval and $X \subset \mathbb{K}$. If $\gamma : [\alpha, \beta] \to X$ is continuously differentiable, we call γ a **path** (in X) and $[\alpha, \beta]$ the **parameter interval** of γ. Furthermore,

$$\gamma^* := \gamma([\alpha, \beta]) = W(\gamma)$$

   is called the **trace** of γ. Furthermore, $\gamma(\alpha)$ is called the **starting point** and $\gamma(\beta)$ the **end point** of γ and the path is **closed,** if the starting point and end point are the same. Finally, the path $\gamma_- : [\alpha, \beta] \to \mathbb{K}$, defined by

$$\gamma_-(t) := \gamma(\alpha + \beta - t) \qquad (t \in [\alpha, \beta]),$$

---

[5] One can show using the Wallis product (see for example [4, Theorem 2.2.1]) that the limit is $\sqrt{2\pi}$. This is the Stirling's formula.

is called the **reverse path** of γ.

2. If $\gamma : [\alpha, \beta] \to \mathbb{K}$ is a path, we define for $f \in C(\gamma^*)$ the **path integral** (or briefly **integral**) of f along γ by

$$\int_\gamma f := \int_\gamma f(\zeta)d\zeta := \int_\alpha^\beta (f \circ \gamma)\gamma'.$$

Furthermore, we call

$$L(\gamma) := \int_\alpha^\beta |\gamma'|$$

the **length** of γ.

Remark 3.5.2 Let γ be a path in $\mathbb{K}$. From the corresponding results for Riemann integrals from Theorem 3.3.12, the linearity of the path integral, i.e., the linearity of the mapping

$$C(\gamma^*) \ni f \mapsto \int_\gamma f \in \mathbb{C},$$

and the linearity of the mapping

$$C(\gamma^*) \ni f \mapsto \int_\alpha^\beta (f \circ \gamma)|\gamma'| \in \mathbb{C}$$

as well as the non-negativity and thus also the monotonicity of the second, but not the first, are derived. Since $|f|$ is continuous and γ^* is compact, $\max_{\gamma^*} |f|$ exists. This results from Theorem 3.3.12 and $|f \circ \gamma| \leq \max_{\gamma^*} |f|$

$$\left| \int_\gamma f \right| \leq \int_\alpha^\beta |f \circ \gamma| \cdot |\gamma'| \leq L(\gamma) \max_{\gamma^*} |f|.$$

Finally, from Theorem 3.3.6.2 we obtain

$$\int_{\gamma_-} f = - \int_\gamma f.$$

Remark and Definition 3.5.3

1. For $a \in \mathbb{C}$ and $\rho > 0$ we define $k_\rho(a) : [-\pi, \pi] \to \mathbb{C}$ by

$$k_\rho(a)(t) := a + \rho e^{it} \qquad (t \in [-\pi, \pi]).$$

Then $K_\rho(a) = (k_\rho(a))^*$ is the circle with radius ρ with center a and for continuous f on $K_\rho(a)$ is

$$\int_{k_\rho(a)} f = \int_{-\pi}^\pi f(a + \rho e^{it})i\rho e^{it}dt.$$

Thus, for all $\rho > 0$

$$\int_{k_\rho(a)} \frac{d\zeta}{\zeta - a} = \int_{-\pi}^{\pi} (\rho e^{it})^{-1} i\rho e^{it}\, dt = i\int_{-\pi}^{\pi} dt = 2\pi i.$$

Also,

$$L(k_\rho(a)) = \int_{-\pi}^{\pi} \rho\, dt = 2\pi\rho.$$

2. If $a, b \in \mathbb{C}$, then for continuous $f : [a, b] \to \mathbb{C}$

$$\int_{s_a^b} f = \int_a^b f = (b - a) \int_0^1 (f(a + (b - a)t)\, dt.$$

Theorem 3.5.4 *Let γ be a path in $\mathbb{K}$.*

1. *If $f_n \in C(\gamma^*)$ for $n \in \mathbb{N}$ and the sequence (f_n) converges uniformly on γ^* to f, then*

$$\int_\gamma f = \lim_{n \to \infty} \int_\gamma f_n.$$

2. *If $g_n \in C(\gamma^*)$ for $n \geq m$ and the series $\sum_{\nu=m}^\infty g_\nu$ converges uniformly on γ^*, then*

$$\int_\gamma \sum_{\nu=m}^\infty g_\nu = \sum_{\nu=m}^\infty \int_\gamma g_\nu.$$

Proof. From the uniform convergence, it follows first that $f \in C(\gamma^*)$. In addition, with Remark 3.5.2

$$\left| \int_\gamma f - \int_\gamma f_n \right| \leq L(\gamma) \|f - f_n\|_{\infty, \gamma^*} \to 0 \qquad (n \to \infty).$$

The 2nd statement follows from 1. by applying to the partial sum sequence. □

Theorem 3.5.5 (Fundamental theorem of calculus for paths) *Let $X \subset \mathbb{K}$ and $f \in C(X)$. If there exists an antiderivative F to f on X, then*

$$\int_\gamma f = F(b) - F(a) =: F(x)\big|_a^b$$

for all paths γ in X with starting point a and endpoint b, in particular

$$\int_\gamma f = 0$$

for all closed paths in X.

Proof. Let $I = [\alpha, \beta]$ be the parameter interval of γ. According to the chain rule, $F \circ \gamma$ is an antiderivative of $(f \circ \gamma)\gamma'$ on I. This results from the fundamental theorem of calculus

$$\int_\gamma f = \int_\alpha^\beta (f \circ \gamma)\gamma' = F(\gamma(\beta)) - F(\gamma(\alpha)) = F(b) - F(a).$$ □

Example 3.5.6 Let $z \in \mathbb{C}$ and γ be any path in $\mathbb{C} \setminus \{z\}$ with starting point a and endpoint b. Then for $m \in \mathbb{Z}, m \neq -1$

$$\int_\gamma (\zeta - z)^m d\zeta = \frac{1}{m+1}(\zeta - z)^{m+1}\Big|_a^b = \frac{1}{m+1}\left((b-z)^{m+1} - (a-z)^{m+1}\right).$$

In particular, for every closed path γ in $\mathbb{C} \setminus \{z\}$

$$\int_\gamma (\zeta - z)^m d\zeta = 0.$$

On the other hand, according to Remark 3.5.3 for $\rho > 0$

$$\int_{k_\rho(z)} \frac{d\zeta}{\zeta - z} = 2\pi i.$$

According to Theorem 3.5.5, the function $\zeta \mapsto 1/(\zeta - z)$ on $K_\rho(z)$ has no antiderivative!

Definition 3.5.7 Let $X \subset \mathbb{K}$ be open, $M \subset \mathbb{C}$ and $\varphi : X \times M \to \mathbb{C}$. If the functions $\phi_\zeta : X \to \mathbb{C}$ with

$$\phi_\zeta(x) := \varphi(x, \zeta) \quad (x \in X)$$

are differentiable for all $\zeta \in M$ on X, we define $D_1\varphi : X \times M \to \mathbb{C}$ by

$$(D_1\varphi)(x, \zeta) := (\phi_\zeta)'(x) \qquad ((x, \zeta) \in X \times M).$$

The following theorem about the smoothness of parameter integrals, i.e., integrals that depend on a parameter, plays a central role in the following.

Theorem 3.5.8 (Parameter integrals) *Let* $X \subset \mathbb{K}$ *and* γ *be a path in* $\mathbb{C}$. *Furthermore, let* $\varphi : X \times \gamma^* \to \mathbb{C}$ *be continuous. If* $\Phi : X \to \mathbb{C}$ *is defined by*

$$\Phi(x) := \int_\gamma \varphi(x, \zeta)\, d\zeta \qquad (x \in X),$$

then the following holds:

1. *If X is open or closed, then Φ is continuous.*
2. *If X is open and $D_1\varphi : X \times \gamma^* \to \mathbb{C}$ is continuous, then Φ is continuously differentiable with*

$$\Phi'(x) = \int_\gamma D_1\varphi(x, \zeta)\, d\zeta \qquad (x \in X).$$

Proof. Let $x \in X$ be fixed. If X is open, we choose $r > 0$ such that $B_r(x) \subset X$, and if X is closed, we choose $r > 0$ arbitrarily. Then $K := B_r(x) \cap X$ is compact.

1. Since $K \times \gamma^* \subset \mathbb{K} \times \mathbb{C}$ is compact, φ is uniformly continuous on $K \times \gamma^*$. If $\varepsilon > 0$ is given, there exists a $0 < \delta (\leq r)$ such that

$$|\varphi(x + h, \zeta) - \varphi(x, \zeta)| < \varepsilon \qquad (|h| < \delta,\ \zeta \in \gamma^*).$$

Then for $|h| < \delta$

$$|\Phi(x + h) - \Phi(x)| = \left| \int_\gamma \varphi(x + h, \zeta) - \varphi(x, \zeta)\, d\zeta \right| \leq \varepsilon L(\gamma).$$

Therefore, Φ is continuous at the point x.

2. We set for $\zeta \in \gamma^*$

$$\psi_\zeta(h) := \varphi(x + h, \zeta) - D_1\varphi(x, \zeta) \cdot h \qquad (|h| < r).$$

Since $D_1\varphi$ is continuous, ψ_ζ is continuously differentiable on $U_r(0)$, and thus according to Remark 3.3.20

$$\varphi(x + h, \zeta) - \varphi(x, \zeta) - D_1\varphi(x, \zeta) \cdot h = \psi_\zeta(h) - \psi_\zeta(0) = \int_0^h \psi_\zeta' \qquad (|h| < r).$$

Now let $\varepsilon > 0$ be given. Since $D_1\varphi$ is continuous on $K \times \gamma^*$ with $\psi_\zeta'(0) = 0$, as in 1. there exists a $\delta > 0$ such that $|\psi\zeta(u)| < \varepsilon$ for $|u| < \delta$ and $\zeta \in \gamma^*$. From this follows

$$|\psi_\zeta(h) - \psi_\zeta(0)| < \varepsilon |h| \qquad (|h| < \delta,\ \zeta \in \gamma^*)$$

and thus for $|h| < \delta$

$$\left| \int_\gamma \varphi(x + h, \zeta) - \varphi(x, \zeta) - D_1\varphi(x, \zeta) \cdot h\, d\zeta \right| \leq \varepsilon |h| L(\gamma).$$

Therefore, for $0 < |h| < \delta$

$$\left| \frac{\Phi(x + h) - \Phi(x)}{h} - \int_\gamma D_1\varphi(x, \zeta) d\zeta \right| \leq \varepsilon L(\gamma).$$

Since $\varepsilon > 0$ was arbitrary, Φ is differentiable at x with $\Phi'(x) = \int_\gamma D_1\varphi(x, \zeta)\, d\zeta$, and by 1. then also continuously differentiable. $\square$

According to the fundamental theorem of calculus, the calculation of integrals becomes simple if one knows suitable antiderivatives. While according to the fundamental theorem, continuous functions on intervals always have antiderivatives, the question of their existence on more general sets in $\mathbb{C}$ is no longer easy to answer. As an application of Theorem 3.5.8, one obtains the existence of antiderivatives for continuously differentiable functions f on star-shaped open sets.

Theorem 3.5.9 *Let $X \subset \mathbb{K}$ be open and star-shaped with respect to a and $f : X \to \mathbb{C}$ continuously differentiable. Then by*

$$F(x) := \int_a^x f \qquad (x \in X)$$

an antiderivative F to f on X is defined.

Proof. Without loss of generality, let $a = 0$. Then

$$F(x) = \int_0^x f = x \int_0^1 f(tx)\, dt = \int_0^1 \varphi(x, t)\, dt \qquad (x \in X)$$

with $\varphi(x, t) := xf(tx)$ for $(x, t) \in X \times [0, 1]$. It follows that

$$D_1 \varphi(x, t) = f(tx) + xf'(tx)t \qquad (x \in X,\ t \in [0, 1]).$$

Since f is continuously differentiable on X, the right side is continuous on $X \times [0, 1]$. According to Theorem 3.5.8, F is continuously differentiable on X with

$$F'(x) = \int_0^1 D_1 \varphi(x, t)\, dt = \int_0^1 f(tx)\, dt + \int_0^1 t \cdot xf'(tx)\, dt$$

$$= \int_0^1 f(tx)\, dt + \left(t \cdot f(tx)\right)\big|_0^1 - \int_0^1 f(tx)\, dt = f(x).$$

Therefore, F is an antiderivative to f on X.   □

Example 3.5.10
1. The set $\Omega := \mathbb{C} \setminus [1, \infty)$ is open and star-shaped with respect to $a = 0$. Since f with $f(z) := 1/(1 - z)$ for $z \in \Omega$ is holomorphic in Ω,

$$F(z) := \int_0^z \frac{d\zeta}{1 - \zeta} \qquad (z \in \Omega)$$

defines an antiderivative of f on Ω with $F(0) = 0$ and thus a holomorphic continuation of $x \mapsto \ln(1/(1 - x))$ on Ω (see Example 3.2.10). If $z \in \mathbb{S} \setminus \{1\}$, then the logarithmic series $\sum_{\nu=1}^{\infty} z^\nu/\nu$ converges according to the Dirichlet criterion (Exercise 2.3.32.2). Therefore, with Example 3.2.10 from the Abel's limit theorem follows

$$\sum_{\nu=1}^{\infty} \frac{z^\nu}{\nu} = \int_0^z \frac{d\zeta}{1 - \zeta} \qquad (z \in \overline{\mathbb{D}} \setminus \{1\}).$$

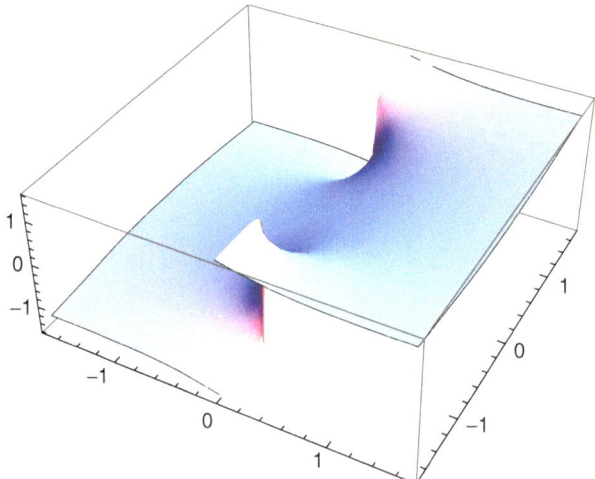

Fig. 3.4 Re(F)

2. The set $\Omega := \mathbb{C} \setminus \{it : t \in (-\infty, -1] \cup [1, \infty)\}$ is open and star-shaped with respect to $a = 0$. Since f with $f(z) = 1/(1 + z^2)$ for $z \in \Omega$ is holomorphic in Ω,

$$F(z) := \int_0^z \frac{d\zeta}{1 + \zeta^2} \qquad (z \in \Omega)$$

defines an antiderivative to f on Ω with $F(0) = 0$ and thus a holomorphic continuation of arctan on Ω. Fig. 3.4 shows the real part of the function F. The real arctangent is recognized as the intersection of the graph with the plane $\{z \in \mathbb{C} : \mathrm{Im}(z) = 0\}$.

As a further application of Theorem 3.5.8 we prove

Theorem 3.5.11 (Error integral) *It holds*

$$\mathrm{erf}(\infty^-) = \frac{2}{\sqrt{\pi}} \int_0^\infty e^{-t^2} \, dt = 1.$$

Proof. We consider the function $\Phi : [0, \infty) \to [0, \infty)$, defined by

$$\Phi(x) := \int_0^1 e^{-x^2(1+t^2)} \frac{dt}{1 + t^2} \qquad (x \geq 0).$$

From $e^{-x^2 t^2} \leq 1$ for $t \in [0, 1]$ follows $\Phi(x) \leq e^{-x^2} \to 0$ $(x \to \infty)$. Furthermore, according to Theorem 3.5.8 with $f(t) := e^{-t^2}$ for $t \in \mathbb{R}$ and $Vf = V_0 f$

$$\Phi'(x) = -2e^{-x^2} \int_0^1 x e^{-(xt)^2} \, dt = -2e^{-x^2} \int_0^x e^{-s^2} \, ds = -((Vf)^2)'(x) \quad (x \geq 0) .$$

So, $(Vf)^2 + \Phi$ is constant on $[0, \infty)$. From

$$((Vf)^2 + \Phi)(0) = \int_0^1 \frac{dt}{1+t^2} = \arctan(1) = \pi/4$$

follows $(Vf)^2(x) = \pi/4 - \Phi(x) \to \pi/4 \; (x \to \infty)$ and thus

$$\int_0^\infty e^{-t^2} \, dt = Vf\big|_0^\infty = \sqrt{\pi}/2.$$

$\square$

Exercises 3.5.12

1. Let $\rho : [-\pi, \pi] \to [0, \infty)$ be continuously differentiable with $\rho(-\pi) = \rho(\pi)$. Furthermore, let $\gamma : [-\pi, \pi] \to \mathbb{C}$ be defined by

$$\gamma(t) := \rho(t)\, e^{it} \qquad (t \in [-\pi, \pi]).$$

   Show:

$$\frac{1}{i} \int_\gamma \bar{\zeta}\, d\zeta = \int_{-\pi}^\pi \rho^2(t)\, dt.$$

2. Consider that the (continuous) function $z \mapsto \bar{z}$ has no antiderivative on $\mathbb{C}$.
3. Prove: There is no differentiable function $f : \mathbb{C}^* \to \mathbb{C}$ with $\exp(f(z)) = z$ for $z \neq 0$.
   Hint: Show that any such function f would be an antiderivative to $z \mapsto 1/z$ on $\mathbb{C}^*$.
4. For $a, b, c \in \mathbb{C}$ we call $\delta(a, b, c) := (s_a^b, s_b^c, s_c^a)$ the oriented triangle with the corners a, b, c and $\Delta(a, b, c) := [a, b] \cup [b, c] \cup [c, a]$ the triangle with corners a, b, c. Show:
   a) If $X \subset \mathbb{K}$ is open, $f : X \to \mathbb{C}$ is continuous and there exists an antiderivative F for f on X, then

$$\int_{\delta(a,b,c)} f := \int_a^b f + \int_b^c f + \int_c^a f = 0$$

   holds for all a, b, c with $\Delta(a, b, c) \subset X$.
   b) If $f \in C(\mathbb{D})$ is such that $\int_{\delta(a,b,c)} f = 0$ for all $a, b, c \in \mathbb{D}$ holds,[6] then by

$$F(z) := \int_0^z f \qquad (z \in \mathbb{D})$$

---

[6] The *Lemma of Goursat-Pringsheim* shows that this condition is fulfilled for differentiable f. Proofs can be found in [2, 3, 5–7, 9–11], all introductions to complex analysis which are recommended. The statement of the lemma is extremely remarkable, as it implies, as we will see later, that every function differentiable on an open set in $\mathbb{C}$ is already "automatically" continuously differentiable, i.e., holomorphic. In fact, holomorphy is usually defined as complex differentiability.

an antiderivative F for f on $\mathbb{D}$ is defined.

Hint: The proof of the fundamental theorem on integral functions can essentially be transferred.

5. (Neile Parabola) A path γ in $\mathbb{C}$ is called smooth if γ' has no zeros. The path $\gamma : [-1, 1] \to \mathbb{C}$ is defined by

$$\gamma(t) = t^2 + it^3 \qquad (t \in [-1, 1]).$$

Show that γ is not smooth, sketch γ^* and calculate the length $L(\gamma)$.

6. (Incomplete Gamma Functions) Let $\Omega := \{z \in \mathbb{C} : \text{Re}(z) > 0\}$ be the right half-plane in $\mathbb{C}$ and $\Gamma : \Omega \to \mathbb{C}$ the Gamma function. Show: If $f_n : \Omega \to \mathbb{C}$ is defined by

$$f_n(z) := \int_{1/n}^{n} e^{-t} t^{z-1} \, dt \qquad (z \in \Omega)$$

for $n \in \mathbb{N}$, then

a) For all $\alpha > 0$, the sequence (f_n) converges uniformly on the half-plane $\{z \in \mathbb{C} : \text{Re}(z) \geq \alpha\}$ towards Γ.

b) For all $n \in \mathbb{N}, f_n \in H(\Omega)$.

c) Γ is continuous on Ω.

7. Show using Theorem 3.5.11 : $\Gamma(1/2) = \sqrt{\pi}$.

3.6   Concepts II: Differentiation and Integration

Hardly any other topic is as closely associated with analysis as the differential and integral calculus, often also referred to as infinitesimal calculus, which essentially goes back to Leibniz and Newton.

While our approach to differentiation is based on the classic approach of difference quotients, i.e., quotients of the form

$$\frac{\tau_a f(h)}{h} = \frac{f(x) - f(a)}{x - a} \, ,$$

when generalizing to the higher-dimensional case, we face the problem of not being able to form corresponding quotients, as we cannot rely on a field structure. The approach of local approximation by an affine-linear function, already hinted at with the *decomposition formula*, carries us much further. This approach can be easily transferred to more general situations:

If $(V, |\cdot|)$ and E are Banach spaces over the common field $\mathbb{K}$, if X is an open subset of V and $f : X \to E$, then f is called Fréchet-differentiable at the point $a \in X$, if a linear mapping $A = A_{f,a} : V \to E$ bounded on the unit ball in V and a function $\varepsilon = \varepsilon_{f,a} : U \to E$ defined and decaying at 0 on a neighborhood U of 0 exist such that

$$(\tau_a f)(h) = f(a + h) - f(a) = A(h) + |h| \cdot \varepsilon(h) \quad (h \in U)$$

holds, in other words, if $|\cdot|^{-1}(\tau_a f - A)$ decays at 0.

It is easy to see that the linear mapping A is uniquely determined in the case of differentiability of f at a and one calls $f'(a) := A$ the Fréchet derivative or total derivative of f at a.[7] In the context of the already considered scalar case $V = \mathbb{K}$ and $E = \mathbb{C}$ (as a vector space over $\mathbb{K}$), it should be noted that the constant $c \in \mathbb{C}$ in the decomposition formula is identified with the linear mapping $A : \mathbb{K} \to \mathbb{C}$, defined by $A(h) := h \cdot c$.

In the case of $\mathbb{K} = \mathbb{R}$ and $V = E = \mathbb{C}$ we thus have two competing derivative concepts: on the one hand, the complex derivative, defined via the limit of the difference quotients, and on the other hand, the just defined Fréchet derivative. If $X \subset \mathbb{C}$ is open and if $f : X \to \mathbb{C}$ is complex differentiable at the point a, so $\tau_a f(h)/h \to c$ for $h \to 0$, then

$$\frac{1}{|h|}|\tau_a f(h) - hc| = |\frac{\tau_a f(h)}{h} - c| \to 0 \quad (h \to 0)$$

follows immediately and thus the real Fréchet differentiability at a. On the other hand, while Fréchet differentiability implies the existence of partial derivatives, it does not yet imply complex differentiability. For this, the Cauchy-Riemann equation must also be satisfied—a massive additional requirement. In fact, the worlds of holomorphic (i.e., complex-differentiable) and real-differentiable functions differ quite significantly. In the next chapter, we will see that holomorphic functions are always analytic!

The Cauchy-Riemann equation, which is usually formulated in the form

$$\bar{\partial} f = 0$$

with the differential operator

$$\bar{\partial} := \big(\partial_{(1,0)} + i\partial_{(0,1)}\big)/2$$

can be understood as a prototype of a (homogeneous) linear partial differential equation. Typical for solutions of corresponding equations are smoothness properties, under suitable conditions up to analyticity.

More than in the case of differentiation, the concept of integration faces a bewildering variety. In addition to the concept of the Riemann integral for regulated functions as introduced by us, one finds for example the classical Riemann integral, and this often in the variant of the Darboux integral, as well as the more advanced Lebesgue integral, which is usually developed within the framework of a general measure and integration theory. It is reassuring that certain consistencies exist among the mentioned variants: regulated functions on compact intervals are Riemann integrable and Riemann integrable functions are in turn Lebesgue integrable—and above all, the value of the integral is the same in each case. In addition, absolutely integrable functions on general (non-compact) intervals are also Lebesgue integrable with the same value.

---

[7] More on this can be found in the literature on multidimensional analysis, such as in [8].

Of fundamental importance for analysis is the intimate relationship between derivatives and integrals as "antiderivatives" that first becomes clearly apparent in the fundamental theorem of calculus. The up and down of differentiation and integration can be described as the pulsating heart of analysis.

It turns out that the existence of antiderivatives is crucial and by no means self-evident. While continuous functions on intervals always have antiderivatives, their existence on sets in the complex plane is a significant condition. This is made plausible by the fundamental theorem for path integrals. If antiderivatives exist, path integrals depend only on the initial and final points, i.e., they are independent of the path's course.

References

1. Beutelspacher, A., Petri, B.: Der goldene Schnitt, 2nd ed. Bibliographisches Institut, Mannheim (1995)
2. Bornemann, F.: Funktionentheorie, 2nd ed. Birkhäuser, Basel (2016)
3. Conway, J.B.: Functions of One Complex Variable, 2nd ed. Springer, New York (1978)
4. Endl, K., Luh, W.: Analysis II, 7th ed. Aula-Verlag, Wiesbaden (1989)
5. Endl, K., Luh, W.: Analysis III, 6th ed. Aula-Verlag, Wiesbaden (1987)
6. Gamelin, T.W.: Complex Analysis. Springer, New York/Berlin (2001)
7. Narasimhan, R., Nievergelt, Y.: Complex Analysis in One Variable. Birkhäuser, Boston (2001)
8. Pöschel, J.: Etwas mehr Analysis. Springer Spektrum, Wiesbaden (2014)
9. Pöschel, J.: Noch mehr Analysis. Springer Spektrum, Wiesbaden (2015)
10. Remmert, R.: Funktionentheorie I. Springer, Berlin (1984)
11. Rudin, W.: Real and Complex Analysis, 3rd ed. McGraw-Hill, New York (1987)

This chapter begins with the examination of Cauchy integrals on the unit circle, marking the start of the original complex analysis. The Cauchy integral formula for the unit circle states that functions in the disc algebra on the open unit disc $\mathbb{D}$ coincide with their Cauchy integral. Since Cauchy integrals are analytical functions, the analyticity of holomorphic functions arises magically, almost by itself. This provides the key to open the door to the fascinating world of holomorphic functions. A first glance into this world is immediately taken.

In order to describe the local behavior of holomorphic functions at isolated singularities, it is necessary to extend the concept of Taylor series to Laurent series. The interpretation of a Laurent series as a (suitably scaled) Fourier series allows for a comparatively direct approach that does not require further topological tools. At the same time, this involves an introduction to Fourier analysis.

In connection with the characterization of poles, the opportunity is finally taken to introduce meromorphic functions and to bring the Riemann sphere into the game.

4.1   Cauchy Integral Formula and Applications

From now on, we will mainly consider functions on open sets in the complex plane. The following theorem provides an important class of analytical functions.

Theorem 4.1.1 *Let γ be a path in $\mathbb{C}$ and let $C_\gamma f : \mathbb{C} \setminus \gamma^* \to \mathbb{C}$ for $f \in C(\gamma^*)$ defined by*

$$(C_\gamma f)(z) := \frac{1}{2\pi i} \int_\gamma \frac{f(\zeta)}{\zeta - z} \, d\zeta \qquad (z \in \mathbb{C} \setminus \gamma^*).$$

J. Müller, *Concepts of Function Theory*, Mathematics Study Resources 12,
https://doi.org/10.1007/978-3-662-69115-1_4

If $a \in \mathbb{C} \setminus \gamma^*$ and $R := \mathrm{dist}(a, \gamma^*)$, then

$$(C_\gamma f)(a + h) = \sum_{\nu=0}^{\infty} c_\nu h^\nu \qquad (h \in U_R(0))$$

with

$$(C_\gamma f)^{(k)}(a) = k! c_k = \frac{k!}{2\pi i} \int_\gamma \frac{f(\zeta)}{(\zeta - a)^{k+1}} \, d\zeta \qquad (k \in \mathbb{N}_0). \qquad (4.1.1)$$

Proof Since γ^* is compact, $R > 0$. From

$$\left| \frac{h}{\zeta - a} \right| \le \frac{|h|}{R} < 1$$

for all $h \in U_R(0)$ and all $\zeta \in \gamma^*$, it follows from the Weierstrass criterion that the geometric series

$$\sum_{\nu=0}^{\infty} \frac{h^\nu}{(\zeta - a)^{\nu+1}} = \frac{1}{\zeta - a} \frac{1}{1 - \frac{h}{\zeta-a}} = \frac{1}{\zeta - a - h}$$

for each fixed $h \in U_R(0)$ converges uniformly on γ^*. Thus, by interchanging summation and integration (see Theorem 3.5.4) for $h \in U_R(0)$ we obtain

$$(C_\gamma f)(a + h) = \frac{1}{2\pi i} \int_\gamma \frac{f(\zeta)}{\zeta - a - h} \, d\zeta = \frac{1}{2\pi i} \int_\gamma f(\zeta) \sum_{\nu=0}^{\infty} \frac{h^\nu}{(\zeta - a)^{\nu+1}} \, d\zeta = \sum_{\nu=0}^{\infty} c_\nu h^\nu .$$

From Theorem 3.2.4 the first equality in (4.1.1) follows. □

Remark and Definition 4.1.2 The function from Theorem 4.1.1 is called **Cauchy integral** of f (with respect to γ). In particular, from Theorem 4.1.1, it follows that $C_\gamma f$ is analytic in $\mathbb{C} \setminus \gamma^*$.

Remark and Definition 4.1.3 We specifically consider $\gamma = k_1(0)$ and the Cauchy integral $Cf := C_{k_1(0)} f$ of f with respect to $k_1(0)$, so

$$(Cf)(z) = \frac{1}{2\pi i} \int_{k_1(0)} \frac{f(\zeta)}{\zeta - z} \, d\zeta \qquad (z \in \mathbb{C} \setminus \mathbb{S}).$$

Then $(Cf)(z) = \sum_{\nu=0}^{\infty} c_\nu z^\nu$ for $z \in \mathbb{D}$ with

$$c_k = \frac{1}{2\pi i} \int_{k_1(0)} \frac{f(\zeta)}{\zeta^{k+1}} \, d\zeta = \frac{1}{2\pi} \int_{-\pi}^{\pi} f(e^{it}) e^{-ikt} \, dt.$$

If specifically $f = 1$, then $c_0 = 1$ according to Remark 3.5.3 and $c_k = 0$ for $k \in \mathbb{N}$ according to Example 3.5.6. So $(C1)(z) = c_0 = 1$ holds for all $z \in \mathbb{D}$, that is, C "reproduces" the constant function 1 in $\mathbb{D}$. Furthermore, according to the reverse triangle inequality $|\zeta - a| \geq 1 - r$ for $|a| \leq r$ and $\zeta \in \mathbb{S}$. This results for $0 \leq r < 1$ in (4.1.1) the important **Cauchy inequality**:

$$|(Cf)^{(k)}(a)| \leq \frac{k!}{(1-r)^{k+1}} \max_{\mathbb{S}} |f| \qquad (a \in B_r(0),\ k \in \mathbb{N}_0). \qquad (4.1.2)$$

Since analytic functions are infinitely differentiable, $C^{\omega}(\Omega) \subset H(\Omega)$. We now want to show that every holomorphic function is already analytic, in other words, that

$$H(\Omega) = C^{\omega}(\Omega)$$

holds—a kind of mathematical miracle. The Cauchy integral formula, which we will now derive in a first version for circles, will be crucial for this.

We write for compact sets $K \subset \mathbb{C}$

$$A(K) := \{f : K \to \mathbb{C} : f \text{ continuous and } f|_{K^{\circ}} \text{ holomorphic}\}.$$

An important role for further considerations is played by the following auxiliary result, in which K is the closed "ring area" between the unit circle $\mathbb{S}$ and the circle $K_r(a(1-r))$ is (compare Fig. 4.1).

Theorem 4.1.4 *Let* $a \in \mathbb{D}$, $0 \leq r < 1$ *and* $K := \overline{\mathbb{D}} \setminus U_r(a(1-r))$. *If* $f \in A(K)$, *then*

$$(Cf)(a) = \frac{1}{2\pi i} \int_{k_1(0)} \frac{f(a(1-r) + r\zeta)}{\zeta - a}\, d\zeta.$$

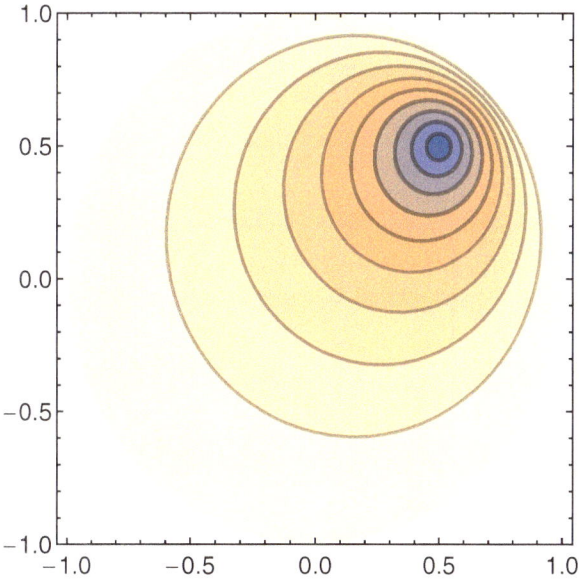

Fig. 4.1   Circles $a(1-r) + r\mathbb{S} = K_r(a(1-r))$ for $a = (1+i)/2$ and $0 < r < 1$

Proof We define $\varphi : [r, 1] \times \mathbb{S}$ by

$$\varphi(\lambda, \zeta) := \frac{f(a + \lambda(\zeta - a))}{\zeta - a} \qquad (\lambda \in [r, 1], \ \zeta \in \mathbb{S})$$

and $\Phi : [r, 1] \to \mathbb{C}$ by

$$\Phi(\lambda) := \int_{k_1(0)} \varphi(\lambda, \zeta) \, d\zeta \qquad (\lambda \in [r, 1]).$$

According to Theorem 3.5.8 Φ is continuous on $[r, 1]$. Since for every $\lambda \in (r, 1)$ the function $\zeta \mapsto f(a + \lambda(\zeta - a))/\lambda$ is an antiderivative of $\zeta \mapsto f(a + \lambda(\zeta - a))$ on $\mathbb{S}$, the fundamental theorem of calculus for path integrals applies and again with Theorem 3.5.8

$$\Phi'(\lambda) = \int_{k_1(0)} D_1 \varphi(\lambda, \zeta) \, d\zeta = \int_{k_1(0)} f'\big(a + \lambda(\zeta - a)\big) \, d\zeta = 0 \quad (r < \lambda < 1).$$

Therefore, Φ is constant on $[r, 1]$ and consequently

$$2\pi i(Cf)(a) = \int_{k_1(0)} \frac{f(\zeta)}{\zeta - a} d\zeta = \Phi(1) = \Phi(r) = \int_{k_1(0)} \frac{f(a(1 - r) + r\zeta)}{\zeta - a} d\zeta.$$

$\square$

We now consider the simple and important case of the closed unit disk $K = \overline{\mathbb{D}} = \{z : |z| \leq 1\}$. The algebra $A(\overline{\mathbb{D}})$ is also called the **disk algebra**.

Theorem 4.1.5 (Cauchy's integral formula for the unit circle) *For $f \in A(\overline{\mathbb{D}})$ holds*

$$C(f)|_{\mathbb{D}} = f|_{\mathbb{D}},$$

thus

$$\frac{1}{2\pi i} \int_{k_1(0)} \frac{f(\zeta)}{\zeta - z} d\zeta = f(z) \qquad (z \in \mathbb{D}).$$

Proof Let $z \in \mathbb{D}$ be fixed. By applying Theorem 4.1.4 with $a = z$ and $r = 0$ and Remark 4.1.3, it follows that $(Cf)(z) = f(z) \cdot (C1)(z) = f(z)$. $\square$

Remark 4.1.6 Theorem 4.1.5 shows in particular that for $f \in A(\overline{\mathbb{D}})$ the function values in $\mathbb{D}$ can be calculated by integration from the boundary values. With Remark 4.1.3, it further follows that $f(z) = \sum_{\nu=0}^{\infty} c_\nu z^\nu$ holds for all $z \in \mathbb{D}$, with

$$\frac{f^{(k)}(0)}{k!} = c_k = \frac{1}{2\pi i} \int_{k_1(0)} \frac{f(\zeta)}{\zeta^{k+1}} \, d\zeta.$$

By suitable scaling, as already indicated, the analyticity of holomorphic functions is obtained. In the proof and in the following, we use that according to the definition of the path integral

$$\int_{k_\rho(a)} f(\zeta)\,d\zeta = \rho \int_{k_1(0)} f(a+\rho\tau)\,d\tau \tag{4.1.3}$$

holds for continuous functions on general circles $K_\rho(a)$.

Theorem 4.1.7 *Let $\Omega \subset \mathbb{C}$ be open and $f \in H(\Omega)$. Then for all $a \in \Omega$ with $R :=$ dist$(a, \partial\Omega)$*

$$f(a+h) = \sum_{\nu=0}^{\infty} \frac{f^{(\nu)}(a)}{\nu!}\, h^\nu \qquad (h \in U_R(0)).$$

*In particular, $f \in C^\omega(\Omega)$. Moreover, for $0 < \rho < R$, one has the **Cauchy integral formula for general circles***

$$f(z) = \frac{1}{2\pi i} \int_{k_\rho(a)} \frac{f(\zeta)}{\zeta - z}\,d\zeta \qquad (z \in U_\rho(a)).$$

Proof Let $a \in \Omega$ and $\rho < R$. We define $g \in A(\overline{\mathbb{D}})$ by

$$g(w) := f(a + \rho w) \qquad (|w| \le 1).$$

Then $f^{(\nu)}(a) = g^\nu(0)/\rho^\nu$ for $\nu \in \mathbb{N}_0$. With Remark 4.1.6 we get

$$f(a+h) = g\left(\frac{h}{\rho}\right) = \sum_{\nu=0}^{\infty} \frac{g^{(\nu)}(0)}{\nu!}\frac{h^\nu}{\rho^\nu} = \sum_{\nu=0}^{\infty}\frac{f^{(\nu)}(a)}{\nu!}h^\nu \quad (h \in U_\rho(0)).$$

Since $\rho < R$ was arbitrary, the representation holds for $|h| < R$ and since $a \in \Omega$ was arbitrary, f is analytic in Ω. Moreover, for $w \in \mathbb{D}$ and $z = a + \rho w$ according to Theorem 4.1.5

$$2\pi i f(z) = 2\pi i\, g(w) = \int_{k_1(0)} \frac{g(\tau)}{\tau - w}\,d\tau = \int_{k_1(0)} \frac{f(a+\rho\tau)}{a + \rho\tau - z}\rho\,d\tau = \int_{k_\rho(a)} \frac{f(\zeta)}{\zeta - z}\,d\zeta.$$

$\square$

Remark 4.1.8 Let $\Omega \subset \mathbb{C}$ be open and $f \in H(\Omega)$. If $\rho < R := $ dist$(a, \partial\Omega)$, then from the Cauchy integral formula for $z = a$ we obtain the important **mean value formula**

$$f(a) = \frac{1}{2\pi} \int_{-\pi}^{\pi} f(a + \rho e^{it})\,dt.$$

Thus: The function value at the center of the circle is obtained as the "integral mean" of the function values at the edge of the circle. Since the Cauchy integral $C_{k_\rho(a)}f$ according to Remark 3.2.9 in conjunction with the fundamental theorem of calculus for path integrals (Theorem 3.5.5) on $\mathbb{C} \setminus B_\rho(a)$ has the value 0, also

$$(f \cdot \mathbb{1}_{U_\rho(a)})(z) = \frac{1}{2\pi i} \int_{k_\rho(a)} \frac{f(\zeta)}{\zeta - z}\,d\zeta \quad (z \in \Omega \setminus K_\rho(a)).$$

With Theorem 4.1.1, the Cauchy integral formula for all derivatives of f is also obtained, thus

$$f^{(k)}(z) = (C_{k_\rho(a)}f)^{(k)}(z) = \frac{k!}{2\pi i} \int_{k_\rho(a)} \frac{f(\zeta)}{(\zeta - z)^{k+1}} \, d\zeta \quad (z \in U_\rho(a), \, k \in \mathbb{N}).$$

Remark and Definition 4.1.9 A function f that is holomorphic in $\mathbb{C}$ is called an **entire function**. In particular, polynomials and exp, sin, cos are entire functions. If f is entire, then with $c_k := f^{(k)}(0)/k!$ according to Theorem 4.1.7 (for $a = 0$ and with z instead of h)

$$f(z) = \sum_{\nu=0}^{\infty} c_\nu z^\nu \quad (z \in \mathbb{C}).$$

As a first important implication, we obtain

Theorem 4.1.10 (Liouville) *If f is entire and bounded, then f is constant.*

Proof By assumption, there exists an $M > 0$ with $|f(z)| \leq M$ for all $z \in \mathbb{C}$. Thus, for $k \in \mathbb{N}$ and $R > 0$ according to Remark 4.1.8 and with $L(k_R(0)) = 2\pi R$

$$|f^{(k)}(0)| = \frac{k!}{2\pi} \Big| \int_{k_R(0)} \frac{f(\zeta)}{\zeta^{k+1}} d\zeta \Big| \leq \frac{k!M}{R^k}.$$

From $1/R^k \to 0$ for $R \to \infty$ it follows $f^{(k)}(0) = 0$ and thus $f(z) = c_0 = f(0)$ for all $z \in \mathbb{C}$. □

Example 4.1.11 If $f = \cos$, then f is unbounded according to Liouville's theorem. In fact, for $t \in \mathbb{R}$

$$\cos(it) = \frac{1}{2}(e^{-t} + e^t) = \cosh(t) \to \infty \quad (t \to \pm\infty).$$

In particular, the complex cosine function—unlike the real one—is unbounded.

Remark 4.1.12 As an application of Liouville's theorem, a very short proof of the Fundamental Theorem of Algebra emerges: Let $p : \mathbb{C} \to \mathbb{C}$ be a non-constant polynomial. Suppose p has no root. Then $1/p$ is an entire function. According to Liouville's theorem, there exists a sequence (z_n) in $\mathbb{C}$ with $|1/p(z_n)| \to \infty$. Since $1/p$ is bounded on all compact subsets of $\mathbb{C}$, it follows that $|z_n| \to \infty$. However, this contradicts $1/p(z) \to 0$ for $|z| \to \infty$ (see Example 2.1.17).

Theorem 4.1.13 (local maximum principle) *Let $\Omega \subset \mathbb{C}$ be open and $f \in H(\Omega)$. If $a \in \Omega$ is a maximum point of $|f|$, then f is constant around a.*

Proof By assumption, there exists an $r > 0$ with

$$|f(z)| \leq |f(a)| \quad \text{for all } z \in U_r(a).$$

Assume there exists a $w \in U_r(a)$ with $|f(w)| < |f(a)|$. If $\rho = |w - a|$, then due to the continuity of $t \mapsto |f(a + \rho e^{it})|$ on $[-\pi, \pi]$ and $|f(a + \rho e^{it})| \leq |f(a)|$ with Exercise 3.3.24.7

$$\frac{1}{2\pi} \int_{-\pi}^{\pi} |f(a + \rho e^{it})| dt < |f(a)| \frac{1}{2\pi} \int_{-\pi}^{\pi} dt = |f(a)|,$$

thus with the mean value theorem

$$|f(a)| \leq \frac{1}{2\pi} \int_{-\pi}^{\pi} |f(a + \rho e^{it})| dt < |f(a)| .$$

Contradiction! Therefore, $|f|$ is constant on $U_r(a)$. It follows that *also f* is constant around a (Exercise 4.1.23.7).   □

Theorem 4.1.14 (Maximum Principle; negative form) *Let $G \subset \mathbb{C}$ be a domain and $f \in H(G)$. If $|f|$ has a maximum point, then f is constant.*

Proof If $|f|$ has a local maximum at a, then f is constant around a according to the local maximum principle. Since f is holomorphic and thus analytic in G, f is constant according to the identity theorem.   □

Remark 4.1.15 If $G \subset \mathbb{C}$ is a domain and $f \in H(G)$, then of course for all zeros z_0 of f

$$|f(z_0)| = 0 \leq |f(z)| \qquad (z \in G),$$

that is, zeros of f are minimum points of $|f|$. But if f is free of zeros and not constant, then the function $|f|$ also has no minimum points, as can be seen immediately by applying Theorem 4.1.14 to the function $1/f$.

Taking into account the fact that continuous functions become maximal on compact sets, we obtain

Theorem 4.1.16 (Maximum Principle; positive form) *Let $G \subset \mathbb{C}$ be a bounded domain and $f \in A(\overline{G})$. Then there exists an $a \in \partial G$ with*

$$|f(a)| = \max_{\overline{G}} |f|.$$

Proof Since G is bounded, $\overline{G} = G \cup \partial G$ is compact. Since with f also $|f|$ is continuous on $\overline{G}$, $|f|$ becomes maximal on $\overline{G}$. If f is constant, the claim is clear. If f is not constant, then $|f|$ has no maximal point on G according to Theorem 4.1.14, so there exists an $a \in \partial G$ as claimed.   □

Remark and Definition 4.1.17 Let $A \subset \mathbb{C}$ and $f : A \to \mathbb{C}$ be continuous. If $r \geq 0$ and $K_r(0) \subset A$, we set

$$M(r, f) := \max_{K_r(0)} |f|.$$

If f is holomorphic in $U_R(0)$, then with the maximum principle (positive form) for all $0 \leq r < R$

$$M(r, f) = \max_{B_r(0)} |f|,$$

and thus in particular $r \mapsto M(r, f)$ is monotonically increasing. From the negative form it follows that $r \mapsto M(r, f)$ is strictly increasing, if f is not constant. If f is entire and not constant, then according to Liouville's theorem $M(r, f) \to \infty$ for $r \to \infty$.

Example 4.1.18 We consider $f = \cos$ on $\mathbb{C}$. If $|z| = r$, then

$$|\cos z| = \left| \sum_{v=0}^{\infty} \frac{(-1)^v z^{2v}}{(2v)!} \right| \leq \sum_{v=0}^{\infty} \frac{r^{2v}}{(2v)!} = \cosh(r) = \cos(ir).$$

Thus, $M(r, f) = \cosh(r)$.

Remark and Definition 4.1.19 Let (X, d_X), (Y, d_Y) be metric spaces and $f_n : X \to Y$. If $a \in X$, the sequence (f_n) is called **uniformly convergent around** $a \in X$, if a neighborhood U of a exists such that $(f_n|_U)$ converges uniformly on U. If this is the case for all $a \in X$, it is said that (f_n) is **locally uniformly convergent** (on X). If $X \subset \mathbb{K}$ is open and are $f_n, f : X \to Y$, then $f_n \to f$ is locally uniformly convergent on X if and only if (f_n) converges uniformly on all compact subsets towards f (Exercise 4.1.23.10).

Remark 4.1.20 Power series are locally uniformly convergent on their circle of convergence according to Remark 2.6.12.

According to Theorem 2.6.6, continuity is transferred to the limit function in the case of locally uniform convergence. The same applies to holomorphy:

Theorem 4.1.21 Let $\Omega \subset \mathbb{C}$ be open and let $f_n : \Omega \to \mathbb{C}$ be holomorphic functions. Furthermore, let $f_n \to f$ converge locally uniformly on Ω. Then f is also holomorphic in Ω, and it holds for all $k \in \mathbb{N}$

$$f_n^{(k)} \to f^{(k)} \qquad (n \to \infty)$$

locally uniformly on Ω.

Proof If $a \in \Omega$, there exists a $\rho > 0$ with $f_n \to f$ uniformly on $B_\rho(a)$. We consider $g_n, g : \overline{\mathbb{D}} \to \mathbb{C}$ with

$$g_n(w) := f_n(a + \rho w) \quad \text{and} \quad g(w) := f(a + \rho w).$$

Then $g_n \to g$ uniformly on $\overline{\mathbb{D}}$. By assumption, $g_n \in A(\overline{\mathbb{D}})$. Moreover, g is continuous on $\mathbb{S}$. From

$$g_n(w) = \frac{1}{2\pi i} \int_{k_1(0)} \frac{g_n(\zeta)}{\zeta - w} d\zeta \;\rightarrow\; \frac{1}{2\pi i} \int_{k_1(0)} \frac{g(\zeta)}{\zeta - w} d\zeta = (Cg)(w) \quad (n \to \infty)$$

for $w \in \mathbb{D}$ follows $g|_{\mathbb{D}} = (Cg)|_{\mathbb{D}}$. Thus, g is analytic in $\mathbb{D}$ and thus f is holomorphic in $U_\rho(a)$. Further, with the Cauchy inequality (4.1.2) for $0 < r < 1$ and $k \in \mathbb{N}_0$

$$\max_{B_r(0)} |g^{(k)} - g_n^{(k)}| \le \frac{k!}{(1-r)^{k+1}} \max_S |g - g_n| \to 0 \quad (n \to \infty).$$

Therefore, $g_n^{(k)} \to g^{(k)}$ $(n \to \infty)$ uniformly on $B_r(0)$. From $f_n^{(k)}(a + \rho w) = \rho^{-k} g_n^{(k)}(w)$ for $w \in \mathbb{D}$ it follows that $f_n^{(k)} \to f^{(k)}$ $(n \to \infty)$ uniformly on $B_{r\rho}(a)$.   □

Example 4.1.22 In Example 2.6.14 we have defined the Riemann zeta function $\zeta : \Omega \to \mathbb{C}$ on $\Omega := \{z \in \mathbb{C} : \operatorname{Re} z > 1\}$ by

$$\zeta(z) := \sum_{\nu=1}^{\infty} \nu^{-z} = \sum_{\nu=1}^{\infty} e^{-z \ln \nu} \quad (\operatorname{Re} z > 1).$$

The partial sum sequence $s_n(z) = \sum_{\nu=1}^{n} \nu^{-z}$ converges locally uniformly on Ω according to Example 2.6.14. Since the partial sums are entire functions, ζ is holomorphic in Ω according to Theorem 4.1.21.

Exercises 4.1.23

1. Show: If $f \in A(\overline{\mathbb{D}})$ and $\sum_{\nu=0}^{\infty} |c_k| < \infty$, where $c_k = f^{(k)}(0)/k!$, then $f(z) = \sum_{\nu=0}^{\infty} c_\nu z^\nu$ for all $z \in \overline{\mathbb{D}}$ with uniform convergence on $\overline{\mathbb{D}}$.

2. Let $\Omega \subset \mathbb{C}$ be open. Show that $H(\Omega)$ is an algebra and that for $f \in H(\Omega)$ and $g \in H(U)$, where $U \supset f(\Omega)$ is open, also $g \circ f \in H(\Omega)$.

3. Let $\Omega \subset \mathbb{C}$ and $a \in \Omega$. Further let $f, g \in H(\Omega)$ with $n_f(a)$, $n_g(a) < \infty$. Show:
   a) $n_{f \cdot g}(a) = n_f(a) + n_g(a)$.
   b) If $n_f(a) \ge n_g(a)$, then there exists $\lim_{z \to a} f(z)/g(z)$ with

$$\lim_{z \to a} \frac{f(z)}{g(z)} = \begin{cases} 0, & \text{if } n_f(a) > n_g(a) \\ f^{(n)}(a)/g^{(n)}(a), & \text{if } n_f(a) = n_g(a) =: n \end{cases}.$$

4. Let $G \subset \mathbb{C}$ be a domain. Show that the ring $(H(G), +, \cdot)$ is free of zero divisors, that is, from $f \cdot g = 0$ follows $f = 0$ or $g = 0$. Does this also apply to arbitrary open sets and the ring $(H(\Omega), +, \cdot)$?

5. For two sequences $u = (u_n)$ and $v = (v_n)$ in $\mathbb{K}^{\mathbb{N}_0}$ the **convolution** $u * v \in \mathbb{K}^{\mathbb{N}_0}$ is defined by

$$(u * v)_n := \sum_{k=0}^{n} u_k v_{n-k} \quad (n \in \mathbb{N}_0).$$

Show: If $\Omega \subset \mathbb{C}$ is open and $f, g \in H(\Omega)$, then for $a \in \Omega$

$$(fg)(a + h) = \sum_{n=0}^{\infty}(u * v)_n h^n \qquad (|h| < \mathrm{dist}(a, \partial\Omega))$$

with $u_n = f^{(n)}(a)/n!$ and $v_n := g^{(n)}(a)/n!$.
Hint: Use the Leibniz formula.

6. Let $f : \mathbb{C}^* \to \mathbb{C}$ be defined by

$$f(z) = \sin(1/z) \qquad (z \in \mathbb{C}^*).$$

Show that f is holomorphic in $\mathbb{C}^*$ and that 0 is an accumulation point of $Z(f)$ in $\mathbb{C}$. Why does this not contradict Theorem 3.2.27?

7. Let $\Omega \subset \mathbb{C}$ be open and $f : \Omega \to \mathbb{C}$ holomorphic. Show that f is locally constant if one of the following conditions is met:
   a) $\bar{f}$ (defined by $\bar{f}(z) = \overline{f(z)}$ for $z \in \Omega$) is holomorphic.
      Hint: Use Remark 3.1.33.
   b) $|f|$ is locally constant.
      Hint: $|f|^2 = f \cdot \bar{f}$.

8. Calculate $M(r, \sin)$ for $r \geq 0$.

9. Let f be an entire function. Show: If there exists an $n \in \mathbb{N}$ with

$$M(r, f)/r^n \to 0 \quad (r \to \infty),$$

then f is a polynomial of degree $\leq n - 1$.

10. Let $X \subset \mathbb{K}$ be open, (Y, d) a metric space and $f_n, f : X \to Y$. Show that (f_n) converges locally uniformly on X to f if and only if (f_n) converges uniformly on every compact subset of X to f.

11. Let $G \subset \mathbb{C}$ be a bounded region and $f \in A(\overline{G})$. Show: If f is zero-free, then there exists a $z_0 \in \partial G$ with

$$|f(z_0)| = \min_{\overline{G}} |f|.$$

12. Let $\Omega := \{z \in \mathbb{C} : \mathrm{Re}(z) > 0\}$ and let (f_n) be the sequence from Exercise 3.5.12.6. Show:
    a) (f_n) converges locally uniformly on Ω.
    b) The gamma function Γ is holomorphic in Ω.

4.2   Fourier and Laurent Series

In the previous sections, we dealt with Taylor series or power series. If $f \in A(\overline{\mathbb{D}})$, then according to Remark 4.1.6

$$f(z) = \sum_{\nu=0}^{\infty} c_\nu z^\nu \qquad (z \in \mathbb{D}),$$

where the Taylor coefficients $c_k = c_k(f, 0)$ have the representation

$$c_k = \frac{1}{2\pi i} \int_{k_1(0)} \frac{f(\zeta)}{\zeta^{k+1}} d\zeta = \frac{1}{2\pi} \int_{-\pi}^{\pi} f(e^{it}) e^{-ikt} dt$$

If $\sum_{\nu=0}^{\infty} |c_\nu| < \infty$, then the series also converges uniformly on the unit circle $\mathbb{S}$ with

$f(\zeta) = \sum_{\nu=0}^{\infty} c_\nu \zeta^\nu$ for $\zeta \in \mathbb{S}$ (Exercise 4.1.23.1).

We now want to examine series developments of this type, which have the significant advantage that no derivatives are needed.

Remark and Definition 4.2.1 If $f : \mathbb{S} \to \mathbb{C}$ is such that $t \mapsto f(e^{it})$ is a regulated function on $[-\pi, \pi]$, we say f is a **regulated function** on $\mathbb{S}$ and write

$$\int f \, dm := \int f(\zeta) dm(\zeta) := \frac{1}{2\pi} \int_{-\pi}^{\pi} f(e^{it}) dt.$$

For continuous f, this also implies

$$\int f \, dm = \frac{1}{2\pi i} \int_{k_1(0)} \frac{f(\zeta)}{\zeta} d\zeta$$

and

$$(Cf)(z) = \int \frac{f(\zeta)}{1 - z\bar{\zeta}} dm(\zeta) \quad (z \in \mathbb{D}).$$

Definition 4.2.2 Let V be a linear space over $\mathbb{K}$.

1. A mapping $\langle \cdot, \cdot \rangle : V \times V \to \mathbb{K}$ is called a **scalar product** (on V) (or **inner product** (on V)), if the following conditions apply:
   (S1) For all $v \in V \setminus \{0\}$, $\langle v, v \rangle > 0$.
   (S2) For all $u, v \in V$, $\langle u, v \rangle = \overline{\langle v, u \rangle}$ applies.
   (S3) For all $v \in V$, the mapping $u \mapsto \langle u, v \rangle$ is linear.
   A linear space with scalar product $(V, \langle \cdot, \cdot \rangle)$ is called a **unitary space**. If V is unitary, then by

$$||v|| := \sqrt{\langle v, v \rangle}$$

   a norm on V is defined (Exercise 4.2.20.1), the so-called **induced norm**.
2. Let $V = (V, \langle \cdot, \cdot \rangle)$ be a unitary space and let $(v_\alpha)_{\alpha \in I}$ be a family of vectors from $V \setminus \{0\}$. Then $(v_\alpha)_{\alpha \in I}$ is called an **orthogonal system**, if

$$\langle v_\alpha, v_\beta \rangle = 0$$

   for all $\alpha, \beta \in I$ with $\alpha \neq \beta$. An orthogonal system is called an **orthonormal system**, if additionally $||v_\alpha|| = 1$ for all $\alpha \in I$ is fulfilled.

In the following, we write for tuples $(a_n)_{n \in \mathbb{Z}}$ in $\mathbb{C}$

$$\sum_{\nu=-\infty}^{\infty} a_\nu := \lim_{n \to \infty} \sum_{\nu=-n}^{n} a_\nu$$

in the case of the existence of the limit on the right side.

Remark 4.2.3 By

$$\langle f, g \rangle := \int f\overline{g}\, dm \qquad (f, g \in C(\mathbb{S}))$$

a scalar product is defined on $C(\mathbb{S})$ (see Exercise 3.3.24.7 for (S1)). For the induced norm we write $\|\cdot\|_2$. If $e_k : \mathbb{S} \to \mathbb{C}$ for $k \in \mathbb{Z}$ is defined by

$$e_k(z) := z^k \qquad (z \in \mathbb{S}),$$

then (compare Remark 4.1.3)

$$\langle e_j, e_k \rangle = \int e_j \overline{e_k}\, dm = \int \zeta^{j-k}\, dm(\zeta) = \frac{1}{2\pi i} \int_{k_1(0)} \zeta^{j-k-1} d\zeta = \begin{cases} 0, & \text{if } k \neq j \\ 1, & \text{if } k = j \end{cases}.$$

Thus, $(e_k)_{k\in\mathbb{Z}}$ is an orthonormal system in $C(\mathbb{S})$. If

$$f(z) = \sum_{\nu=-\infty}^{\infty} c_\nu z^\nu = \lim_{n\to\infty} \sum_{\nu=-n}^{n} c_\nu z^\nu \qquad (z \in \mathbb{S})$$

with uniform convergence on $\mathbb{S}$, then $f \in C(\mathbb{S})$ and $c_k = \langle f, e_k \rangle$ for all k.

Because: Due to the uniform convergence, with Theorem 3.5.4

$$\langle f, e_k \rangle = \int \left(\sum_{\nu=-\infty}^{\infty} c_\nu e_\nu \right) \overline{e_k}\, dm = \sum_{\nu=-\infty}^{\infty} c_\nu \int e_\nu \overline{e_k}\, dm = c_k.$$

Definition 4.2.4 Let $f \in C(\mathbb{S})$. Then for $k \in \mathbb{Z}$

$$\widehat{f}(k) := \langle f, e_k \rangle = \int f\overline{e_k}\, dm = \frac{1}{2\pi} \int_{-\pi}^{\pi} f(e^{it})e^{-ikt}\, dt = \frac{1}{2\pi i} \int_{k_1(0)} \frac{f(\zeta)}{\zeta^{k+1}} d\zeta$$

is called the k-th **Fourier coefficient** of f and the mapping

$$C(\mathbb{S}) \ni f \mapsto \widehat{f} \in \mathbb{C}^{\mathbb{Z}}$$

Fourier transformation. For $n \in \mathbb{N}_0$ further $s_n f$ with

$$(s_n f)(z) := \sum_{\nu=-n}^{n} \widehat{f}(\nu) z^\nu = \sum_{\nu=-n}^{n} \langle f, e_\nu \rangle e_\nu(z) \qquad (z \in \mathbb{S})$$

is called the n-th **Fourier partial sum** of f and $(s_n f)_{n\in\mathbb{N}_0}$ **Fourier series** of f.

Remark 4.2.5 If $f \in A(\overline{\mathbb{D}})$, then $\widehat{f}(k) = c_k(f, 0) = f^{(k)}(0)/k!$ for $k \in \mathbb{N}_0$, that is, for $k \geq 0$ the Fourier coefficients and the Taylor coefficients coincide. Also, in this case, $\widehat{f}(-k) = 0$ for $k > 0$, as can be seen from Remark 4.1.6 with $z \mapsto f(z)z^k$ instead of f.

Example 4.2.6 We consider $f : \mathbb{S} \to \mathbb{R}$ with

$$f(e^{it}) := t^2 \qquad (t \in (-\pi, \pi]).$$

Then we get

$$\widehat{f}(0) = \frac{1}{2\pi} \int_{-\pi}^{\pi} t^2 \, dt = \frac{\pi^2}{3}$$

and with twice partial integration

$$\widehat{f}(k) = \frac{1}{2\pi} \int_{-\pi}^{\pi} t^2 e^{-ikt} \, dt = (-1)^k \frac{2}{k^2}$$

for $k \in \mathbb{Z}$, $k \neq 0$. Thus, for $z = e^{it}$ the Fourier series is given by

$$\sum_{\nu=-\infty}^{\infty} \widehat{f}(\nu) z^\nu = \frac{\pi^2}{3} + \sum_{\nu=1}^{\infty} (-1)^\nu \frac{2}{\nu^2} (z^\nu + z^{-\nu}) = \frac{\pi^2}{3} + 4 \sum_{\nu=1}^{\infty} \frac{(-1)^\nu}{\nu^2} \cos(\nu t).$$

The convergence of the series is uniform according to the Weierstrass criterion on $\mathbb{S}$.

Remark 4.2.7 The Fourier transformation $C(\mathbb{S}) \ni f \mapsto \widehat{f} \in \mathbb{C}^{\mathbb{Z}}$ is a linear mapping. We want to show that the mapping is injective, that is, functions $f \in C(\mathbb{S})$ are completely determined by the sequence of Fourier coefficients. To this end, we prove that

$$\|f - s_n f\|_2 \to 0 \qquad (n \to \infty) \tag{4.2.1}$$

holds for all $f \in C(\mathbb{S})$, that is, $(s_n f)$ converges "in the mean square" to f. We set

$$T_n := \text{span}\{e_k : k \in \{-n, \ldots, n\}\}.$$

An element from T_n, thus a linear combination of the e_{-n}, ..., e_n, is called a **trigonometric polynomial**[1] of degree $\leq n$. Since $(e_k)_{k \in \mathbb{Z}}$ is an orthonormal system, it follows (Exercise 4.2.20.2)

$$\|f - s_n f\|_2 = \min_{p \in T_n} \|f - p\|_2 = \text{dist}(f, T_n).$$

So: $s_n f \in T_n$ is the best approximation from T_n to f with respect to the $\|\cdot\|_2$-norm. In particular, $p = s_n p$ for all $p \in T_n$. To prove (4.2.1), it is therefore sufficient to show that a sequence (p_n) with $p_n \in T_n$ and

$$\|f - p_n\|_2 \to 0 \quad (n \to \infty)$$

---

[1] With $z = e^{it}$ for $t \in (-\pi, \pi]$ is $e_k(z) = e^{ikt} = \cos(kt) + i \sin(kt)$, and thus trigonometric polynomials are linear combinations of the trigonometric functions $\cos(kt)$ and $\sin(kt)$.

exists. Since for all $f \in C(\mathbb{S})$

$$\|f\|_2 = \left(\int |f|^2 \, dm \right)^{1/2} \le \|f\|_\infty$$

holds, it is sufficient to show that $p_n \in T_n$ exist with

$$\|f - p_n\|_\infty \to 0 \qquad (n \to \infty),$$

in other words, the set of trigonometric polynomials $\bigcup_{n\in\mathbb{N}} T_n$ is dense in the space $(C(\mathbb{S}), \|\cdot\|_\infty)$.

Remark and Definition 4.2.8 Let $f, g \in C(\mathbb{S})$. We define the **convolution** $f * g : \mathbb{S} \to \mathbb{C}$ of f and g by

$$(f * g)(z) := \int f(z\bar{\zeta}) g(\zeta) \, dm(\zeta) \qquad (z \in \mathbb{S}).$$

Then $f * g$ is continuous on $\mathbb{S}$ (Exercise 4.2.20.6) with

$$f * g = g * f.$$

Specifically, if $p \in T_n$, then $p = s_n p$ and thus

$$(f * p)(z) = \sum_{\nu=-n}^{n} \widehat{p}(\nu) z^\nu \int \bar{\zeta}^\nu f(\zeta) \, dm(\zeta) = \sum_{\nu=-n}^{n} \widehat{p}(\nu) \widehat{f}(\nu) z^\nu \qquad (z \in \mathbb{S}).$$

Therefore, $f * p \in T_n$ and

$$\widehat{f * p} = \widehat{f} \cdot \widehat{p}.$$

If $A \subset \mathbb{S}$ is such that $\mathbb{1}_A = \mathbb{1}_{A,\mathbb{S}}$ is a regular function on $\mathbb{S}$, then for all regulated functions f on $\mathbb{S}$, $f\mathbb{1}_A$ and $f\mathbb{1}_{\mathbb{S}\setminus A}$ are also regulated functions (cf. Exercise 3.3.24.2). With

$$\int_A f \, dm := \int f\mathbb{1}_A \, dm$$

the mapping $f \mapsto \int_A f \, dm$ is linear and nonnegative with

$$\int f \, dm = \int_A f \, dm + \int_{\mathbb{S}\setminus A} f \, dm.$$

This particularly applies to $A = B_\delta := U_{\delta,\mathbb{S}}(1)$ with $\delta > 0$.

Theorem 4.2.9 (Approximate Unit) *For $n \in \mathbb{N}$ let $Q_n \in T_n$ with $Q_n \ge 0$ on $\mathbb{S}$ and $\int Q_n \, dm = 1$. If for all $\delta > 0$*

$$\int_{\mathbb{S}\setminus B_\delta} Q_n \, dm \to 0 \qquad (n \to \infty), \tag{4.2.2}$$

then the sequence $(f * Q_n)$ converges for all $f \in C(\mathbb{S})$ uniformly on $\mathbb{S}$ to f.

Proof Let $\varepsilon > 0$ be given. Since f is continuous on the compact set $\mathbb{S}$, f is uniformly continuous. Since $|z\bar{\zeta} - z| = |\zeta - 1|$ holds for all $z, \zeta \in \mathbb{S}$, there exists a $\delta > 0$ such that

$$|f(z\bar{\zeta}) - f(z)| < \varepsilon$$

for all $z \in \mathbb{S}$ and $\zeta \in B_\delta$. From $Q_n \geq 0$ and $\int Q_n \, dm = 1$ it follows for $z \in \mathbb{S}$ and $n \in \mathbb{N}$ initially

$$
\begin{aligned}
|(f * Q_n)(z) - f(z)| &= \left| \int (f(z\bar{\zeta}) - f(z)) Q_n(\zeta) \, dm(\zeta) \right| \\
&\leq \int |f(z\bar{\zeta}) - f(z)| Q_n(\zeta) \, dm(\zeta) \\
&\leq \int_{B_\delta} \varepsilon Q_n(\zeta) \, dm(\zeta) + \int_{\mathbb{S}\backslash B_\delta} 2\|f\|_\infty Q_n(\zeta) \, dm(\zeta),
\end{aligned}
$$

and thus also

$$\|f * Q_n - f\|_\infty \leq \varepsilon + 2\|f\|_\infty \int_{\mathbb{S}\backslash B_\delta} Q_n \, dm.$$

With (4.2.2) it follows

$$\|f - f * Q_n\|_\infty \leq 2\varepsilon$$

for n sufficiently large. $\square$

Remark 4.2.10 We call a sequence (Q_n) as in Theorem 4.2.9 a sequence of good kernels. Naturally, the question arises about the existence of good kernels. An obvious idea is to consider the partial sums $s_n f$ as a convolution with suitable kernels and then apply the above theorem. In fact, for $f \in C(\mathbb{S})$ according to Remark 4.2.8, the partial sums $s_n f$ can be written as $s_n f = f * D_n$, where

$$D_n := \sum_{\nu=-n}^{n} e_\nu$$

denotes the n-th **Dirichlet kernel** (Fig. 4.2). As we will show in the last chapter, the partial sums $s_n f$ do not even converge pointwise on $\mathbb{S}$ for all $f \in C(\mathbb{S})$. In particular, the sequence of Dirichlet kernels cannot represent a sequence of good kernels.

Instead, we consider the arithmetic means of the D_n. It holds that

$$\sum_{\nu=0}^{n} D_\nu = \sum_{\nu=0}^{n} \sum_{\mu=-\nu}^{\nu} e_\mu = \sum_{\mu=-n}^{n} e_\mu \sum_{\nu=|\mu|}^{n} 1 = \sum_{\mu=-n}^{n} (n+1-|\mu|)e_\mu.$$

So

$$\frac{1}{n+1} \sum_{\nu=0}^{n} s_\nu f = \frac{1}{n+1} \sum_{\nu=0}^{n} (f * D_\nu) = f * F_n,$$

Fig. 4.2 Dirichlet kernel D_5
(blue)

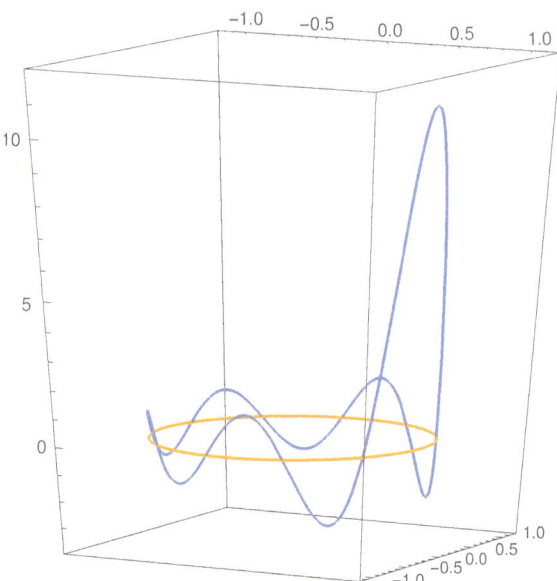

where

$$F_n := \sum_{\nu=-n}^{n} \left(1 - \frac{|\nu|}{n+1}\right) e_\nu \qquad (n \in \mathbb{N})$$

denotes the n-th **Fejér kernel** (Fig. 4.3). The sequence (F_n) actually proves to be a sequence of good kernels:

Firstly, $F_n \in T_n$ and

$$\int F_n \, dm = \sum_{\nu=-n}^{n} \left(1 - \frac{|\nu|}{n+1}\right) \int e_\nu \, dm = 1.$$

Furthermore, for $z \in \mathbb{S}$

$$\frac{1}{n+1} \Big| \sum_{j=0}^{n} z^j \Big|^2 = \frac{1}{n+1} \Big(\sum_{j=0}^{n} z^j \Big) \Big(\sum_{j=0}^{n} \bar{z}^j \Big) =$$

$$= \frac{1}{n+1} \sum_{j,k=0}^{n} z^{j-k} = \frac{1}{n+1} \sum_{\nu=-n}^{n} (n+1-|\nu|) z^\nu = F_n(z),$$

thus $F_n \geq 0$ and for $z \in \mathbb{S} \setminus B_\delta$

$$F_n(z) = \frac{1}{n+1} \Big| \sum_{j=0}^{n} z^j \Big|^2 = \frac{1}{n+1} \Big| \frac{z^{n+1}-1}{z-1} \Big|^2 \leq \frac{1}{n+1} \cdot \frac{4}{\delta^2} \to 0 \quad (n \to \infty).$$

Fig. 4.3 Fejér kernel F_5 (blue)

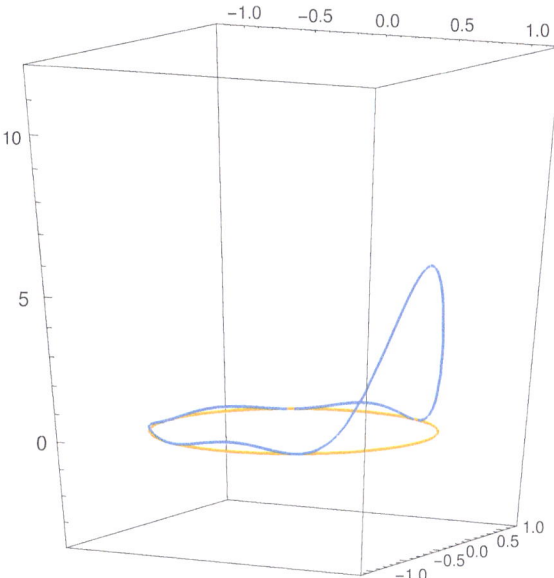

Therefore, $F_n \to 0$ uniformly on $\mathbb{S} \setminus B_\delta$. Thus, in particular, (4.2.2) is also fulfilled.

As an important consequence, we obtain

Theorem 4.2.11 (Fejér) *For all $f \in C(\mathbb{S})$*

$$\frac{1}{n+1} \sum_{\nu=0}^{n} s_\nu f \to f \quad (n \to \infty) \ uniformly \ on \ \mathbb{S}.$$

Proof The assertion follows directly from Theorem 4.2.9 and Remark 4.2.10.   □

As a consequence of Theorem 4.2.11, we obtain

Theorem 4.2.12 *Let $f \in C(\mathbb{S})$. Then the following holds*
1. $\|f - s_n f\|_2 \to 0 \ (n \to \infty)$.

2. *(**Parseval's equation**)* $\|f\|_2^2 = \displaystyle\sum_{\nu=-\infty}^{\infty} |\widehat{f}(\nu)|^2$.

3. *The Fourier transformation $f \mapsto \widehat{f}$ is injective.*

4. *If $(s_n f)$ converges uniformly on $\mathbb{S}$, then $s_n f \to f$, thus $f = \displaystyle\sum_{\nu=-\infty}^{\infty} \widehat{f}(\nu) e_\nu$.*

Proof 1. Since $f * F_n$ is a trigonometric polynomial of degree $\leq n$, it follows from Remark 4.2.7

$$\|f - s_n f\|_2 \leq \|f - f * F_n\|_2 \leq \|f - f * F_n\|_\infty \to 0 \qquad (n \to \infty).$$

2. It holds that $s_n f \in T_n$ and $\langle f - s_n f, p \rangle = 0$ for all $p \in T_n$. Thus, from the Pythagorean theorem (Exercise 4.2.20.2)

$$\|f\|_2^2 = \|f - s_n f\|_2^2 + \|s_n f\|_2^2.$$

By 1. $\|f - s_n f\|_2^2 \to 0$ holds $(n \to \infty)$, so $\|s_n f\|_2^2 \to \|f\|_2^2$ follows $(n \to \infty)$. Also, again with the Pythagorean theorem,

$$\|s_n f\|_2^2 = \| \sum_{v=-n}^{n} \widehat{f}(v) e_v \|_2^2 = \sum_{v=-n}^{n} |\widehat{f}(v)|^2 \|e_v\|_2^2 = \sum_{v=-n}^{n} |\widehat{f}(v)|^2.$$

Thus, 2. follows.

3. If $f, g \in C(\mathbb{S})$ with $\widehat{f} = \widehat{g}$, then $\|f - g\|_2 = 0$ according to 2. so $f = g$.

4. Due to the uniform convergence, $g : \mathbb{S} \to \mathbb{C}$ with

$$g := \lim_{n \to \infty} s_n f = \sum_{v=-\infty}^{\infty} \widehat{f}(v) e_v$$

is continuous on $\mathbb{S}$. Moreover, according to Remark 4.2.3, $\widehat{g} = \widehat{f}$ then holds. According to 3. $f = g$.   □

Remark 4.2.13 From Parseval's equation, it follows that for $f \in C(\mathbb{S})$ in particular $\widehat{f}(k) \to 0$ holds for $k \to \pm\infty$.

Example 4.2.14 From Example 4.2.6 and Theorem 4.2.12.4 we obtain

$$t^2 = \frac{\pi^2}{3} + 4 \sum_{v=1}^{\infty} \frac{(-1)^v}{v^2} \cos(vt)$$

for all $t \in [-\pi, \pi]$. In particular, for $t = \pi$ we get

$$\zeta(2) = \sum_{v=1}^{\infty} \frac{1}{v^2} = \frac{\pi^2}{6}$$

and for $t = 0$

$$\sum_{v=1}^{\infty} \frac{(-1)^{v-1}}{v^2} = \frac{\pi^2}{12}.$$

Remark 4.2.15 If $f \in A(\overline{\mathbb{D}} \setminus U_r(0))$ for some $r \in [0, 1)$, then from Theorem 4.1.4 with $a = 0$

$$\frac{1}{2\pi i} \int_{k_1(0)} f(\zeta) \frac{d\zeta}{\zeta} = (Cf)(0) = \frac{1}{2\pi i} \int_{k_1(0)} f(r\zeta) \frac{d\zeta}{\zeta}.$$

Applying this to $z \mapsto f(z) z^{-k}$ for $k \in \mathbb{Z}$, in the case $r > 0$

$$\widehat{f}(k) = \frac{1}{2\pi i} \int_{k_1(0)} f(r\zeta)(r\zeta)^{-k} \frac{d\zeta}{\zeta} = r^{-k} \int f(r\zeta) \overline{\zeta}^k \, dm(\zeta) \quad (k \in \mathbb{Z})$$

and thus

$$|\widehat{f}(k)| \le r^{-k} \max_{K_r(0)} |f| \qquad (k \in \mathbb{Z}).$$

If $r = 0$, this applies for negative k.

Correspondingly, in the case of $f \in A(B_R(0) \setminus \mathbb{D})$ for $R > 1$ by applying to $z \mapsto f(Rz)(Rz)^{-k}$

$$\widehat{f}(k) = \frac{1}{2\pi i} \int_{k_1(0)} f(R\zeta)(R\zeta)^{-k} \frac{d\zeta}{\zeta} = \frac{1}{R^k} \int f(R\zeta)\overline{\zeta}^{k} \, dm(\zeta) \quad (k \in \mathbb{Z})$$

and thus

$$|\widehat{f}(k)| \le R^{-k} \max_{K_R(0)} |f| \qquad (k \in \mathbb{Z}).$$

If $0 \le r < R \le \infty$ and $a \in \mathbb{C}$, then

$$V_{r,R}(a) := \{z \in \mathbb{C} : r < |z - a| < R\} = U_R(a) \setminus B_r(a)$$

is the open **annulus** with center a, inner radius r and outer radius R. In particular, then $V_{0,R}(a) = U_R^*(a)$ and $V_{0,\infty}(0) = U_\infty^*(0) = \mathbb{C}^*$.

Remark 4.2.16 Let $f \in C(\mathbb{S})$ and $c_k := \widehat{f}(k)$ for $k \in \mathbb{Z}$. Further, let

$$0 \le r < 1 < R \le \infty.$$

Then the following holds:

1. If f is a restriction of a function that is continuous on $U_R(0) \setminus \mathbb{D}$ and holomorphic in the interior, then f is for every $R' \in (1, R)$ a restriction of a function in $A(B_{R'}(0) \setminus \mathbb{D})$. From Remark 4.2.15, it follows that the power series $\sum_{\nu=0}^{\infty} c_\nu z^\nu$ has a radius of convergence $\ge R$. Therefore, $\sum_{\nu=0}^{\infty} c_\nu z^\nu$ converges locally uniformly on $U_R(0)$.

2. If f is a restriction of a function that is continuous on $\overline{\mathbb{D}} \setminus B_r(0)$ and holomorphic in the interior, then f is for every $r' \in (r, 1)$ a restriction of a function in $A(\overline{\mathbb{D}} \setminus U_{r'}(0))$. From Remark 4.2.15, it follows that the power series $\sum_{\mu=1}^{\infty} c_{-\mu} w^\mu$ has a radius of convergence $\ge 1/r$. Therefore, the series $\sum_{\mu=1}^{\infty} c_{-\mu} z^{-\mu}$ converges locally uniformly on $\mathbb{C} \setminus B_r(0)$.

If 1. and 2. are fulfilled, then

$$g(z) := \sum_{\nu=-\infty}^{\infty} c_\nu z^\nu = \sum_{\nu=0}^{\infty} c_\nu z^\nu + \sum_{\mu=1}^{\infty} c_{-\mu} z^{-\mu} \qquad (z \in V_{r,R}(0))$$

exists with local uniform convergence on $V_{r,R}(0)$, and therefore also with uniform convergence on all compact subsets of $V_{r,R}(0)$. In particular, uniform convergence on $\mathbb{S}$ is given, and from Theorem 4.2.12.4 it follows that $g|_{\mathbb{S}} = f$. Since g is also holomorphic in $V_{r,R}(0)$, it can be seen that f is actually the restriction of a function that is holomorphic on $V_{r,R}(0)$.

By suitable scaling, it follows

Theorem 4.2.17 *Let $0 \leq r < R \leq \infty$, $a \in \mathbb{C}$ and $f \in H(V_{r,R}(a))$. Then there exists exactly one tuple $(c_k)_{k\in\mathbb{Z}} = (c_{k,r,R}(f,a))_{k\in\mathbb{Z}}$ such that*

$$f(a+h) = \sum_{\nu=-\infty}^{\infty} c_\nu h^\nu$$

with locally uniform convergence on $V_{r,R}(0)$. Here,

$$c_k = \frac{1}{\rho^k} \int f(a+\rho\tau)\bar{\tau}^k \, dm(\tau) = \frac{1}{2\pi i} \int_{k_\rho(a)} \frac{f(\zeta)}{(\zeta-a)^{k+1}} \, d\zeta$$

for any $r < \rho < R$.

Proof

1. Let $\sigma \in (r, R)$ be fixed. We consider $g \in H(V_{r/\sigma,R/\sigma}(0))$, defined by

$$g(w) := f(a+\sigma w) \qquad (w \in V_{r/\sigma,R/\sigma}(0)).$$

According to Remark 4.2.16,

$$f(a+h) = g\left(\frac{h}{\sigma}\right) = \sum_{\nu=-\infty}^{\infty} \frac{\widehat{g}(\nu)}{\sigma^\nu} h^\nu$$

holds for $|h| = \sigma$, where the series on the right side converges locally uniformly on $V_{r,R}(0)$. Moreover, for $k \in \mathbb{Z}$ and $r < \rho < R$ with (4.1.3) and Remark 4.2.15,

$$\frac{1}{2\pi i} \int_{k_\rho(a)} \frac{f(\zeta)}{(\zeta-a)^{k+1}} \, d\zeta = \frac{1}{\rho^k} \int f(a+\rho\tau)\bar{\tau}^k \, dm(\tau)$$

$$= \frac{1}{\rho^k} \int g(\rho\tau/\sigma)\bar{\tau}^k \, dm(\tau) = \frac{\widehat{g}(k)}{\sigma^k},$$

where the right side is independent of ρ! Since $\sigma \in (r, R)$ was arbitrary, the claimed series representation follows.

2. If $f(a+h) = \sum_{\nu=-\infty}^{\infty} b_\nu h^\nu$ locally uniform in $V_{r,R}(0)$, then for $\rho \in (r, R)$

$$c_k = \rho^{-k} \int f(a+\rho\tau)\bar{\tau}^k \, dm(\tau) = \sum_{\nu=-\infty}^{\infty} b_\nu \rho^{\nu-k} \int \tau^{\nu-k} \, dm(\tau) = b_k \qquad (k \in \mathbb{Z}).$$

Thus, there is only one sequence (c_k) as in 1.   □

Definition 4.2.18 Under the conditions of the preceding theorem, $c_k = c_{k,r,R}(f, a)$ is called the k-th **Laurent coefficient** of f and $\sum_{\nu=-\infty}^{\infty} c_\nu h^\nu$ is the **Laurent series** (or also **Laurent expansion**) of f with respect to a and the pair of radii (r, R).

Example 4.2.19

1. We consider the function $f : \mathbb{C} \setminus \{1\} \to \mathbb{C}$ with

$$f(z) = \frac{1}{z - 1} \qquad (z \neq 1).$$

Then, the Laurent development of f with respect to 0 and the pair of radii $(1, \infty)$ is given by

$$f(z) = \frac{1}{z} \frac{1}{1 - 1/z} = \sum_{\mu=1}^{\infty} z^{-\mu} = \sum_{\nu=-\infty}^{-1} z^\nu \qquad (|z| > 1).$$

2. Let $f : \mathbb{C} \setminus \{1, 2\} \to \mathbb{C}$ be defined by

$$f(z) = \frac{1}{z - 1} + \frac{1}{2 - z} \qquad (z \neq 1, 2).$$

Then, the Laurent development of f with respect to 0 and the pair of radii $(1, 2)$ is given by

$$f(z) = \sum_{\mu=1}^{\infty} z^{-\mu} + \sum_{\nu=0}^{\infty} \frac{z^\nu}{2^{\nu+1}} = \sum_{\nu=-\infty}^{-1} z^\nu + \sum_{\nu=0}^{\infty} \frac{z^\nu}{2^{\nu+1}} \qquad (1 < |z| < 2).$$

Exercises 4.2.20

1. Let $(V, \langle \cdot, \cdot \rangle)$ be a unitary space. Prove:
   a) (**Cauchy-Schwarz Inequality**) For all $u, v \in V$, the following holds

$$|\langle u, v \rangle|^2 \leq \langle u, u \rangle \cdot \langle v, v \rangle.$$

   b) A norm on V is defined by $||v|| := \sqrt{\langle v, v \rangle}$.
2. Let $(V, \langle \cdot, \cdot \rangle)$ be a unitary space and $||\cdot||$ the induced norm. Show:
   a) (**Pythagoras**) If $(u_\alpha)_{\alpha \in I}$ is an orthogonal system and $J \subset I$ is finite, then

$$\left\| \sum_{j \in J} u_j \right\|^2 = \sum_{j \in J} \|u_j\|^2.$$

   b) If $(v_j)_{j \in J}$ is a finite orthonormal system and $u \in V$, then

$$\left\langle u - \sum_{j \in J} \langle u, v_j \rangle v_j, v_k \right\rangle = 0$$

   holds for all $k \in J$ and

$$||u - v|| \geq \left\| u - \sum_{j \in J} \langle u, v_j \rangle v_j \right\|$$

   holds for all $v \in \mathrm{span}\{v_j : j \in J\}$.

3. Let $f \in C(\mathbb{S})$ with

$$\sum_{\nu=-\infty}^{\infty} |\widehat{f}(\nu)| < \infty.$$

Show that the Fourier series of f converges uniformly to f on $\mathbb{S}$.

4. Verify that the Fourier coefficients of $f : \mathbb{S} \to \mathbb{R}$, defined by

$$f(e^{it}) := \frac{\pi}{2} - |t| \qquad (t \in (-\pi, \pi]),$$

are given by

$$\widehat{f}(k) = \begin{cases} \frac{2}{\pi k^2}, & \text{for } k \text{ odd} \\ 0, & \text{for } k \text{ even} \end{cases}$$

and prove that

$$f(e^{it}) = \frac{4}{\pi} \sum_{\mu=0}^{\infty} \frac{\cos((2\mu + 1)t)}{(2\mu + 1)^2}$$

holds with uniform convergence on $\mathbb{S}$.

5. Calculate $\frac{1}{n+1} \sum_{\nu=0}^{n} s_\nu f$ for $f : \mathbb{S} \to \mathbb{C}$ with $f(z) = z$.

6. Let $f, g \in C(\mathbb{S})$. Show:
   a) $f * g \in C(\mathbb{S})$,
   b) $f * g = g * f$.

7. Prove:
   a) For all $f, g \in C(\mathbb{S})$ is $\widehat{f * g} = \widehat{f} \cdot \widehat{g}$.
   b) $(C(\mathbb{S}), *)$ is an abelian semigroup.
      Hint: Use the injectivity of the Fourier transformation.

8. Determine the Laurent coefficients of $f : \mathbb{C} \setminus \{\pm i, 0\} \to \mathbb{C}$ with

$$f(z) = \frac{1}{z(1 + z^2)} \qquad (z \in \mathbb{C} \setminus \{\pm i, 0\})$$

with respect to 0 and the radius pairs $(0, 1)$ and $(1, \infty)$.

4.3   Isolated Singularities

Often one is interested in the behavior of holomorphic functions when approaching boundary points of the domain. The simplest case of such a boundary point is that of an isolated point, which we will now examine in more detail.

Remark and Definition 4.3.1 Let $\Omega \subset \mathbb{C}$ be an open set and $a \in \Omega$. If $f \in H(\Omega \setminus \{a\})$, then a is called an **isolated singularity** of f. If we set $R := \operatorname{dist}(a, \partial\Omega)$, then f has, according to Theorem 4.2.17, exactly one Laurent expansion

$$f(a + h) = \sum_{\nu=-\infty}^{\infty} c_\nu(a)h^\nu = \sum_{\mu=1}^{\infty} c_{-\mu}(a)h^{-\mu} + \sum_{\nu=0}^{\infty} c_\nu(a)h^\nu$$

with respect to a and with radius pair $(0, R)$, where $c_k(a) := c_{k,r,R}(f, a)$. We then briefly speak of the Laurent expansion of f with respect to a. In addition, the power series $\sum_{\nu=0}^{\infty} c_\nu(a)h^\nu$ is called the **regular part** and the series $\sum_{\mu=1}^{\infty} c_{-\mu}(a)h^{-\mu}$ the **principal part** of the Laurent expansion. Then a is called

1. **removable singularity**, if $c_{-\mu}(a) = 0$ for all $\mu \in \mathbb{N}$,
2. **pole** of **order** $p \in \mathbb{N}$, if $c_{-p}(a) \neq 0$ and $c_{-\mu} = 0$ for all $\mu > p$.
3. **essential singularity**, if $c_{-\mu}(a) \neq 0$ for infinitely many $\mu \in \mathbb{N}$.

Remark and Definition 4.3.2 If f has a pole of order p at a, we set

$$n_f(a) := -p.$$

If one considers that the above Laurent expansion can be understood as a generalization of the power series expansion with center a of a function holomorphic in Ω, one can see that the definition of the order $n_f(a)$ as negative pole order is a natural extension of the original.

Example 4.3.3
1. Let $f \in H(\mathbb{C}^*)$ be defined by

$$f(z) = \frac{e^z - 1}{z} \qquad (z \in \mathbb{C}^*).$$

Here,

$$f(z) = \sum_{\nu=0}^{\infty} \frac{z^\nu}{(\nu + 1)!} \qquad (z \in \mathbb{C}^*)$$

holds with locally uniform convergence. Therefore, $c_{-\mu}(0) = 0$ for all $\mu \in \mathbb{N}$ and f has a removable singularity at 0.
2. For $p \in \mathbb{N}$ let $f \in H(\mathbb{C}^*)$ be defined by

$$f(z) := e^z z^{-p} \qquad (z \in \mathbb{C}^*).$$

Then

$$f(z) = \sum_{k=0}^{\infty} \frac{1}{k!} z^{k-p} = \sum_{\nu=-p}^{\infty} \frac{1}{(\nu + p)!} z^\nu \qquad (z \in \mathbb{C}^*)$$

holds with locally uniform convergence, so $c_{-p}(0) = 1$ and $c_{-\mu}(0) = 0$ for $\mu > p$. Thus, f at 0 has a pole of order p.
3. Let $f \in H(\mathbb{C}^*)$ be defined by

$$f(z) = e^{1/z} \qquad (z \in \mathbb{C}^*).$$

Then

$$f(z) = 1 + \sum_{\mu=1}^{\infty} \frac{1}{\mu!} z^{-\mu} \qquad (z \in \mathbb{C}^*)$$

holds with locally uniform convergence. Therefore, here $c_{-\mu}(0) = 1/\mu! \neq 0$ for all $\mu \in \mathbb{N}$. Consequently, f at 0 has an essential singularity.

We now want to derive characterizations for all three types of isolated singularities.

Theorem 4.3.4 (Riemann's Removable Singularity Theorem) *Let $\Omega \subset \mathbb{C}$ be open and $f \in H(\Omega \setminus \{a\})$. Then the following are equivalent*

a) *f has a removable singularity at a.*
b) *f can be extended by $f(a) := c_0(a)$ to a holomorphic function f on Ω.*
c) *There exists a neighborhood U of a such that f is bounded on $U \setminus \{a\}$.*

Proof a) $\Rightarrow$ b): If a is a removable singularity of f, then

$$\tilde{f}(z) := \sum_{\nu=0}^{\infty} c_\nu(a)(z-a)^\nu$$

is holomorphic on $U_R(a)$, where $R = \mathrm{dist}(a, \partial\Omega)$, and it holds

$$\tilde{f}(z) = f(z) \qquad (z \in U_R^*(a)).$$

The statement b) $\Rightarrow$ c) is clear.
c) $\Rightarrow$ a): Let $\delta > 0$ and $M \geq 0$ be such that $|f(z)| \leq M$ for $z \in U_\delta^*(a)$. According to Theorem 4.2.17, for $k \in \mathbb{N}$

$$|c_{-k}(a)| \leq \rho^k \int |f(a + \rho\tau)| dm(\tau) \leq M\rho^k \qquad (0 < \rho < \delta).$$

From $M\rho^k \to 0$ for $\rho \to 0^+$ it follows $c_{-k}(a) = 0$.
The following characterization applies for poles of order p.   □

Theorem 4.3.5 *Let $\Omega \subset \mathbb{C}$ be open, $a \in \Omega$ and $f \in H(\Omega \setminus \{a\})$. Then the following are equivalent:*

a) *f has a pole of order p at a.*
b) *There exists an open neighborhood V of 0 and a function $g \in H(V)$ with $g(0) \neq 0$ and*

$$f(a + h) = g(h) h^{-p} \qquad (h \in V \setminus \{0\}).$$

c) *There exists an open neighborhood U of a such that f has no zeros in U and $(1/f)|_{U \setminus \{a\}}$ can be extended by $(1/f)(a) := 0$ to a holomorphic function on U with a zero of order p at a.*

Proof a) $\Rightarrow$ b): Let $R > 0$ with $U_R(a) \subset \Omega$ and $g(h) := h^p f(a + h)$ for $0 < |h| < R$. If

$$f(a + h) = \sum_{\nu=-p}^{\infty} c_\nu(a) h^\nu$$

is the Laurent expansion with respect to a, then $c_{-p}(a) \neq 0$ and thus

$$g(h) = \sum_{\nu=-p}^{\infty} c_\nu(a) h^{\nu+p} = \sum_{\nu=0}^{\infty} c_{\nu-p}(a) h^\nu$$

for $0 < |h| < R$. Therefore, g can be holomorphically extended by $g(0) := c_{-p}(a) \neq 0$ to $V := U_R(0)$. By definition,

$$f(a + h) = g(h) h^{-p} \quad (h \in U_R^*(0)).$$

b) $\Rightarrow$ c): According to the assumption, there exists an $r > 0$ with $g(h) \neq 0$ for $h \in U_r(0)$, so

$$(1/f)(a + h) = h^p (1/g)(h) \quad (h \in U_r^*(0)).$$

With $U := U_r(a)$ is $1/f$ (defined and) holomorphic on U and according to Theorem 3.2.17 has a zero of order p at a.

c) $\Rightarrow$ a): According to the assumption and Theorem 3.2.17

$$(1/f)(a + h) = h^p g(h) \quad (h \in U_r(0))$$

with a zero-free function $g \in H(U_r(0))$. Therefore,

$$f(a + h) = h^{-p}(1/g)(h) \quad (h \in U_r(0)).$$

Since $1/g$ is holomorphic in $U(0)$, $1/g$ has a power series representation

$$(1/g)(h) = \sum_{\nu=0}^{\infty} b_\nu h^\nu \quad (h \in U_r(0)).$$

Thus,

$$f(a + h) = \sum_{\nu=0}^{\infty} b_\nu h^{\nu-p} = \sum_{\nu=-p}^{\infty} b_{\nu+p} h^\nu$$

with locally uniform convergence in $U_r^*(0)$ and $c_{-p}(a) = b_0 = 1/g(0) \neq 0$. □

As a consequence, we obtain

Theorem 4.3.6 *Let $\Omega \subset \mathbb{C}$ be open, $a \in \Omega$ and $f \in H(\Omega \setminus \{a\})$. Then the following are equivalent*

a) *f has a pole at a (of any positive order).*
b) *It holds $|f(z)| \to \infty$ for $z \to a$.*

Proof If f has a pole at a, say of order p, then with g as in Theorem 4.3.5 (since $g(0) \neq 0$)

$$|f(a+h)| = \frac{|g(h)|}{|h|^p} \to \infty \qquad (h \to 0).$$

Conversely, if

$$|f(a+h)| \to \infty \qquad (h \to 0),$$

then there exists an open neighborhood U of a with $f(z) \neq 0$ in $U \setminus \{a\}$ and

$$\lim_{h \to 0} (1/f)(a+h) = 0,$$

thus $1/f$ is holomorphically extendable at the point a with a zero, say of order p. According to Theorem 4.3.5, f has a pole of order p. $\qquad\square$

Remark and Definition 4.3.7 If f has a removable singularity at a point a, then f can be uniquely extended to a function that is holomorphic at a. We also want to assign a value to functions at pole points. To do this, we extend the complex numbers by a point, which we call ∞, and set

$$\mathbb{C}_\infty := \mathbb{C} \cup \{\infty\}.$$

Considering the above characterizations of poles, we agree

$$1/\infty := 0 \quad \text{and} \quad 1/0 := \infty.$$

Now if $\Omega \subset \mathbb{C}$ is open, then $f : \Omega \to \mathbb{C}_\infty$ is called **meromorphic** (in Ω), if for all $a \in \Omega$ an open neighborhood U of a exists such that $f|_U$ or $(1/f)|_U$ is holomorphic. Note that according to this definition, meromorphic functions can also be constant with value ∞ around points a.[2] We set

$$M(\Omega) := M(\Omega, \mathbb{C}_\infty) := \{f : \Omega \to \mathbb{C}_\infty : f \text{ meromorphic in } \Omega\}.$$

According to Theorem 4.3.5, f has a pole at the point $a \in \Omega$ (and the function value $f(a) = \infty$) if and only if $1/f$ has a zero of finite order at a. Here, $n_{1/f}(a) = -n_f(a)$. We write $P(f)$ for the set of poles of f (and again $Z(f)$ for the set of zeros). For $f : \Omega \to \mathbb{C}_\infty$, $f \in M(\Omega)$ if and only if $1/f \in M(\Omega)$. Here, $P(1/f) = Z(f)$, if f is not constant with value 0 or ∞ at any point.

Remark and Definition 4.3.8 We extend the addition on $\mathbb{C}$ to a mapping on $(\mathbb{C}_\infty \times \mathbb{C}_\infty) \setminus \{(\infty, \infty)\}$ such that ∞ acts absorbingly with respect to $\mathbb{C}$, thus

---

[2] Most often, the case of locally constant functions with value ∞ is not included in the definition of meromorphic functions. This has advantages from an algebraic point of view (the functions that are meromorphic in this sense form the quotient field of the ring of functions that are holomorphic in G; see for example [1]). However, for our purposes, the extended definition proves to be more practical.

$$\infty + z := z + \infty := \infty \quad (z \in \mathbb{C}),$$

and the multiplication to a mapping on $(\mathbb{C}_\infty \times \mathbb{C}_\infty) \setminus \{(0, \infty), (\infty, 0)\}$ such that ∞ is absorbing with respect to $\mathbb{C}^* \cup \{\infty\}$, thus

$$z \cdot \infty = \infty \cdot z := \infty \quad (z \in \mathbb{C}^* \cup \{\infty\}).$$

Therefore, if $f, g \in M(\Omega)$ are not locally constant at any point with value 0 or ∞, then $f \cdot g$ and $f/g \in M(\Omega)$ are also in the following sense: The functions $f \cdot g$ and f/g, which are defined pointwise on Ω at least apart from the zero and pole points of f and g, have uniquely determined meromorphic continuations on Ω, which are then also denoted by $f \cdot g$ or f/g (cf. Exercise 4.3.15.2). Also, in this sense sums of two such functions are meromorphic.

Example 4.3.9
1. It holds $Z(\sin) = \pi\mathbb{Z}$ and $Z(\cos) = \pi(\mathbb{Z} + 1/2)$. Therefore,

$$\tan = \sin \cdot (1/\cos) \quad \text{and} \quad \cot = 1/\tan$$

   are defined as $\mathbb{C}_\infty$-valued functions pointwise on $\mathbb{C}$ and are meromorphic in $\mathbb{C}$. Since all zeros of sin and cos are of first order, tan has zeros of first order at the positions $k\pi$ and cot at $(k + 1/2)\pi$ for $k \in \mathbb{Z}$. Therefore, $\cot = 1/\tan$ has poles of first order at $k\pi$ and $\tan = 1/\cot$ has poles of first order at $(k + 1/2)\pi$.
2. Let $f \in H(\mathbb{C}^*)$ be defined by

$$f(z) = \sin(1/z) \quad (z \in \mathbb{C}^*).$$

   Then f and $1/f$ are meromorphic in $\mathbb{C}^*$ with

$$P(1/f) = Z(f) = \{1/(k\pi) : k \in \mathbb{Z}^*\}.$$

Remark and Definition 4.3.10 We now want to provide $\mathbb{C}_\infty$ with a natural metric so that meromorphic functions become continuous mappings. For this purpose, let

$$S^2 := K_{1/2, \mathbb{R}^3}(0, 0, 1/2)$$

be the surface of the ball with center $(0, 0, 1/2)$ and radius $1/2$ in $(\mathbb{R}^3, || \cdot ||_2)$. Then, by

$$\varphi(z) := \frac{1}{1 + |z|^2} (\operatorname{Re} z, \operatorname{Im} z, |z|^2) \quad (z \in \mathbb{C})$$

an bijective mapping from $\mathbb{C}$ to $S^2 \setminus \{(0, 0, 1)\}$ is defined (Exercise 4.3.15.4). Geometrically, the point $\varphi(z)$ results as the intersection of the sphere S^2 with the line segment between the points $(0, 0, 1)$, i.e., the "north pole" of the sphere, and the point $(s, t, 0)$ (see Fig. 4.4). With $\varphi(\infty) := (0, 0, 1)$, $\varphi : \mathbb{C}_\infty \to S^2$ is bijective. The inverse mapping, called **stereographic projection**, is given by

$$\varphi^{-1}(\xi, \eta, \zeta) = \begin{cases} \frac{\xi}{1-\zeta} + i\frac{\eta}{1-\zeta}, & \text{if } \zeta \neq 1 \\ \infty, & \text{if } \zeta = 1 \end{cases}$$

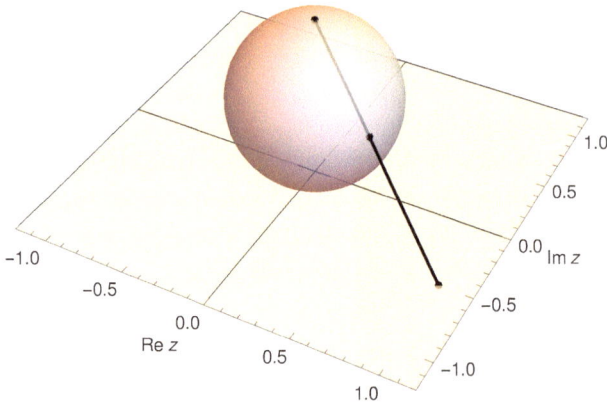

Fig. 4.4 Stereographic projection for $z = 1 - i/2$.

(again Exercise 4.3.15.4). Thus, a metric on $\mathbb{C}_\infty$ is defined by

$$\chi(z, w) := \|\varphi(z) - \varphi(w)\|_2 \qquad (z, w \in \mathbb{C}_\infty)$$

Figure 4.4 illustrates the distance $\chi(1 - i/2, \infty)$ as the length of the part of the line segment from the north pole to the point $(1, -1/2, 0)$ that lies within the sphere.

The metric χ is called **chordal metric**. Furthermore, one then speaks of the **Riemann sphere** $\mathbb{C}_\infty$, since in addition to the metric structure according to Remark 4.3.8 there is also an arithmetic structure on $\mathbb{C}_\infty$ originating from $\mathbb{C}$ (with appropriate restrictions). One can calculate (Exercise 4.3.15.5) that for $z \in \mathbb{C}$

$$\chi(z, w) = \begin{cases} \dfrac{|z-w|}{\sqrt{(1+|z|^2)(1+|w|^2)}}, & \text{if } w \in \mathbb{C} \\[2mm] \dfrac{1}{\sqrt{1+|z|^2}}, & \text{if } w = \infty \end{cases}$$

holds. If (w_n) is a sequence in $\mathbb{C}$, then $\chi(w_n, \infty) \to 0$ exactly when $|w_n| \to \infty$, and $\chi(w_n, w) \to 0$ for $w \in \mathbb{C}$ exactly when $|w_n - w| \to 0$. If $f : \Omega \to \mathbb{C}_\infty$ is meromorphic, then f is continuous, also at poles.

Theorem 4.3.11 *The metric space* $(\mathbb{C}_\infty, \chi)$ *is compact.*

Proof The mapping $\varphi : (\mathbb{C}_\infty, \chi) \to (S^2, d_{S^2})$ is an isometry. Thus, according to Theorem 2.5.10 (applied to the inverse function φ^{-1}), the compactness of S^2 transfers to $\mathbb{C}_\infty$.   □

Finally, we will characterize the behavior of functions near essential singularities. We will later learn a significant strengthening with the Picard theorem.

Theorem 4.3.12 (Casorati-Weierstraß) *Let* $\Omega \subset \mathbb{C}$ *be open and* $f \in H(\Omega \setminus \{a\})$. *Then the following are equivalent:*

a) f has an essential singularity at a.
b) For all open neighborhoods U of a in Ω, $f(U \setminus \{a\})$ is dense in $\mathbb{C}$.
c) For every $w \in \mathbb{C}$ there exists a sequence (z_n) in $\Omega \setminus \{a\}$ with $z_n \to a$ and $f(z_n) \to w$ for $n \to \infty$.

Proof a) $\Rightarrow$ b): Assume there exists an open neighborhood U of a such that $f(U \setminus \{a\})$ is not dense in $\mathbb{C}$. Then there exist a $w \in \mathbb{C}$ and a $\delta > 0$ such that $|f(z) - w| \geq \delta$ for all $z \in U \setminus \{a\}$. We define $g : U \setminus \{a\} \to \mathbb{C}$ by

$$g(z) := \frac{1}{f(z) - w} \qquad (z \in U \setminus \{a\}).$$

Then $g \in H(U \setminus \{a\})$ with $|g(z)| \leq 1/\delta$ for all $z \in U$, $z \neq a$. Thus, g has a removable singularity at a according to Theorem 4.3.4 (we also write g for the continuation). If $g(a) \neq 0$, then f has a removable singularity at a, and if $g(a) = 0$, then f has a pole at a. Contradiction!

b) $\Rightarrow$ c): If $w \in \mathbb{C}$, then for every $n \in \mathbb{N}$ there exists a z_n with $0 < |z_n - a| < 1/n$ and $|f(z_n) - w| < 1/n$. Thus, $z_n \to a$ and $f(z_n) \to w$ as $n \to \infty$.

c) $\Rightarrow$ a): If condition c) holds, then f is unbounded in every neighborhood of a, and certainly $|f(z)| \to \infty$ does not hold for $z \to a$. Consequently, f has neither a removable singularity nor a pole at a (Theorem 4.3.4 or 4.3.6). Therefore, f has an essential singularity at a. □

Example 4.3.13 If $f(z) = e^{1/z}$ for $z \in \mathbb{C}^*$, then f has an essential singularity at 0. For $w \in \mathbb{C}^*$ and $w = re^{i\varphi}$ with $\varphi \in (-\pi, \pi]$ the sequence

$$z_n = \left(\ln r + i(\varphi + 2n\pi) \right)^{-1}$$

satisfies $z_n \to 0$ $(n \to \infty)$ and

$$f(z_n) = e^{\ln r + i(\varphi + 2n\pi)} = re^{i\varphi} = w.$$

Therefore, $f(z_n) = w$ for all $n \in \mathbb{N}$ and thus $f(U \setminus \{0\}) = \mathbb{C}^*$ for all neighborhoods U of 0. In this example, therefore, every $w \neq 0$ is actually taken infinitely often as a value in every (punctured) neighborhood of 0, much stronger than guaranteed by the Casorati-Weierstrass theorem.

Remark 4.3.14 An entire function f is called **transcendental**, if f is not a polynomial. By transferring the theorem of Casorati-Weierstrass, one can see: If f is transcendental, then for every $w \in \mathbb{C}$ there exists a sequence (z_n) in $\mathbb{C}$ with $|z_n| \to \infty$ and $f(z_n) \to w$ $(n \to \infty)$.

Because: Let $f(z) = \sum_{\nu=0}^{\infty} c_\nu z^\nu$ be $(z \in \mathbb{C})$. Then $g : \mathbb{C}^* \to \mathbb{C}$ with $g(z) := f(1/z)$ for $z \in \mathbb{C}^*$ has the Laurent expansion

$$g(z) = c_0 + \sum_{\mu=1}^{\infty} c_\mu z^{-\mu} \qquad (z \in \mathbb{C}^*).$$

Since $c_\mu \neq 0$ for infinitely many μ (remark: f is not a polynomial), g has an essential singularity at 0. Therefore, according to Theorem 4.3.12, for every $w \in \mathbb{C}$ there exists a sequence (ζ_n) in $\mathbb{C}^*$ with $\zeta_n \to 0$ and $g(\zeta_n) \to w$ $(n \to \infty)$. The sequence (z_n) with $z_n = 1/\zeta_n$ then fulfills $|z_n| \to \infty$ and $f(z_n) \to w$ $(n \to \infty)$.

Exercises 4.3.15

1. Determine the types of isolated singularities of the following functions:
   a) $f(z) = \sin(z)/z$ for $z \neq 0$,
   b) $f(z) = \sin(1/z)$ for $z \neq 0$,
   c) $f(z) = 1/\sin(z)$ for $z \notin \pi\mathbb{Z}$,
   d) $f(z) = 1/\sin(1/z)$ for $1/z \notin \pi\mathbb{Z}^*$ and $z \neq 0$.
2. Let $\Omega \subset \mathbb{C}$ be open and f, $g \in M(\Omega)$ not constant at any point with value 0 or value ∞. Show:
   a) With $A := (P(f) \cap Z(g)) \cup (Z(f) \cap P(g))$,

   $$(f \cdot g)(z) := f(z)g(z) \quad (z \in \Omega \setminus A)$$

   defines a function $f \cdot g \in M(\Omega \setminus A)$ which has removable singularities or poles at all points $a \in A$ and can therefore be uniquely extended to a meromorphic function in Ω.
   b) With $A := (Z(f) \cap Z(g)) \cup (P(f) \cap P(g))$,

   $$(f/g)(z) := f(z) \cdot (1/g(z)) \quad (z \in \Omega \setminus A)$$

   defines a function $f/g \in M(\Omega \setminus A)$ which has removable singularities or poles at all points $a \in A$ and can therefore be uniquely extended to a meromorphic function in Ω.
   c) For the correspondingly extended functions $f \cdot g$ and f/g,

   $$n_{f \cdot g} = n_f + n_g \quad \text{and} \quad n_{f/g} = n_f - n_g.$$

3. Let $f, g : \mathbb{C} \to \mathbb{C}$ be defined by $f(z) = 1 - \cos z$ and $g(z) = z \cdot \sin z$ for $z \in \mathbb{C}$. Determine $Z(f/g)$ and $P(f/g)$ as well as $\lim\limits_{h \to 0} f(h)/g(h)$.
4. a) Let $(\xi, \eta, \zeta) \in \mathbb{R}^3$. Show that (ξ, η, ζ) is in S^2 exactly when

   $$\xi^2 + \eta^2 = \zeta(1 - \zeta)$$

   holds.
   b) Show: By

   $$\varphi(z) := \frac{1}{|z|^2 + 1} \, (\mathrm{Re}\, z, \mathrm{Im}\, z, |z|^2) \quad (z \in \mathbb{C})$$

   a bijective mapping from $\mathbb{C}$ to $S^2 \setminus \{(0, 0, 1)\}$ is defined with

   $$\varphi^{-1}(\xi, \eta, \zeta) = \frac{\xi}{1 - \zeta} + i \, \frac{\eta}{1 - \zeta} \quad ((\xi, \eta, \zeta) \in S^2, \zeta \neq 1)$$

and

$$\varphi\left(\frac{1}{z}\right) = (0, 0, 1) - \varphi(-\bar{z}) \qquad (z \in \mathbb{C}^*).$$

5. Prove:
   a) For $z \in \mathbb{C}$,

$$\chi(z, w) = \begin{cases} \dfrac{|z-w|}{\sqrt{(1+|z|^2)(1+|w|^2)}}, & \text{if } w \in \mathbb{C} \\ \dfrac{1}{\sqrt{1+|z|^2}}, & \text{if } w = \infty \end{cases}.$$

   Hint: Use statement a) from the previous Exercise.
   b) For $z, w \in \mathbb{C}_\infty$, $\chi(1/z, 1/w) = \chi(z, w)$.
6. Let $f : \mathbb{C}^* \to \mathbb{C}$ be defined by

$$f(z) = \cosh\left(\frac{1}{z}\right) = \frac{1}{2}\left(e^{1/z} + e^{-1/z}\right) \qquad (z \neq 0).$$

Show: For all open neighborhoods U of 0, $f\left(U \setminus \{0\}\right) = \mathbb{C}$.

4.4 Concepts III: Approximation and Series Developments

In Sect. 4.2 we considered the linear space $C(\mathbb{S})$ of continuous functions on $\mathbb{S}$. If one equips $C(\mathbb{S})$ with the supremum norm $\|\cdot\|_\infty$, one obtains a Banach space. With the $\|\cdot\|_2$-norm, $C(\mathbb{S})$ becomes a unitary space, although in this case the induced metric is not complete. If one considers the convolution $*$ as another operation on $C(\mathbb{S})$, then $C(\mathbb{S})$ becomes a (commutative) algebra. The question arises as to the existence of a unit element, i.e., a function $e \in C(\mathbb{S})$ with

$$f * e = f \qquad (f \in C(\mathbb{S})).$$

If such an e existed, then

$$\hat{e}(k)\hat{e}_k(k) = (\widehat{e * e_k})(k) = \hat{e}_k(k) = 1 \qquad (k \in \mathbb{Z})$$

and thus $\hat{e} = 1_{\mathbb{C}^{\mathbb{Z}}}$, i.e., $s_n e = D_n$. This contradicts Remark 4.2.13, and consequently, no unit element exists.

At this point, the concept of approximate units comes into play. In general, these are families in function spaces that are, in a certain sense, approximately neutral with respect to an operation (in the above example, the convolution on $C(\mathbb{S})$). We have seen that approximate units exist in $C(\mathbb{S})$, such as the family (F_n) of Fejér kernels.

Why are we interested in such approximate units? With $f * F_n$ we have found a sequence of trigonometric polynomials and thus very simple functions, namely linear combinations of the powers

$$e_k(z) = z^k \qquad (k \in \mathbb{Z}),$$

which approximate a given continuous function f on the unit circle $\mathbb{S}$ with increasing n and thus model it in a way. So, the functions e_k can be seen as a kind of building blocks of continuous functions on $\mathbb{S}$: The closure of the linear span—here with respect to the supremum norm on $\mathbb{S}$—is the space $C(\mathbb{S})$.

If we consider the $\|\cdot\|_2$-norm instead of the supremum norm, the situation becomes even more favorable: For any continuous functions, the sequence of Fourier partial sums $(s_n f)$ itself converges to f. We have seen similar phenomena in other situations: Analytic functions can be represented locally as their Taylor series, i.e., in the form

$$f(x + h) = \sum_{\nu=0}^{\infty} \frac{f^{(\nu)}(x)}{\nu!} h^\nu .$$

Again, we see a series expansion in monomials h^k, but now for h in a suitable neighborhood of zero and $k \in \mathbb{N}_0$. Moreover, the convergence here is of much higher quality, namely with geometric speed, and the closer h approaches zero, the better. In the case of functions in the disc algebra, the Taylor and Fourier series coincide. As we will show in the last chapter with the Runge theorems, monomials can be seen as building blocks of holomorphic functions in very general situations: If you have open sets without "holes", polynomials, i.e., linear combinations of the monomials h^k, are sufficient for local uniform approximation, in the case of general open sets rational functions, i.e., quotients of polynomials.

If we no longer deal with analytic functions, the Taylor theorem shows that—depending on the smoothness of the function—approximation by Taylor polynomials up to the corresponding polynomial degree is possible. The Taylor theorem plays an absolutely central role in higher-dimensional real analysis.

One can raise the question of why one is interested in statements about approximations or series expansions.

First of all, of course, the constructive aspect is central: Suitable sections of series expansions or related approximations are in many cases easily accessible for recursive and thus efficient numerical calculation. No computer "knows" the elementary functions, for example. It depends on calculating approximations in one way or another. It gives a good feeling when you can theoretically approximate with any given error tolerance.

Another aspect is more of a structural nature: By approximation, statements about the building blocks can possibly be transferred to the approximated objects. We have learned examples at various points: Continuity is transferred in the case of local uniform convergence from the approximates to the limit functions; the same applies to holomorphy. Integral formulas can often be transferred by limit transitions. The introduction of the Riemann integrals is based on a typical approximation process. The building blocks in this case are functions whose integrals are given in a very simple and "natural" way, namely indicator functions of intervals and the step functions resulting from them by linear combination. We have seen how properties of the integrals of step functions are transferred by (uniform) approximation to regulated functions and thus in particular to continuous functions. Not only the calculation of seemingly self-evident and supposedly self-explanatory mathematical parameters such as areas or path lengths is based on corresponding approximations, but—much more dramatically—even their existence.

Reference

1. Remmert, R.: Funktionentheorie I. Springer, Berlin (1984)

According to the fundamental theorem of calculus, integrals over closed paths vanish if an antiderivative exists. With the Cauchy theorem, a far-reaching generalization of this statement and—closely related to this—also of the Cauchy integral formula is proved. As a consequence, the residue theorem is obtained, according to which the local behavior of holomorphic functions at their isolated singularities determines the integrals over cycles. In the residue theorem, global and local theory merge, the wonder world of function theory reveals itself in all its beauty.

The residue theorem is used—unsurprisingly—in the calculation of integrals. Some simple examples give an impression of the power and elegance of this method. Another application of the residue theorem lies in determining the number of zeros, as shown in Rouché's theorem. By localizing, central statements about the mapping behavior of holomorphic functions are obtained.

5.1   Cauchy Theorem and Applications

For the further, it makes sense to suitably extend the definition of path integrals. We start with further results and terms from topology.

Remark and Definition 5.1.1

1. Let (X, d) be a metric space. For $x \in X$,

$$G(x) = G_X(x) := \bigcup \{A \subset X : x \in A \text{ and } A \text{ connected}\}$$

   is called **connected component** or shortly **component** of X with respect to x. According to Remark 3.2.25, $G(x)$ is connected and according to Exercise 5.1.20.1 for $x, y \in X$ either $G(x) = G(y)$ or $G(x) \cap G(y) = \emptyset$. Thus, $(G(x))_{x \in X}$ is a decomposition of X into maximal connected subsets. If (Y, d_Y) is another

J. Müller, *Concepts of Function Theory*, Mathematics Study Resources 12,
https://doi.org/10.1007/978-3-662-69115-1_5

metric space, then according to Remark 3.2.21 a locally constant function
$f : X \to Y$ is constant on each component of X.

2. If $X \subset \mathbb{K}$ is open, then $G_X(x)$ is also open for all $x \in X$ (Exercise 5.1.20.2).
   Thus, each component of X is open and therefore a domain. In addition, X has
   at most countably many components (again Exercise 5.1.20.2).

Remark and Definition 5.1.2

1. Let I be a finite set, $M \subset \mathbb{C}$ and

$$\gamma_\iota : [\alpha_\iota, \beta_\iota] \to M$$

for $\iota \in I$ paths with starting points a_ι and end points b_ι. The tuple $\gamma := (\gamma_\iota)_{\iota \in I}$
is then called a **chain** (in M) and $\gamma^* := \bigcup_{\iota \in I} \gamma_\iota^*$ the **trace** of γ. If a bijective
mapping $\sigma : \{1, ..., n\} \to I$ exists such that for $j = 1, ..., n-1$ the end points
$b_{\sigma(j)}$ of $\gamma_{\sigma(j)}$ coincide with the starting points $a_{\sigma(j+1)}$ of $\gamma_{\sigma(j+1)}$, we speak of again
of a **path** (in M). From Theorem 3.2.23 and Remark 3.2.25 it follows that the
trace γ^* of a path is connected. Furthermore, the path γ is called **closed**, if
additionally $a_{\sigma(1)} = b_{\sigma(n)}$ holds (this condition is independent of the choice of
σ). Moreover, we call $a_{\sigma(1)}$ **starting point** and $b_{\sigma(n)}$ **end point** of γ.[1] Finally,
we set $\gamma_- := ((\gamma_\iota)_-)_{\iota \in I}$.

Remark and Definition 5.1.3
Let $M \subset \mathbb{C}$ be a set. Then M is called **path-connected**, if for all points $x, y \in M$ a path γ in M exists with starting point x and
endpoint y.

Theorem 5.1.4
Let $M \subset \mathbb{C}$. Then the following holds

1. *If M is path-connected, then M is also connected.*
2. *If M is open and connected, then M is also path-connected.*

Proof.

1. Let $a \in M$ be fixed. Then for every $z \in M$ a path $\gamma(z)$ in M exists with starting
   point a and endpoint z. Thus, $M = \bigcup_{z \in M} \gamma(z)^*$. Since $\gamma(z)^*$ is connected and
   $a \in \bigcap_{z \in M} \gamma(z)^*$ holds, M is connected according to Remark 3.2.25.
2. Let $a \in M$ be fixed and A the set of all $z \in M$ such that a path $\gamma(z)$ in M exists
   with endpoint z and starting point a. If $z \in A$, then there exists a $\delta > 0$ with
   $U_\delta(z) \subset M$. If $w \in U_\delta(z)$, then $(\gamma(z), s_z^w)$ is a path in M with starting point a and
   endpoint w. Thus, A is open in M. The same consideration yields the closedness
   of A in M. Since $A \neq \emptyset$ (Remark: $a \in A$), it follows that $A = M$.

---

[1] Note that for closed paths, the starting and end points are not unique—which is quite natural.

Remark and Definition 5.1.5 If γ is a chain, we define for $f \in C(\gamma^*)$

$$\int_\gamma f = \int_\gamma f(\zeta)d\zeta := \sum_{\iota \in I} \int_{\gamma_\iota} f$$

and

$$L(\gamma) := \sum_{\iota \in I} L(\gamma_\iota) .$$

Again, $L(\gamma)$ is called the **length** of γ. Directly from the respective definition, with Remark 3.5.2

$$\left| \int_\gamma f \right| \le \max_{\gamma^*} |f| \cdot L(\gamma) .$$

With the following theorem, we connect to the fundamental theorem of calculus for path integrals.

Theorem 5.1.6 *Let there be $G \subset \mathbb{C}$ a domain and $f : G \to \mathbb{C}$ continuous.*

1. *If F is an antiderivative to f in G, then*

$$\int_\gamma f = F(b) - F(a)$$

   holds for arbitrary paths in G with starting point a and endpoint b.
2. *A primitive function F to f in G exists exactly when for all closed paths γ in G*

$$\int_\gamma f = 0$$

   holds.

Proof.

1. If F is an antiderivative of f and $\gamma = (\gamma_\iota)_{\iota \in I}$ is a path in G with starting point a and endpoint b, then according to the fundamental theorem calculus for path integrals and with σ as in Remark 5.1.2

$$\int_\gamma f = \sum_{\iota \in I} \int_{\gamma_\iota} f = \sum_{j=1}^n \left(F(b_{\sigma(j)}) - F(a_{\sigma(j)}) \right) = F(b) - F(a).$$

   In particular, the integral vanishes when γ is closed.
2. By 1., we only need to show the reverse direction of 2. For this, let $a \in G$ be fixed. According to Theorem 5.1.4, G is path-connected. If $\gamma(z)$ denotes any path in G with starting point a and endpoint z, then by

$$F(z) := \int_{\gamma(z)} f \qquad (z \in G),$$

a function $F : G \to \mathbb{C}$ is defined.

It is important to note: The value of the integral is independent of the choice of the path $\gamma(z)$, because if $\tilde{\gamma}(z)$ is another such path, then $\gamma := (\gamma(z), \tilde{\gamma}(z)_-)$ is a closed path and therefore, according to the assumption,

$$0 = \int_\gamma f = \int_{\gamma(z)} f - \int_{\tilde{\gamma}(z)} f.$$

If $z \in G$ and $U_r(z) \subset G$, then for $|h| < r$ with $\gamma := (\gamma(z), s_z^{z+h}, \gamma(z+h)_-)$

$$0 = \int_\gamma f = \int_{\gamma(z)} f + \int_z^{z+h} f - \int_{\gamma(z+h)} f = F(z) + \int_z^{z+h} f - F(z+h).$$

Therefore, $F(z+h) - F(z) = \int_z^{z+h} f$ follows and thus

$$\left| \frac{F(z+h) - F(z)}{h} - f(z) \right| = \left| \frac{1}{h} \int_z^{z+h} \big(f(\zeta) - f(z)\big)\, d\zeta \right| \leq \max_{[z, z+h]} |f - f(z)| \to 0$$

for $h \to 0$. Therefore, F is differentiable at z with $F'(z) = f(z)$. Since $z \in G$ was arbitrary, F is an antiderivative of f.

We now consider more general Cauchy integrals.

Remark and Definition 5.1.7 Let $\gamma = (\gamma_\iota)_{\iota \in I}$ be a chain and $f : \gamma^* \to \mathbb{C}$ continuous. Then the function $C_\gamma f : \mathbb{C} \setminus \gamma^* \to \mathbb{C}$, defined by

$$(C_\gamma f)(z) := \frac{1}{2\pi i} \int_\gamma \frac{f(\zeta)}{\zeta - z}\, d\zeta \qquad (z \notin \gamma^*)$$

is called the **Cauchy integral** of f with respect to γ. In the case of a path γ, the Cauchy integral generalizes to one from Remark 4.1.2. According to Theorem 4.1.1,

$$C_\gamma f = \sum_{\iota \in I} C_\gamma f \in H(\mathbb{C} \setminus \gamma^*).$$

Furthermore, for $z \notin \gamma^*$

$$|(C_\gamma f)(z)| \leq \frac{1}{2\pi} \max_{\gamma^*} |f| \cdot L(\gamma) / \mathrm{dist}(z, \gamma^*) \to 0 \qquad (|z| \to \infty).$$

Remark and Definition 5.1.8 A chain $\gamma = (\gamma_\iota)_{\iota \in I}$ we call a **cycle**, if a decomposition $(I_\kappa)_{\kappa \in M}$ of I exists such that $(\gamma_\iota)_{\iota \in I_\kappa}$ for all $\kappa \in M$ is a closed path. Furthermore, for cycles γ

$$\mathrm{ind}_\gamma(z) := (C_\gamma 1)(z) = \frac{1}{2\pi i} \int_\gamma \frac{d\zeta}{\zeta - z}$$

is called the **index** (or also **winding number**) of z with respect to γ. According to Remark 4.1.8, for $a \in \mathbb{C}$ and $\rho > 0$

$$\text{ind}_{k_\rho(a)}(z) = \mathbb{1}_{U_\rho(a)}(z) \qquad (z \in \mathbb{C} \setminus K_\rho(a)).$$

For compact sets $K \subset \mathbb{C}$, the open set $\mathbb{C} \setminus K$ has exactly one unbounded component.

Theorem 5.1.9 *Let γ be a cycle. Then* ind_γ *is constant on each component of $\mathbb{C} \setminus \gamma^*$ and an integer, thus* $\text{ind}_\gamma(\mathbb{C} \setminus \gamma^*) \subset \mathbb{Z}$. *Moreover,* ind_γ *has the value 0 on the unbounded component of $\mathbb{C} \setminus \gamma^*$.*

Proof.

1. Let's first assume $\gamma : [\alpha, \beta] \to \mathbb{C}$ is a path. Then

$$2\pi i \, \text{ind}_\gamma(z) = \int_\alpha^\beta \frac{\gamma'(s)}{\gamma(s) - z} \, ds \quad (z \in \mathbb{C} \setminus \gamma^*).$$

For $z \in \mathbb{C} \setminus \gamma^*$ we define $\varphi = \varphi_z : [\alpha, \beta] \to \mathbb{C}$ by

$$\varphi(t) := \exp\left(\int_\alpha^t \frac{\gamma'(s)}{\gamma(s) - z} ds \right) \quad (t \in [\alpha, \beta]).$$

With the chain rule and the main theorem on integral functions, we get

$$\varphi' = \varphi \frac{\gamma'}{\gamma - z}$$

on $[\alpha, \beta]$. According to the quotient rule,

$$\left(\frac{\varphi}{\gamma - z} \right)' = \frac{\varphi'(\gamma - z) - \varphi\gamma'}{(\gamma - z)^2} = 0$$

on $[\alpha, \beta]$. Thus, there exists a constant c with

$$\varphi(t) = c(\gamma(t) - z) \quad (t \in [\alpha, \beta]).$$

From $\varphi(\alpha) = 1$, we get $c = 1/(\gamma(\alpha) - z)$, thus

$$\varphi(t) = \frac{\gamma(t) - z}{\gamma(\alpha) - z} \quad (t \in [\alpha, \beta]).$$

2. We show that $\text{ind}_\gamma(z)$ is an integer for all $z \in \mathbb{C} \setminus \gamma^*$. It is sufficient to prove the assertion for closed paths $\gamma = (\gamma_\iota)_{\iota \in I}$. For this, let $\varphi_\iota = \varphi_{\iota,z}$ be as in 1. with γ_ι instead of γ. Since γ is a closed path, with 1.

$$\exp(2\pi i \, \text{ind}_\gamma(z)) = \prod_{\iota \in I} \exp(2\pi i \, \text{ind}\gamma_\iota(z)) = \prod_{\iota \in I} \varphi_\iota(\beta_\iota) = \prod_{\iota \in I} \frac{\gamma_\iota(\beta_\iota) - z}{\gamma_\iota(\alpha_\iota) - z} = 1$$

and thus $\text{ind}_\gamma(z) \in \mathbb{Z}$.

3. Let G be a component of $\mathbb{C} \setminus \gamma^*$. Since G is connected and ind_γ is continuous and integral on G, ind_γ is constant in G according to Theorem 3.2.24. Moreover, according to Remark 5.1.7

$$\text{ind}_\gamma(z) \to 0 \qquad (|z| \to \infty).$$

Thus, $|\text{ind}_\gamma(z)| < 1$ for $|z|$ sufficiently large. Since ind_γ is constant on the unbounded component of $\mathbb{C} \setminus \gamma^*$, $\text{ind}_\gamma(z) = 0$ there.

Definition 5.1.10 Let γ be a cycle. Then

$$\text{Int}(\gamma) := \{z \in \mathbb{C} \setminus \gamma^* : \text{ind}_\gamma(z) \neq 0\}$$

is called the **interior** of γ and

$$\text{Ext}(\gamma) := \{z \in \mathbb{C} \setminus \gamma^* : \text{ind}_\gamma(z) = 0\}$$

the **exterior** of γ. If $\Omega \subset \mathbb{C}$ is open and γ is a cycle in Ω, then γ is called **nullhomologous** in Ω or Ω-**nullhomologous**, if $\text{ind}_\gamma(z) = 0$ for all $z \in \Omega^c = \mathbb{C} \setminus \Omega$, that is, if $\Omega^c \subset \text{Ext}(\gamma)$ holds.[2]

The following theorem, which essentially has a real variable character, is used in the proof of the subsequent Cauchy theorem.

Theorem 5.1.11 *Let $X \subset \mathbb{K}$ be open and $f : X \to \mathbb{C}$ twice continuously differentiable. Furthermore, let $g : X \times X \to \mathbb{C}$ be defined by*

$$g(x, y) := \begin{cases} \frac{f(y) - f(x)}{y - x}, & x \neq y \\ f'(x), & x = y \end{cases}.$$

Then $D_1 g$ is (existent and) continuous on $X \times X$.

Proof. Since differentiability and continuity are local properties, we can assume without loss of generality that X is convex (then $[u, v] \subset X$ for all $u, v \in X$).

1. We show:

$$D_1 g(x, y) = \begin{cases} \frac{f(y) - f(x) - f'(x)(y - x)}{(y - x)^2}, & y \neq x \\ \frac{f''(x)}{2}, & y = x \end{cases}.$$

Because: If $(x, y) \in X \times X$ with $y \neq x$, then with the quotient rule

$$D_1 g(x, y) = \frac{1}{(y - x)^2} (f(y) - f(x) - f'(x)(y - x)).$$

If $y = x$, then according to the Taylor theorem (Theorem 3.3.21) for $0 \neq h \in X - x$

---

[2] More generally, two cycles γ and $\widetilde{\gamma}$ are homologous in Ω, if the cycle $(\gamma, \widetilde{\gamma}_-)$ is nullhomologous in Ω.

$$\frac{1}{h}(g(x+h,x) - g(x,x)) = \frac{1}{h^2}(f(x+h) - f(x) - f'(x)h) = \int_0^1 (1-t)f''(x+th)\,dt.$$

From the continuity of f'' at x it follows

$$\int_0^1 (1-t)f''(x+th)\,dt \to f''(x)\int_0^1 (1-t)dt = \frac{f''(x)}{2} \qquad (h \to 0).$$

2. We show: $D_1 g$ is continuous on $X \times X$.
   Because: The continuity of $D_1 g$ is clear at all points (x,y) with $x \neq y$. Let $a \in X$ and $\varepsilon > 0$. As in 1., for $x, y \in X$ with $y \neq x$

$$D_1 g(x,y) = \frac{1}{(y-x)^2}(f(y) - f(x) - f'(x)(y-x)) = \int_0^1 (1-t)f''(x+t(y-x))\,dt.$$

From the continuity of f'' at a and $\int_0^1 (1-t)\,dt = 1/2$ the existence of a neighborhood U of (a,a) follows with

$$\left| \int_0^1 (1-t)f''(x+t(y-x))\,dt - \frac{f''(a)}{2} \right| < \varepsilon \quad \text{and} \quad |f''(x) - f''(a)| < \epsilon$$

for $(x,y) \in U$. Thus, according to 1.,

$$|D_1 g(x,y) - D_1 g(a,a)| = |D_1 g(x,y) - f''(a)/2| < \varepsilon$$

for $(x,y) \in U$ and thus the continuity of $D_1 g$ at (a,a).

Remark 5.1.12 An argument analogous (simpler) to the proof of Theorem 5.1.11 using Taylor's theorem for $n = 0$ instead of $n = 1$ also provides the continuity of g, even for (once) continuously differentiable f (Exercise 5.1.20.4).

With this, we can prove a central result for global function theory:

Theorem 5.1.13 (Cauchytheorem) Let $\Omega \subset \mathbb{C}$ open. If γ is a cycle in Ω, the following statements are equivalent:

a) γ is Ω-nullhomologous.
b) For all $f \in H(\Omega)$ is $f \cdot \mathrm{ind}_\gamma|_{\Omega \setminus \gamma^*} = C_\gamma f|_{\Omega \setminus \gamma^*}$, thus

$$f(z) \cdot \mathrm{ind}_\gamma(z) = \frac{1}{2\pi i} \int_\gamma \frac{f(\zeta)}{\zeta - z}\,d\zeta \qquad (z \in \Omega \setminus \gamma^*).$$

c) For all $f \in H(\Omega)$ is

$$\int_\gamma f = 0.$$

Proof. a) $\Rightarrow$ b) Let g be as in Theorem 5.1.11. Since g is continuous, a function $\Phi : \Omega \to \mathbb{C}$ is defined by

$$\Phi(z) := \frac{1}{2\pi i} \int_\gamma g(z,\zeta)d\zeta = \frac{1}{2\pi i} \sum_{\iota \in I} \int_{\gamma_\iota} g(z,\zeta)\,d\zeta \qquad (z \in \Omega).$$

Since also $D_1 g$ is continuous, Φ is holomorphic in Ω according to Theorem 3.5.8. For $z \in \Omega \setminus \gamma^*$

$$\Phi(z) = \frac{1}{2\pi i} \int_\gamma \frac{f(\zeta)-f(z)}{\zeta - z}\,d\zeta = (C_\gamma f)(z) - f(z)\mathrm{ind}_\gamma(z).$$

So it is enough to show $\Phi = 0$.

Firstly, $C_\gamma f$ is holomorphic in $\mathbb{C} \setminus \gamma^*$ according to Remark 5.1.7 with $\Phi(z) = (C_\gamma f)(z)$ for all $z \in \Omega \cap \mathrm{Ext}(\gamma)$. Furthermore, according to the assumption, $\partial\Omega \subset \Omega^c \subset \mathrm{Ext}(\gamma)$. Thus, an entire function F is defined by

$$F(z) := \begin{cases} \Phi(z), & z \in \Omega \\ (C_\gamma f)(z), & z \in \Omega^c \end{cases}.$$

From $F(z) = (C_\gamma f)(z)$ for $z \in \mathrm{Ext}(\gamma)$, it also follows, again with Remark 5.1.7,

$$F(z) = (C_\gamma f)(z) \to 0 \qquad (|z| \to \infty).$$

With the Liouville theorem, it follows that F is constant and thus also $F = 0$ in $\mathbb{C}$. Therefore, for all $z \in \Omega$,

$$\Phi(z) = F(z) = 0.$$

b) $\Rightarrow$ c) Let $a \in \Omega \setminus \gamma^*$. Then, with b) applied to $f_a : \Omega \to \mathbb{C}$ with $f_a(z) = (z-a)f(z)$,

$$0 = 2\pi i f_a(a) \cdot \mathrm{ind}_\gamma(a) = \int_\gamma \frac{f_a(\zeta)}{\zeta - a}\,d\zeta = \int_\gamma f.$$

c) $\Rightarrow$ a) Follows from $\zeta \mapsto \frac{1}{\zeta - z} \in H(\Omega)$ for all $z \notin \Omega$.

Remark and Definition 5.1.14

1. Let $\Omega \subset \mathbb{C}$ be an open set. Every bounded component of $\mathbb{C} \setminus \Omega$ is called a **hole** of Ω. If γ is a cycle in Ω such that no hole of Ω lies in $\mathrm{Int}(\gamma)$, then γ is nullhomologous in Ω.

   Because: ind_γ is continuous and integer on $\mathbb{C} \setminus \gamma^*$, and in particular on $\mathbb{C} \setminus \Omega$. Since ind_γ is constant on each component A of $\mathbb{C} \setminus \Omega$, a component either lies entirely in $\mathrm{Int}(\gamma)$ or entirely in $\mathrm{Ext}(\gamma)$. Therefore, all bounded components (if they exist) lie in $\mathrm{Ext}(\gamma)$. From $\mathrm{ind}_\gamma(z) \to 0$ for $|z| \to \infty$, it follows that $\mathrm{ind}_\gamma = 0$ also on all unbounded components.

2. A domain G is called **simply connected**, if G has no holes. Then, according to 1., every cycle in G and in particular every closed path in G is nullhomologous in G.

Theorem 5.1.15 *Let G be a simply connected domain and $f \in H(G)$.*

1. (***Cauchy's Integral Formula*** *and* ***Cauchy's Integral Theorem***) *For all closed paths γ in G, the following holds*

$$f(z) \cdot \operatorname{ind}_\gamma(z) = \frac{1}{2\pi i} \int_\gamma \frac{f(\zeta)}{\zeta - z} \, d\zeta \qquad (z \in G \setminus \gamma^*)$$

and

$$\int_\gamma f = 0.$$

2. *The function f has an antiderivative in G.*

Proof. Since every closed path in G is nullhomologous in G, 1. follows from the Cauchy theorem. With Theorem 5.1.6, 2. also follows.

As a first application, we want to deal with the question of the existence of logarithms and general powers in $\mathbb{C}$. In the context of introducing elementary functions, we defined the real logarithm function as the inverse of the real exponential function. The fact that the exponential function is no longer injective in the complex plane suggests that the situation here will become more complicated.

Theorem 5.1.16 *Let $G \subset \mathbb{C}$ be a domain and $g \in H(G)$ without zeros.*

1. *If G is simply connected, then there exists a primitive function f to g'/g in G with $e^f = g$.*
2. *If $f_1, f_2 : G \to \mathbb{C}$ are continuous, then $e^{f_1} = e^{f_2}$ holds if and only if some $k \in \mathbb{Z}$ exists with*
$$f_1(z) = f_2(z) + 2k\pi i \qquad (z \in G).$$

Proof.

1. Since G is simply connected, according to Theorem 5.1.15, a function $f \in H(G)$ exists with $f' = g'/g$. Here, f can be chosen such that for some $a \in G$ and $g(a) = re^{i\theta}$ additionally $f(a) = \ln(r) + i\theta$ holds (if necessary, add a suitable constant to f). It follows

$$(ge^{-f})' = g'e^{-f} + ge^{-f}(-f') = 0.$$

According to Remark 3.2.21, a constant c exists with $g(z) = ce^{f(z)}$ for all $z \in G$. From $e^{f(a)} = g(a)$ it follows that $c = 1$ and thus the assertion.
2. If $f_1, f_2 \in C(G)$ with $e^{f_1} = e^{f_2}$, then $e^{f_1 - f_2} = e^{f_1}/e^{f_2} = 1$ in G. Thus,

$$\varphi(z) = f_1(z) - f_2(z) \in 2\pi i \mathbb{Z}$$

for all $z \in G$. Since G is connected and φ is continuous on G, φ is constant on G according to Theorem 3.2.24, that is, there exists a $k \in \mathbb{Z}$ with

$$f_1(z) = f_2(z) + 2k\pi i \qquad (z \in G).$$

The reversal is clear.

Remark and Definition 5.1.17 Let $G \subset \mathbb{C}$ be a domain and $g \in H(G)$ without zeros. Any function $f \in H(G)$ with $e^f = g$ in G is called a **branch of the logarithm** of g in G. If f is such a branch, then f_k with $f_k(z) = f(z) + 2k\pi i$ for every $k \in \mathbb{Z}$ is a branch. According to Theorem 5.1.16.2, all branches are given by these (countably infinitely many) functions. Furthermore, for $m \in \mathbb{N}$, $m \geq 2$ the function $e^{f/m}$ is called a **branch of the m-th root** of g in G (Remark: $(e^{f/m})^m = e^f = g$).

If $G \subset \mathbb{C}$ is simply connected, then according to Theorem 5.1.16 there exist functions $f \in H(G)$ with $e^f = g$, that is, branches of the logarithm of f in G, and therefore also m-th roots of g.

Example 5.1.18 Since the slit plane

$$\mathbb{C}^- := \mathbb{C} \setminus (-\infty, 0]$$

is simply connected, according to Remark 5.1.17 a branch f of the logarithm of $g(z) = z$ exists on $\mathbb{C}^-$ with $f(1) = 0$ (we have essentially already proven the existence in Example 3.5.10).

If $z \in \mathbb{C}^-$, then due to the polar coordinate representation of z with respect to $\alpha = 0$ there is exactly one $\theta \in (-\pi, \pi)$ with $z = |z|e^{i\theta}$. One calls $\arg(z) := \theta$ then the **argument** of z. Since the mapping

$$\mathbb{C}^- \ni z \mapsto (|z|, \arg(z)) \in (0, \infty) \times (-\pi, \pi)$$

is continuous (see Exercise 2.5.18.8),

$$\mathbb{C}^- \ni z \mapsto \ln|z| + i\arg(z) \in \mathbb{C}$$

is continuous with

$$e^{\ln|z|+i\arg(z)} = |z|e^{i\arg(z)} = z \qquad (z \in \mathbb{C}^-).$$

From $f(1) = 0 = \ln(1) + i\arg(1)$ follows $f(z) = \ln|z| + i\arg(z)$ for all $z \in \mathbb{C}^-$ with Theorem 5.1.16. For $z = r > 0$, in particular, $f(r) = \ln r$ results, that is, f extends the real logarithm $r \mapsto \ln r$ holomorphically to $\mathbb{C}^-$. One calls f the **principal branch of the logarithm** (of z) in $\mathbb{C}^-$ and also writes for it

$$f(z) =: \log z \qquad (z \in \mathbb{C}^-).$$

According to Theorem 5.1.16.2, all other branches have the form

$$z \mapsto \log z + 2k\pi i = \ln|z| + i(\arg(z) + 2k\pi)$$

for a $k \in \mathbb{Z}$ (Fig. 5.1).

What about the validity of the functional equation

$$\log(zw) = \log(z) + \log(w)$$

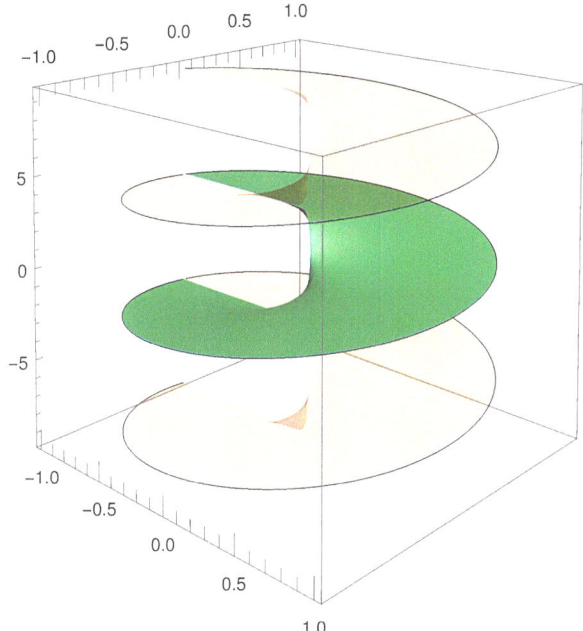

Fig. 5.1 Imaginary parts of three branches of the logarithm with $z \mapsto \arg(z) = \operatorname{Im}(\log(z))$ in green

for $z, w \in \mathbb{C}^-$? If $\arg(z) + \arg(w) \in (-\pi, \pi)$ is true, then $\arg(zw) = \arg(z) + \arg(w)$ and thus also

$$\log(zw) = \log z + \log w.$$

However, if $\arg(z) + \arg(w) > \pi$ is true, then $\arg(zw) = \arg(z) + \arg(w) - 2\pi$, thus

$$\log(zw) = \log z + \log w - 2\pi i.$$

A "correction term" $2\pi i$ is therefore added. If $\varphi + \theta = \pi$, then $\log(zw)$ is not even defined.

Remark and Definition 5.1.19 If $\alpha \in \mathbb{C}$, then we set

$$z^\alpha := e^{\alpha \cdot \log z} \qquad (z \in \mathbb{C}^-).$$

The function $z \mapsto z^\alpha$ is holomorphic in $\mathbb{C}^-$ with derivative $z \mapsto \alpha z^{\alpha-1}$, and for $z \in \mathbb{C}^-$ and $\alpha, \beta \in \mathbb{C}$

$$z^{\alpha+\beta} = z^\alpha z^\beta.$$

However, the problems hinted at above with the functional equation of the logarithm mean that for $z, w \in \mathbb{C}^-$ in general *not* $(zw)^\alpha = z^\alpha w^\alpha$ holds. It can therefore be seen that a too careless handling of complex logarithms and powers can easily lead to false conclusions.

Specifically, if $\alpha = 1/m$ for an $m \in \mathbb{N}$, $m \geq 2$, then one writes for $z \in \mathbb{C}^-$ also $\sqrt[m]{z}$ instead of $z^{1/m}$, and in the case $m = 2$ also briefly $\sqrt{z}$. The function $z \mapsto \sqrt[m]{z}$ is called the **principal branch of the m-th root** of z in $\mathbb{C}^-$ and for $m = 2$ briefly the **principal branch of the root** of z in $\mathbb{C}^-$.

Exercises 5.1.20

1. Let (X, d) be a metric space. Show: An equivalence relation on X is defined by

$$x \sim y :\Leftrightarrow y \in G_X(x).$$

2. Let $X \subset \mathbb{K}$ be open. Show:
   a) Every component $G_X(x)$ of X is open.
   b) X has at most countably many components.
3. Calculate ind_γ for
   a) $\gamma = (k_1(0), k_{1/2}(0))$,
   b) $\gamma = (k_1(0), (k_{1/2}(0))_-)$,
   c) $\gamma = (k_1(0), k_{1/2}(1/2))$.
4. Let $X \subset \mathbb{K}$ be open and $f : X \to \mathbb{C}$ continuously differentiable. Show: If g is as in Theorem 5.1.11, then g is continuous.
5. Let $\Omega \subset \mathbb{C}$ be open, and let $f\,H(\Omega)$. Furthermore, let $a \in \Omega$ and $w_0 := f(a)$, where a is a zero of order m of $f - w_0$. Show: There exist an open neighborhood U of a and a function φ holomorphic in U with $\varphi(a) = 0$ and $\varphi'(a) \neq 0$ and such that

$$f(z) = w_0 + \varphi^m(z) \qquad (z \in U).$$

6. Show: For all $z \in \overline{\mathbb{D}} \setminus \{1\}$ the following holds

$$\sum_{\nu=1}^{\infty} \frac{z^\nu}{\nu} = \log\left(\frac{1}{1-z}\right).$$

Hint: Use example 3.5.10.

7. (Complex binomial series) Let $\alpha \in \mathbb{C}$. Show: For $|z| < 1$ the following holds

$$(1+z)^\alpha = \sum_{\nu=0}^{\infty} \binom{\alpha}{\nu} z^\nu.$$

8. At the end of a letter dated June 17, 1746 to Christian Goldbach, Leonhard Euler writes:

Recently, I have found that this expression $(\sqrt{-1})^{\sqrt{-1}}$ has a real value, which in decimal fractions $= 0.2078795763$, which seems remarkable to me.

What do you think about this?

5.2 Residue Theorem and Applications

We now want to merge the Cauchy theorem with the local theory of isolated singularities.

Remark and Definition 5.2.1 Let $\Omega \subset \mathbb{C}$ be open and $a \in \Omega$. If f is holomorphic in $\Omega \setminus \{a\}$, the (-1)th coefficient $c_{-1}(a)$ of the Laurent development of f with respect to a is called the **residue** of f at the point a. We write

$$\operatorname{res}_f(a) := c_{-1}(a) .$$

According to Theorem 4.2.17

$$\operatorname{res}_f(a) = \rho \int f(a + \rho\tau)\tau \, dm(\tau) = \frac{1}{2\pi i} \int_{k_\rho(a)} f$$

for $0 < \rho < \operatorname{dist}(a, \partial\Omega)$.

Example 5.2.2 (cf. Example 4.3.3)

1. If f has a removable singularity at a, then $\operatorname{res}_f(a) = 0$.
2. If $p \in \mathbb{N}$ and $f(z) = e^z/z^p$ for $z \in \mathbb{C}^*$, then

$$\operatorname{res}_f(0) = \frac{1}{(p-1)!}.$$

3. For

$$f(z) = e^{1/z} = 1 + \sum_{\mu=1}^{\infty} \frac{1}{\mu!} z^{-\mu} \qquad (z \in \mathbb{C}^*)$$

$\operatorname{res}_f(0) = 1$.

Theorem 5.2.3 (Residue Theorem) *Let $\Omega \subset \mathbb{C}$ be open and γ an Ω-nullhomologous cycle. If f is holomorphic in $\Omega \setminus A$ for a discrete set A in Ω and is $\gamma^* \cap A = \emptyset$, then*[3]

$$\frac{1}{2\pi i} \int_\gamma f = \sum_{w \in A} \operatorname{ind}_\gamma(w) \cdot \operatorname{res}_f(w) . \tag{5.2.1}$$

Proof. Let $A_\gamma := A \cap \operatorname{Int}(\gamma)$. Since $\operatorname{Int}(\gamma) \cup \gamma^*$ is compact, A_γ is finite (see Exercise 2.5.18.4). We set $\Omega_\gamma := (\Omega \setminus A) \cup A_\gamma$. By assumption and by definition of A_γ, γ is also Ω_γ-nullhomologous. Without loss of generality, we can assume $A_\gamma \neq \emptyset$ (for $A_\gamma = \emptyset$ the assertion follows from the Cauchy theorem).

---

[3] With the notation introduced in Sect. 1.6, the right side can alternatively be written as $\sum_{w \in \operatorname{Int}(\gamma)} \operatorname{ind}_\gamma(w) \cdot \operatorname{res}_f(w)$ or as $\sum_{w \in \Omega \setminus \gamma^*} \operatorname{ind}_\gamma(w) \cdot \operatorname{res}_f(w)$.

We choose $\delta > 0$ such that $U_\delta(w) \subset \text{Int}(\gamma)$ for all $w \in A_\gamma$ and

$$|w - v| > 2\delta$$

for all $w, v \in A_\gamma$, $w \neq v$ holds. Then f for all $w \in A_\gamma$ according to Theorem 4.2.17 has a Laurent development

$$f(w + h) = \sum_{\nu=-\infty}^{\infty} c_\nu(w) h^\nu \qquad (h \in U_\delta^*(0))$$

with respect to w. The main part

$$\varphi_w(h) := \sum_{\mu=1}^{\infty} a_{-\mu}(w) h^{-\mu}$$

then converges locally uniformly to $\mathbb{C}^*$ (cf. Remark 4.2.16) and in particular uniformly to $\gamma^* - w$. For $\mu > 1$, $\int_\gamma (\zeta - w)^{-\mu} d\zeta = 0$ according to Example 3.5.6. Thus, for $w \in A_\gamma$

$$\int_\gamma \varphi_w(\zeta - w) d\zeta = \sum_{\mu=1}^{\infty} c_{-\mu}(w) \int_\gamma (\zeta - w)^{-\mu} d\zeta = 2\pi i \, \text{ind}_\gamma(w) \cdot \text{res}_f(w).$$

The function $g : \Omega \setminus A \to \mathbb{C}$

$$g(z) := f(z) - \sum_{w \in A_\gamma} \varphi_w(z - w) \qquad (z \in \Omega \setminus A)$$

is holomorphic in $\Omega \setminus A$, and for $w \in A_\gamma$ it holds in $U_\delta^*(0)$

$$g(w + h) = \sum_{\nu=0}^{\infty} a_\nu(w) h^\nu - \sum_{v \in A_\gamma, \, w \neq v} \varphi_v(h + w - v).$$

Since the right side is holomorphic in $U_\delta(0)$, g has a removable singularity at w. Therefore, g can be holomorphically extended to Ω_γ. Since γ is nullhomologous in Ω_γ,

$$\int_\gamma g = 0$$

results from the Cauchy theorem. Consequently,

$$\int_\gamma f = \sum_{w \in A_\gamma} \int_\gamma \varphi_w(\zeta - w) \, d\zeta = 2\pi i \sum_{w \in A_\gamma} \text{ind}_\gamma(w) \cdot \text{res}_f(w).$$

Remark 5.2.4 We highlight various important special cases of the residue theorem.

1. If $a \in \Omega$ and $\rho > 0$ with $B_\rho(a) \subset \Omega$, then in the case $A \cap K_\rho(a) = \emptyset$ with $\gamma = k_\rho(a)$ according to Remark 5.1.8[4]

$$\rho \int f(a + \rho\tau)\tau \, dm(\tau) = \frac{1}{2\pi i} \int_{k_\rho(a)} f = \sum_{w \in A \cap U_\rho(a)} \mathrm{res}_f(w). \qquad (5.2.2)$$

2. If G is a simply connected domain, then (5.2.1) applies to all closed paths γ in G.
3. In the case $A = \emptyset$, (5.2.1) again results in

$$\int_\gamma f = 0$$

for all Ω-nullhomologous γ and all $f \in H(\Omega)$ (see Theorem 5.1.15).

In order to use the residue theorem efficiently, it is important to have techniques for calculating residues available. For poles, the following applies

Theorem 5.2.5 *Let* $\Omega \subset \mathbb{C}$ *be open,* $a \in \Omega$ *and* $f \in H(\Omega \setminus \{a\})$.

1. *If* f *has a pole of order* p *at* a, *then the function* $g \in H(\Omega \setminus \{a\})$ *with* $g(z) := (z - a)^p f(z)$ *has a removable singularity at* a *and*

$$\mathrm{res}_f(a) = \frac{1}{(p-1)!} \lim_{z \to a} g^{(p-1)}(z) = \frac{1}{(p-1)!} g^{(p-1)}(a) \, ,$$

   where in the second equation g *is continuously extended at* a.
2. *If there exist an open neighborhood* U *of* a *and functions* $f_1, f_2 \in H(U)$ *with* $f_2(a) = 0$, $f_2'(a) \neq 0$ *and* $f = f_1/f_2$ *in* $U \setminus \{a\}$, *then*

$$\mathrm{res}_f(a) = \frac{f_1(a)}{f_2'(a)} \, .$$

Proof.

1. It holds for $h \in V_{0,R}(0)$, where $R := \mathrm{dist}(a, \partial\Omega)$,

$$h^p f(a + h) = \sum_{\nu=-p}^{\infty} c_\nu(a) h^{\nu+p} = \sum_{\nu=0}^{\infty} c_{\nu-p}(a) h^\nu = g(a + h) \, .$$

Thus, g has a removable singularity at a and it holds

---

[4] The proof of the residue theorem shows that this simple version does not require the Cauchy theorem.

$$\mathrm{res}_f(a) = c_{-1}(a) = \frac{g^{(p-1)}(a)}{(p-1)!} = \frac{1}{(p-1)!} \lim_{z \to a} g^{(p-1)}(z).$$

2. If $f_1(a) \neq 0$, then by assumption f_2/f_1 has a root of order 1 at a, thus f has a pole of order 1 at a. According to 1.,

$$\mathrm{res}_f(a) = \lim_{z \to a} (z - a)\frac{f_1(z)}{f_2(z)} = \lim_{z \to a} f_1(z) \lim_{z \to a} \frac{1}{\frac{f_2(z)-f_2(a)}{z-a}} = \frac{f_1(a)}{f_2'(a)}.$$

If $f_1(a) = 0$, then f_1/f_2 has a removable singularity at a, since f_2 has a root of order 1 at a (Exercise 4.1.23.3). Thus, $\mathrm{res}_f(a) = 0$.

Example 5.2.6

1. Let $f = \cot = \cos/\sin$. Then, with $f_1 = \cos$ and $f_2 = \sin$ according to Theorem 5.2.5.2

$$f_1(k\pi) = (-1)^k, \qquad f_2(k\pi) = 0, \qquad f_2'(k\pi) = \cos(k\pi) = (-1)^k$$

and thus

$$\mathrm{res}_f(k\pi) = \frac{f_1(k\pi)}{f_2'(k\pi)} = 1 \qquad (k \in \mathbb{Z}).$$

2. Let $f(z) = 1/(1 + z^2)$ for $z \in \mathbb{C} \setminus \{\pm i\}$. With $f_1(z) := 1$ and $f_2(z) := 1 + z^2$ according to Theorem 5.2.5.2

$$\mathrm{res}_f(\pm i) = \pm\frac{1}{2i}.$$

This can also be easily seen by partial fraction decomposition: It holds

$$f(z) = \frac{1}{(z + i)(z - i)} = \frac{1}{2i}\left(\frac{1}{z - i} - \frac{1}{z + i}\right),$$

thus

$$\mathrm{res}_f(i) = \frac{1}{2\pi i} \int_{k_1(i)} f(\zeta)\, d\zeta = \frac{1}{2i}\frac{1}{2\pi i} \int_{k_1(i)} \frac{d\zeta}{\zeta - i} = \frac{1}{2i}$$

(and correspondingly for $-i$).

3. Let $f(z) = 1/(1 + z^2)^2$ for $z \in \mathbb{C} \setminus \{\pm i\}$. Then f has poles of second order at $\pm i$. With

$$g(z) := (z - i)^2 f(z) = \frac{1}{(z + i)^2} \qquad (z \neq \pm i)$$

according to Theorem 5.2.5.1

$$\mathrm{res}_f(i) = \lim_{z \to i} g'(z) = -\lim_{z \to i} \frac{2}{(z + i)^3} = \frac{1}{4i}.$$

Remark 5.2.7 An interesting class of integrals that can possibly be calculated using the residue theorem are integrals of the form

$$\int_{-\pi}^{\pi} f(\cos t)dt \qquad \text{and} \qquad \int_{-\pi}^{\pi} f(\sin t)dt\,,$$

where f is a rational function without poles on $[-1, 1]$. We set for $z \neq 0$

$$j(z) := \frac{1}{2}\left(z + \frac{1}{z}\right), \qquad k(z) := \frac{1}{2i}\left(z - \frac{1}{z}\right).$$

Then f^* and f^{**}, defined by

$$f^*(z) := f(j(z))/z \quad \text{and} \quad f^{**}(z) := f(k(z))/z,$$

are rational functions without poles on $\mathbb{S}$ (cf. Exercise 2.3.32.13 for the case of the Joukowski mapping j; corresponding statements apply for k).

From $\cos t = j(e^{it})$ and $\sin t = k(e^{it})$ for $t \in [-\pi, \pi]$ results with (5.2.2)

$$\frac{1}{2\pi}\int_{-\pi}^{\pi} f(\cos t)dt = \int f^*(\zeta)\zeta\, dm(\zeta) = \sum_{w \in P(f^*) \cap \mathbb{D}} \operatorname{res}_{f^*}(w) \qquad (5.2.3)$$

respectively

$$\frac{1}{2\pi}\int_{-\pi}^{\pi} f(\sin t)dt = \int f^{**}(\zeta)\zeta\, dm(\zeta) = \sum_{w \in P(f^{**}) \cap \mathbb{D}} \operatorname{res}_{f^{**}}(w)\,. \qquad (5.2.4)$$

Example 5.2.8 For $p \in \mathbb{N}$ and $c > 1$ we consider the integral

$$\int_{-\pi}^{\pi} \frac{dt}{(c + \cos t)^p}\,.$$

Here

$$f(u) = \frac{1}{(c + u)^p}\,,$$

is therefore

$$f^*(z) = \frac{1}{z} \cdot \frac{1}{(c + (z + 1/z)/2)^p} = \frac{2^p z^{p-1}}{(z^2 + 2cz + 1)^p} = \frac{2^p z^{p-1}}{(z - w_1)^p (z - w_2)^p}$$

with $w_1 = -c + \sqrt{c^2 - 1} \in (-1, 0)$ and $w_2 = -c - \sqrt{c^2 - 1} < -1$. Thus, from (5.2.3) and Theorem 5.2.5.1 with $g(z) := 2^p z^{p-1}/(z - w_2)^p$ for $z \neq w_2$

$$\int_{-\pi}^{\pi} \frac{dt}{(c + \cos t)^p} = 2\pi \cdot \operatorname{res}_{f^*}(w_1) = \frac{2\pi}{(p-1)!} g^{(p-1)}(w_1)\,.$$

For $p = 1$ we obtain

$$\int_{-\pi}^{\pi} \frac{dt}{c + \cos t} = 2\pi \cdot \frac{2}{w_1 - w_2} = \frac{2\pi}{\sqrt{c^2 - 1}},$$

and for $p = 2$ f

$$\int_{-\pi}^{\pi} \frac{dt}{(c + \cos t)^2} = 2\pi \cdot 4 \cdot \frac{-w_1 - w_2}{(w_1 - w_2)^3} = \frac{2\pi c}{\sqrt{c^2 - 1}^3}.$$

Remark and Definition 5.2.9 For $0 < \rho < 1$ we consider the integral

$$\int_{-\pi}^{\pi} \frac{dt}{1 - 2\rho \cos t + \rho^2}.$$

If

$$f(u) = \frac{1}{1 - 2\rho u + \rho^2}$$

and

$$f^*(z) = \frac{1}{z} f\left(\frac{1}{2}\left(z + \frac{1}{z}\right)\right) = \frac{1}{z} \cdot \frac{1}{1 - \rho(z + 1/z) + \rho^2} = \frac{1}{(z - \rho)(1 - \rho z)},$$

then f^* has the two simple poles $\rho < 1$ and $1/\rho > 1$. Thus, according to (5.2.3) and Theorem 5.2.5.1

$$\int_{-\pi}^{\pi} \frac{dt}{1 - 2\rho \cos t + \rho^2} = 2\pi \cdot \operatorname{res}_{f^*}(\rho) = 2\pi \cdot \lim_{z \to \rho} \frac{1}{1 - \rho z} = \frac{2\pi}{1 - \rho^2}.$$

For the function $P_\rho : U_{1/\rho}(0) \to \mathbb{R}$ with

$$P_\rho(\zeta) := \operatorname{Re}\left(\frac{1 + \rho\zeta}{1 - \rho\zeta}\right) = \frac{1 - \rho^2}{|1 - \rho\zeta|^2} \qquad (|\zeta| < 1/\rho)$$

one has

$$P_\rho(e^{it}) = \frac{1 - \rho^2}{1 - 2\rho \cos t + \rho^2} \qquad (-\pi \le t \le \pi)$$

and thus

$$\int P_\rho \, dm = \frac{1}{2\pi} \int_{-\pi}^{\pi} P_\rho(e^{it}) \, dt = 1.$$

Fig. 5.2 Trace
$\gamma_{10}^* = \tau_{10}^* \cup \sigma_{10}^*$ of the path γ_{10}

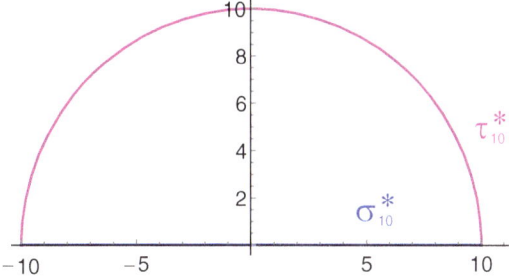

for $0 < \rho < 1$. The function P_ρ is called (ρ-)**Poisson kernel**. We also refer to the family $(P_\rho)_{0<\rho<1}$ as the Poisson kernel.

We will now see that the residue theorem can also be used to calculate improper integrals of the form $\int_{-\infty}^{\infty} f$.[5] As a preparation, we consider suitable closed paths.

Remark 5.2.10 For $R > 0$, let $\sigma_R(t) := t$ for $t \in [-R, R]$ and $\tau_R(t) := Re^{it}$ for $t \in [0, \pi]$ (Fig. 5.2). Then $\gamma_R := (\tau_R, \sigma_R)$ and $-\gamma_R = (-\tau_R, -\sigma_R)$ are closed paths in $\mathbb{C}$. If f is continuous on $K_R(0) \cup [-R, R]$, then $-\sigma_R = (\sigma_R)_-$ and thus

$$\int_{\gamma_R} f + \int_{-\gamma_R} f = \int_{(\gamma_R, -\gamma_R)} f = \int_{(\tau_R, -\tau_R)} f = \int_{k_R(0)} f \,.$$

If $z \in U_R(0) \cap \mathbb{H}$, where $\mathbb{H} = \{z : \text{Im}(z) > 0\}$ denotes the open upper half-plane, then $z \in \text{Ext}(-\gamma_R)$ and consequently

$$\text{ind}_{\gamma_R}(z) = \text{ind}_{\gamma_R}(z) + \text{ind}_{-\gamma_R}(z) = \text{ind}_{k_R(0)}(z) = 1.$$

Remark and Definition 5.2.11 Let $f : \mathbb{R} \to \mathbb{C}$ be continuous. If the limit $\lim_{R\to\infty} \int_{-R}^{R} f$ exists, then the limit is called the **Cauchy principal value** of the integral $\int_{-\infty}^{\infty} f$. From the definition of improper integrals, it follows: If $\int_{-\infty}^{\infty} f$ exists, then the Cauchy principal value exists and the two values coincide. Conversely, the existence of the Cauchy principal value does not generally imply the existence of the improper integral.

With this, we prove:

Theorem 5.2.12 *Let $A \subset \mathbb{H}$ be finite, $\Omega \supset \mathbb{H} \cup \mathbb{R}$ open and $f\, H(\Omega \setminus A)$.*

1. *If $\lim_{R\to\infty} \int_{\tau_R} f$ exists, then the Cauchy principal value of the integral exists $\int_{-\infty}^{\infty} f$ and*

---

[5] The following examples only give a first impression of the enormous potential inherent in the method. A whole range of further applications, also in connection with the calculation of series values, can be found for example in [1–3, 5].

$$\lim_{R \to \infty} \int_{-R}^{R} f = 2\pi i \left(\sum_{w \in A} \mathrm{res}_f(w) \right) - \lim_{R \to \infty} \int_{\tau_R} f .$$

2. *If an $\alpha > 1$ and an $R_0 > 0$ exist such that $R \mapsto R^\alpha \max_{\tau_R^*} |f|$ is bounded on $[R_0, \infty)$, then f is absolutely integrable on $\mathbb{R}$ with*

$$\int_{-\infty}^{\infty} f = 2\pi i \sum_{w \in A} \mathrm{res}_f(w) .$$

Proof.

1. Let $R_0 > 0$ be such that $A \subset U_{R_0}(0)$. For $R \geq R_0$, we consider the closed path γ_R from Remark 5.2.10. From the residue Theorem and Remark 5.2.10 it follows (Remark: $\Omega^c \subset \mathrm{Ext}(\gamma_R)$)

$$\int_{\gamma_R} f = 2\pi i \sum_{w \in A} \mathrm{res}_f(w)$$

for all $R \geq R_0$, thus also

$$\int_{-R}^{R} f = \int_{\gamma_R} f - \int_{\tau_R} f = 2\pi i \left(\sum_{w \in A} \mathrm{res}_f(w) \right) - \int_{\tau_R} f$$

and thus the assertion for $R \to \infty$.

2. Without loss of generality, let R_0 be such that $A \subset U_{R_0}(0)$. By assumption, there exists an $M > 0$ with $\max_{\tau_R^*} |f| \leq MR^{-\alpha}$ for $R \geq R_0$. In particular, $|f(t)| \leq M|t|^{-\alpha}$ for $t \geq R_0$ and $t \leq -R_0$.
   From the existence of the integral $\int_1^\infty t^{-\alpha} dt$ for $\alpha > 1$, the absolute integrability of f on $\mathbb{R}$ follows (Remark: f is continuous on $\mathbb{R}$). In particular, the improper integral $\int_{-\infty}^\infty f$ exists and coincides with the Cauchy principal value. Moreover, for $R \geq R_0$

$$\left| \int_{\tau_R} f \right| \leq \max_{\tau_R^*} |f| \cdot L(\tau_R) \leq \pi MR^{1-\alpha} \to 0 \qquad (R \to \infty).$$

Thus, 2. follows from 1.

Remark 5.2.13 In particular, Theorem 5.2.12.2 can be applied to integrands of the form

$$e^{i\omega t} \frac{p(t)}{q(t)} = \cos(\omega t) \frac{p(t)}{q(t)} + i \sin(\omega t) \frac{p(t)}{q(t)},$$

where $\omega \geq 0$ and p, q are polynomials with $\deg(q) \geq \deg(p) + 2$ and $q(t) \neq 0$ for $t \in \mathbb{R}$.

Because: We consider $f \in H(\mathbb{C} \setminus Z(q))$, defined by

$$f(z) := e^{i\omega z} \frac{p(z)}{q(z)} \qquad (z \notin Z(q)).$$

Then $|e^{i\omega z}| = e^{-\omega \text{Im}(z)} \leq 1$ for $z \in \mathbb{R} \cup \mathbb{H}$, and according to example 2.1.17 $z^2 p(z)/q(z)$ has a limit for $|z| \to \infty$. Thus, the prerequisites of Theorem 5.2.12.2 are met with $A := Z(q) \cap \mathbb{H}$ and $\Omega := (\mathbb{C} \setminus Z(q)) \cup A$.

Example 5.2.14 For $\omega \geq 0$, let $f : \mathbb{C} \setminus \{\pm i\} \to \mathbb{C}$ be defined by

$$f(z) = \frac{e^{i\omega z}}{1 + z^2} \qquad (z \in \mathbb{C} \setminus \{\pm i\}).$$

Then f is as in Remark 5.2.13. Note that

$$\int_{-\infty}^{\infty} \sin(\omega t)/(1 + t^2)\, dt = 0$$

holds, since the integrand is odd. Therefore, according to Theorem 5.2.12.2,

$$\int_{-\infty}^{\infty} \frac{\cos(\omega t)}{1 + t^2}\, dt = \int_{-\infty}^{\infty} \frac{e^{i\omega t}}{1 + t^2}\, dt = 2\pi i \, \text{res}_f(i).$$

Furthermore, (for example, according to Theorem 5.2.5.2 with $f_1(z) = e^{i\omega z}$ and $f_2(z) = 1 + z^2$)

$$\text{res}_f(i) = \frac{e^{-\omega}}{2i},$$

and therefore [6]

$$\int_{-\infty}^{\infty} \frac{\cos(\omega t)}{1 + t^2}\, dt = \pi e^{-\omega}.$$

In particular, this shows how in the case $\omega = 0$ the circle number

$$\pi = \int_{-\infty}^{\infty} \frac{1}{1 + t^2}\, dt$$

results from the residue of $z \mapsto 1/(1 + z^2)$ at the point i (and thus as a circle integral). The value π can be seen in Fig. 3.4 as the jump height of the real part of the complex arctangent at the points $\pm i$. Wonder worlds !

Exercises 5.2.15

1. Let $\Omega \subset \mathbb{C}$ be open, γ nullhomologous in Ω and $f \in H(\Omega)$. Show that the general Cauchy integral formula

---

[6] The function $\omega \mapsto \pi e^{-|\omega|}$ is the Fourier transform of the function $t \mapsto 1/(1 + t^2)$ and thus, up to normalization, the characteristic function of the Cauchy distribution. More on this topic can be found in the literature on Fourier analysis and statistics.

$$\operatorname{ind}_\gamma(z)f(z) = \frac{1}{2\pi i} \int_\gamma \frac{f(\zeta)}{\zeta - z}\, d\zeta \qquad (z \in \Omega \setminus \gamma^*)$$

follows from the residue theorem.

2. Let $A \subset \mathbb{C}$ be finite and let $f \in H(\mathbb{C} \setminus A)$ with

$$z \cdot f(z) \to 0 \qquad (|z| \to \infty).$$

Show:

$$\sum_{w \in A} \operatorname{res}_f(w) = 0.$$

3. Let p and $q \neq 0$ be polynomials. Show: If φ_w for $w \in Z(q)$ are the principal parts of the Laurent development of p/q with respect to w, then there exists a polynomial r with

$$(p/q)(z) = r(z) + \sum_{w \in Z(q)} \varphi_w(z - w) \quad (z \in \mathbb{C} \setminus Z(q)).$$

This additive decomposition of p/q into rational functions, which have (at most) one pole, is called **partial fraction decomposition** of p/q.

4. Calculate the residues of the following functions at the isolated singularities:
   a) $f(z) = z/\sin z$ for $z \in \mathbb{C} \setminus \pi \mathbb{Z}$,
   b) $f(z) = z^2/(1 - \cos z)$ for $z \in \mathbb{C} \setminus 2\pi \mathbb{Z}$.

5. Calculate $\displaystyle\int_{-\pi}^{\pi} \frac{dt}{1 + \sin^2 t}$.

6. Calculate $\displaystyle\int_{-\infty}^{\infty} \frac{dt}{(1 + t^2)^2}$ and $\displaystyle\int_{-\infty}^{\infty} \frac{dt}{1 + t^4}$.

7. a) The entire function $f : \mathbb{C} \to \mathbb{C}$ is defined by

$$f(z) := \begin{cases} (e^{iz} - 1)/z, & \text{if } z \neq 0 \\ i, & \text{if } z = 0 \end{cases}.$$

   Show:

$$\lim_{R \to \infty} \int_{-R}^{R} f = \pi i.$$

   Hint: First consider that $\int_0^\pi e^{-R \sin t} dt \to 0$ for $R \to \infty$.

   b) Show:

$$\int_{-\infty}^{\infty} \frac{\sin t}{t}\, dt = \pi .$$

5.3 Mapping Behavior of Holomorphic Functions

We now come to further applications of the residue theorem, which essentially deal with the local mapping behavior of holomorphic functions.

Theorem 5.3.1 *Let it be $\Omega \subset \mathbb{C}$ open. If f is meromorphic in Ω and not constant at any point with value 0 or ∞, then f'/f is meromorphic in Ω with simple poles at all $a \in P(f) \cup Z(f)$ and*

$$\operatorname{res}_{f'/f}(z) = n_f(z) \qquad (z \in \Omega).$$

Proof. We set $A := Z(f) \cup P(f)$. Then $(f'/f)|_{\Omega \setminus A}$ is holomorphic. Therefore, $n_f(z) = 0 = \operatorname{res}_{f'/f}(z)$ holds for all $z \notin A$.

If a is a zero of f of order $n := n_f(a)$, there exists a holomorphic function g in an open neighborhood U of 0 with $g(h) \neq 0$ in U and

$$f(a + h) = h^n g(h) \qquad (h \in U).$$

Therefore, it follows

$$\frac{f'(a + h)}{f(a + h)} = \frac{n}{h} + \frac{g'(h)}{g(h)} \qquad (h \in U).$$

Thus, f'/f has a pole of order 1 at a, and it holds

$$\operatorname{res}_{f'/f}(a) = n.$$

If a is a pole of order p of f, then $1/f$ has a zero of order p at a. Therefore, $\operatorname{res}_{(1/f)'/(1/f)}(a) = p$ is valid. Furthermore, on $\Omega \setminus A$ it holds

$$(1/f)'/(1/f) = -f'/f \, ,$$

therefore $\operatorname{res}_{f'/f}(a) = -p$.

Theorem 5.3.2 (Rouché)[7] *Let $\Omega \subset \mathbb{C}$ be open and $f, g \in H(\Omega)$. If $a \in \Omega$ and $R > 0$ with $B_R(a) \subset \Omega$ and $|f - g| < |f|$ on $K_R(a)$, then f and g have the same number of zeros in $U_R(a)$ including multiplicities, thus*

$$\sum_{w \in Z(f) \cap U_R(a)} n_f(w) = \sum_{w \in Z(g) \cap U_R(a)} n_g(w).$$

Proof. We consider the function $\varphi \in C(\Omega \times [0, 1])$ with

$$\varphi(z, t) := f(z) + t(g - f)(z) \quad (t \in [0, 1], z \in \Omega).$$

For $t \in [0, 1]$ and $\zeta \in K_R(a)$

---

[7] The Rouché theorem can also be formulated for more general cycles. For our purposes, the simple version, which can be proven without the Cauchy theorem, is sufficient.

$$|\varphi(\zeta,t)| \geq |f(\zeta)| - t|(g-f)(\zeta)| \geq |f(\zeta)| - |(g-f)(\zeta)| > 0,$$

so that the holomorphic function $z \mapsto \varphi(z,t)$ on $K_R(a)$ has no zeros. With

$$\psi(t,\zeta) := D_1\varphi(\zeta,t)/\varphi(\zeta,t) \quad (\zeta \in K_R(a),\, t \in [0,1])$$

according to Theorem 3.5.8 the function $\Phi : [0,1] \to \mathbb{C}$, defined by

$$\Phi(t) := \frac{1}{2\pi i} \int_{k_R(a)} \psi(t,\zeta)\, d\zeta \quad (t \in [0,1]),$$

is continuous. Furthermore, $\Phi([0,1]) \subset \mathbb{N}_0$ according to Theorem 5.3.1 and (5.2.2) and thereby

$$\Phi(0) = \sum_{w \in Z(f) \cap U_R(a)} n_f(w), \qquad \Phi(1) = \sum_{w \in Z(g) \cap U_R(a)} n_g(w).$$

The intermediate value theorem implies that Φ is constant. Therefore, $\Phi(0) = \Phi(1)$.

Example 5.3.3

1. We prove once again the Fundamental Theorem of Algebra in a quantitative version: Let $p(z) = \sum_{\nu=0}^{d} c_\nu z^\nu$ be a polynomial of degree $d \in \mathbb{N}$. If $R > 0$ is such that

$$\sum_{\nu=0}^{d-1} |c_\nu| R^{\nu-d} < |c_d|,$$

then p has exactly d roots including multiplicities in $U_R(0)$, thus

$$\sum_{w \in Z(p) \cap U_R(0)} n_p(w) = d.$$

Because: If $q(z) := c_d z^d$, then for $|z| = R$

$$|p(z) - q(z)| \leq \sum_{\nu=0}^{d-1} |c_\nu| R^\nu < |c_d| R^d = |q(z)|.$$

From Theorem 5.3.2 it follows that $\sum_{w \in Z(p) \cap U_R(0)} n_p(w) = n_q(0) = d$.

2. We consider the equation

$$e^z = 1 + 2z$$

and look for all solutions in $\mathbb{D}$.
Obviously, $z = 0$ is a solution. From

$$|e^z - 1| \leq \sum_{\nu=1}^{\infty} 1/\nu! = e - 1 < 2 \quad (|z| = 1)$$

it follows with $f(z) := 2z$ and $g(z) := 1 + 2z - e^z$ for $|z| = 1$

$$|f(z) - g(z)| = |e^z - 1| < 2 = |f(z)|.$$

Therefore, f and g have the same number of zeros in $\mathbb{D}$, namely one. Consequently, $z = 0$ is the only solution of the equation in $\mathbb{D}$.

We now want to investigate the local mapping behavior of holomorphic functions.

Definition 5.3.4 If (X, d_X) and (Y, d_Y) are metric spaces, then a mapping $f : X \to Y$ is called **open**, if images of open sets are open, that is, if $f(U)$ is open for all open sets $U \subset X$.

Theorem 5.3.5 *Let $\Omega \subset \mathbb{C}$ be open and $f \in H(\Omega)$. Then the following holds*

1. *f is locally injective at the point $a \in \Omega$ if and only if $f'(a) \neq 0$ holds.*
2. *If f is not constant at any point $a \in \Omega$, then f is open.*

Proof. We first show: If f is not constant around $a \in \Omega$, then for every sufficiently small $\rho > 0$ there exists a $\delta > 0$ such that f and $f - w$ for all $w \in U_\delta(0)$ have the same number of zeros in $U_\rho(a)$ including multiplicities.

Without restriction, let $f(a) = 0$ (otherwise consider $f - f(a)$). Furthermore, let $\rho > 0$ be such that $B_\rho(a) \subset \Omega$ and $f(z) \neq 0$ for all $z \in B_\rho(a) \setminus \{a\}$ (such a ρ exists according to Theorem 3.2.17). Since f is continuous on $K_\rho(a)$,

$$\delta := \min_{K_\rho(a)} |f|$$

exists and $\delta > 0$ holds. For $w \in U_\delta(0)$, according to Rouché's theorem, the functions f and $f - w$ have the same number of zeros in $U_\rho(a)$ including multiplicities.

1. If $f'(a) \neq 0$, then $f - f(a)$ has a simple zero at a. Therefore, $f|_{U_\rho(a)}$ is injective for sufficiently small $\rho > 0$ according to the above consideration. Conversely, if f is injective on a neighborhood U of a, then $f - f(a)$ again has a simple zero at a according to the above consideration, and therefore $f'(a) \neq 0$.
2. Let f be not constant at any point. If $a \in \Omega$, then according to the above consideration (applied to $f - f(a)$), for each $w \in U_\delta(0)$ there exists at least one $z \in U_\rho(a)$ with $f(z) = w + f(a)$. Therefore, $U_\delta(f(a)) \subset f(U_\rho(a))$. Since $a \in \Omega$ was arbitrary, $f(U)$ is open for every open set $U \subset \Omega$.

Remark 5.3.6 (Local Invertibility) Let $\Omega \subset \mathbb{C}$ be open and $f \in H(\Omega)$. If $a \in \Omega$ with $f'(a) \neq 0$, then according to Theorem 5.3.5 there exist open neighborhoods U of a in Ω and V of $f(a)$ in $f(\Omega)$ such that $f_U : U \to V$ with $f_U(z) := f(z)$ for $z \in U$ is bijective. Furthermore, $g := (f_U)^{-1} : V \to U$ is continuous, since f is open and thus pre-images of open sets under g are open. According to the inversion rule (Theorem 3.1.12), g is holomorphic with

$$g' = 1/(f' \circ g).$$

Remark 5.3.7 Let $G \subset \mathbb{C}$ be a domain and $f \in H(G)$ not constant.

1. (**Domain preservation**) According to the identity theorem and Theorem 5.3.5, $f(G)$ is open and due to the continuity of f also connected, thus also a domain.
2. For all $a \in G$ and all $r > 0$ with $U_r(a) \subset G$, $f(U_r(a))$ is open. Therefore, in particular, there exists a $w \in f(U_r(a))$ with $|w| > |f(a)|$. Thus, $|f|$ does not have a local maximum at a. This shows that Theorem 5.3.5 includes the maximum principle.

Finally, we study some implications of Rouché's theorem on sequences of functions.

Theorem 5.3.8 *Let $\Omega \subset \mathbb{C}$ be open, $a \in \Omega$ and (f_n) a sequence in $H(\Omega)$ with $f_n \to f$ locally uniformly on Ω. Then the following holds*

1. *If $R > 0$ with $B_R(a) \subset \Omega$ and f has no zeros on $K_R(a)$, then for n sufficiently large f and f_n have the same number of zeros in $U_R(a)$ including multiplicities.*
2. *If f is not constant with value 0 around a, then there exists a $\rho > 0$ such that for all $0 < r < \rho$ the functions f_n for n sufficiently large (depending on r) in $U_r(a)$ have exactly $n_f(a)$ zeros including multiplicities.*

Proof.

1. Since f is continuous on $K := K_R(a)$, there exists

$$\delta := \min_K |f|$$

and $\delta > 0$. Since (f_n) converges uniformly on K to f, there exists an $n_0 > 0$ such that

$$\max_K |f - f_n| < \delta$$

holds for all $n \geq n_0$. The first assertion follows from Rouché's theorem.
2. By assumption and Theorem 3.2.17, 0 is not an accumulation point of $Z(f)$. If $\rho > 0$ with $f(z) \neq 0$ in $B_\rho(a) \setminus \{a\}$, the second assertion follows from the first, applied with r instead of R.

Example 5.3.9 Let

$$f(z) = e^z = \sum_{\nu=0}^{\infty} \frac{z^\nu}{\nu!} \qquad (z \in \mathbb{C}).$$

Then, for the n-th partial sums $s_n(z) = \sum_{\nu=0}^{n} z^\nu/\nu!$ according to Theorem 5.3.8 one has: For all $R > 0$ there exists an $n_0(R)$ such that s_n for all $n \geq n_0(R)$ in $U_R(0)$ has no zero. This means that for each compact subset $K \subset \mathbb{C}$ the zeros, which according to the fundamental theorem of algebra do exist, lie outside of K for sufficiently large n.

Theorem 5.3.10 (Hurwitz) *Let $G \subset \mathbb{C}$ be a domain and (f_n) a sequence of holomorphic functions in G with $f_n \to f$ locally uniform on G.*

1. *If $w \in f(G)$ and f is not constant, then $w \in f_n(G)$ for n sufficiently large.*
2. *If f_n is injective for infinitely many n, then either f is constant or f is injective.*

Proof.

1. Let $a \in G$ with $f(a) = w$, so $n_{f-w}(a) > 0$. By assumption, $f - w$ is not constant. Therefore, according to the identity theorem, $f - w$ is also not constant around a. Thus, $0 \in (f_n - w)(G)$ for n sufficiently large according to Theorem 5.3.8.2.
2. Let f be not constant. We consider $w \in f(G)$ and $a \in G$ with $f(a) = w$. If $z \in G$, $z \neq a$ and are $U := U_\delta(a)$ and $V := U_\delta(z)$ with $\delta := |z - a|/2$, then according to Theorem 5.3.8.2 there exists an n_0 with $0 \in (f_n - w)(U)$ for $n \geq n_0$. Since f_n is injective for infinitely many n, it follows that $w \notin f_n(V)$ for infinitely many n. According to 1., w is then also not in $f(V)$, so in particular $f(z) \neq w$.

Remark 5.3.11 The statements of Hurwitz's theorem may sound a bit more self-evident than they are. If one considers the sequence (f_n) in $C^\omega(\mathbb{R})$ with $f_n(x) = x^2 + 1/n$ for $x \in \mathbb{R}$, then $f_n(x) \to x^2 (n \to \infty)$ uniformly on $\mathbb{R}$. Here, all f_n are without zeros, but the non-constant limit function is not.

Exercise 5.3.12

1. (**Argument Principle**) Let $\Omega \subset \mathbb{C}$ be open and $f \in M(\Omega)$ not constant at any point with value 0 or ∞. Prove: If γ is a null-homologous cycle in Ω with $\gamma^* \cap A = \emptyset$, where $A := Z(f) \cup P(f)$, then

$$\text{ind}_{f \circ \gamma}(0) = \frac{1}{2\pi i} \int_\gamma \frac{f'}{f} = \sum_{w \in A} \text{ind}_\gamma(w) \cdot n_f(w).$$

2. Show:
   a) For $u, v \in \mathbb{C}$ equality holds in the triangle inequality, i.e., $|u - v| = |u| + |v|$, exactly when $0 \in [u, v]$.
   b) The statement of Rouché's theorem also holds under the condition $|f - g| < |f| + |g|$ instead of $|f - g| < |f|$.

3. How many zeros does the polynomial $P(z) = z^7 + z^5 - 8z^3 + 2z + 1$ have in the annulus $V_{1,2}(0)$?

5.4   Concepts IV: Kernels and Integral Representations

In the context of applications of the residue theorem, we have introduced the Poisson kernels P_ρ. We have seen that P_ρ is continuous and positive on $U_{1/\rho}(0)$ and that $\int P_\rho \, dm = 1$ holds. Furthermore, it can be shown that the "mass" of P_ρ for $\rho \to 1$ concentrates at the point 1 in the sense that for every $\delta > 0$

$$\lim_{\rho \to 1^-} \int_{U_{\delta,\mathbb{S}}(1)} P_\rho \, dm = 1$$

is fulfilled—the Poisson kernels prove to be a family of good kernels, just like the Fejér kernels. This shows, as in the theorem about approximate identities, that for all $f \in C(\mathbb{S})$

$$\max_{\mathbb{S}} |f - f * P_\rho| \to 0 \qquad (\rho \to 1^-) \tag{5.4.5}$$

holds. The function $Pf : \mathbb{D} \to \mathbb{C}$ with

$$(Pf)(z) = (Pf)(\rho u) := (f * P_\rho)(u),$$

where $z = \rho u$ denotes the polar form of z, is called the Poisson integral of the function f. From (5.4.5) it follows that Pf provides a continuous extension of f to $\overline{\mathbb{D}}$. If $f \in C(\mathbb{S})$ is real-valued, then

$$(Pf)(\rho u) = \int f(\zeta) P_\rho(u\overline{\zeta}) \, dm(\zeta) = \text{Re}\left(\int f(\zeta) \frac{\zeta + \rho u}{\zeta - \rho u} \, dm(\zeta) \right).$$

Since the integration kernel $(\zeta + z)/(\zeta - z)$ is holomorphic in $\mathbb{C} \setminus \mathbb{S}$ as a function of the variable z, the parameter integral

$$z \mapsto \int f(\zeta) \frac{\zeta + z}{\zeta - z} \, dm(\zeta) \tag{5.4.6}$$

is also holomorphic in $\mathbb{C} \setminus \mathbb{S}$ according to the theorem on the differentiation of parameter integrals. This is an occurrence of a general principle, which roughly states that parameter integrals with kernel functions are at least as smooth as the kernels as a function of the parameter. The integration against the kernel thus typically results in a smoothing, as in the case of the Fejér kernels F_n up to trigonometric polynomials of degree $\leq n$.

From the above considerations, it also follows that Pf for real-valued f is the real part of a holomorphic function in $\mathbb{D}$ and thus a so-called harmonic function, and that Pf as such solves the Dirichlet problem with respect to the boundary function f.[8]

The kernel of all kernels of complex analysis is the Cauchy kernel $1/(\zeta - z)$, holomorphic as a function of z in $\mathbb{C} \setminus B$ for any closed sets $B \subset \mathbb{C}$ and $\zeta \in B$. If, for example, $B = \gamma^*$, where γ is a path or more generally a cycle in $\mathbb{C}$, then the corresponding parameter integral is the Cauchy integral $C_\gamma f$. The analyticity of the Cauchy kernel—quite simply the geometric series—leads to the analyticity of Cauchy integrals. This fortunate circumstance proves to be the basis of all that makes complex analysis what it is and fundamentally distinguishes it from real analysis.

Considering the simplest case $\gamma = k_1(0)$ of the unit circle, the Cauchy integral Cf for continuous f on $\mathbb{S}$ essentially corresponds to the parameter integral (5.4.6), because from

$$\frac{2\zeta}{\zeta - z} = 1 + \frac{\zeta + z}{\zeta - z}$$

follows

$$2(Cf)(z) = \int f(\zeta) \frac{2\zeta}{\zeta - z}\, dm(\zeta) = \hat{f}(0) + \int f(\zeta) \frac{\zeta + z}{\zeta - z}\, dm(\zeta).$$

However, here *not* anymore $(Cf)(\rho u) \to f(u)$ $(\rho \to 1^-)$ applies for arbitrary continuous functions on $\mathbb{S}$; the Cauchy kernel is a so-called singular kernel. Only forming the real part leads to an approximate identity and thus to a good kernel. On the other hand, according to the Cauchy integral formula $(Cf)|_{\mathbb{D}} = f|_{\mathbb{D}}$ holds for functions f in the disc algebra $A(\overline{\mathbb{D}})$. In this sense, the Cauchy kernel is even reproducing for the disc algebra.

In the non-compact situation $B = \mathbb{R}$ the Cauchy integral $C_{\mathbb{R}}f$ is improper and as such no longer defined for arbitrary continuous functions. If f is absolutely integrable, then

$$z \mapsto \frac{1}{\pi i} \int_{-\infty}^{\infty} \frac{f(t)}{t - z}\, dt$$

is again holomorphic in $\mathbb{C} \setminus \mathbb{R}$. The real part of the Cauchy kernel proves to be a good kernel again, with the limit transition for $\mathrm{Im}(z) \to 0$ instead of $\rho \to 1^-$ being considered. The imaginary part, on the other hand, again represents a singular kernel. For continuously differentiable f and $t \in \mathbb{R}$ the integral is defined as the so-called Cauchy principal value (a proof can be found in [4]). The corresponding mapping is the Hilbert transformation, which plays an important role in signal processing, among other things.

---

[8] More details about the Dirichlet problem and harmonic functions, further interesting properties of the Poisson kernel in connection with partial differential equations, as well as an interpretation as Abel summation kernel can be found in [6].

References

1. Bornemann, F.: Funktionentheorie, 2nd ed. Birkhäuser, Basel (2016)
2. Conway, J.B.: Functions of One Complex Variable. 2nd ed. Springer, New York (1978)
3. Gamelin, T.W.: Complex Analysis. Springer, New York (2001)
4. Lax, P.D., Zalcman, L.: Complex Proofs of Real Theorems. American Mathematical Society, Providence (2012)
5. Remmert, R.: Funktionentheorie I. Springer, Berlin (1984)
6. Stein, E.M., Shakarchi, R.: Fourier analysis. An introduction. Princeton Lectures in Analysis, Vol. 1. Princeton University Press, Princeton (2003)

The dynamic behavior of polynomials within the framework of the Fatou-Julia dichotomy is one of the most fascinating topics in complex analysis. An introduction to the theory is the content of the last section of this chapter. The essential theoretical foundation is formed by the spherical normality theorem of Montel, also briefly referred to as Montel's big theorem. As a further consequence of Montel's big theorem, one obtains Picard's big theorem, which in turn is undoubtedly one of the jewels of function theory.

A declared goal of this chapter is the self-contained elaboration of this foundation. The presentation is noticeably tighter compared to the previous chapters. In a first step, Montel's (little) normality theorem for families of holomorphic functions is derived, which follows directly from the Arzelà-Ascoli theorem using the Cauchy inequality. Therefore, the chapter starts with equicontinuous families and related compactness statements. Montel's little theorem is then crucially used in the proof of the Riemann mapping theorem, which implies the conformal equivalence of simply connected strict subdomains of $\mathbb{C}$ and the open disk $\mathbb{D}$.

As an approach to Montel's great theorem, the path via the Zalcman lemma is chosen, a statement about the existence of limit functions for *non-normal* families of meromorphic functions after suitable rescaling. In preparation for this, a suitable distortion theorem for spherical derivatives is derived.

6.1 Normal Families of Continuous Functions

Definition 6.1.1 Let (X, d) be a metric space.

1. For $\varepsilon > 0$, a subset E of X is called $\boldsymbol{\varepsilon}$-**dense** (in X), if

$$X = \bigcup_{x \in E} U_\varepsilon(x).$$

J. Müller, *Concepts of Function Theory*, Mathematics Study Resources 12,
https://doi.org/10.1007/978-3-662-69115-1_6

Fig. 6.1 32nd roots of unity

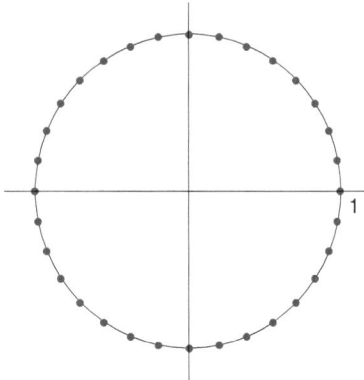

2. The space X is called **precompact**, if for all $\varepsilon > 0$ there exists a finite, ε-dense subset E. If $M \subset X$, then M is called **precompact**, if (M, d_M) is precompact (or $M = \emptyset$). This is the case exactly when for every $\varepsilon > 0$ there exists a finite set $E \subset X$ with $M \subset \bigcup_{x \in E} U_{\varepsilon, X}(x)$ (Exercise 6.1.21.1). In particular, subsets of precompact sets and finite unions of precompact subsets are also precompact.

Example 6.1.2

1. In $\mathbb{R}$ for every d with $0 < d < 2\varepsilon$ the set $d\mathbb{Z}$ is an ε-dense subset. Accordingly, in $\mathbb{C}$ for every d with $0 < d < \sqrt{2}\varepsilon$ the lattice $d\mathbb{Z} + id\mathbb{Z}$ is an ε-dense subset.
2. In $\mathbb{S}$ for every $m \in \mathbb{N}$ with $\pi/m < \varepsilon$ the set of the m-th roots of unity is an ε-dense subset (this follows from 1. and $|e^{it} - e^{is}| \leq |t - s|$ for $t, s \in \mathbb{R}$). In particular, $\mathbb{S}$ is precompact. Figure 6.1 shows the 32nd roots of unity.

Remark 6.1.3 Every precompact metric space (X, d) is separable.

Because: For every $n \in \mathbb{N}$ there exists a finite set E_n with $\bigcup_{x \in E_n} U_{1/n}(x) = X$. We define $A := \bigcup_{n=1}^{\infty} E_n$. Then A is countable according to Theorem 1.4.22. If $a \in X$ and $\varepsilon > 0$ are given, it follows for n with $1/n < \varepsilon$

$$U_\varepsilon(a) \cap A \supset U_{1/n}(a) \cap E_n \neq \emptyset .$$

So $a \in \overline{A}$ and consequently $\overline{A} = X$.

Theorem 6.1.4 *Let (X, d) be a metric space. A set $M \subset X$ is precompact if and only if every sequence in M has a Cauchy subsequence.*

Proof $\Rightarrow$: Let $(x_n)_{n \in \mathbb{N}}$ be a sequence in M. Since M is precompact, there exists a finite set $E_1 \subset M$ with

$$M = \bigcup_{y \in E_1} U_{1/2, M}(y).$$

Since E_1 is finite, there exist a $y_1 \in E_1$ and an infinite set $I_1 \subset \mathbb{N}$ with

$$x_n \in M_1 := U_{1/2,M}(y_1) \quad (n \in I_1).$$

In addition, diam$(M_1) \leq 1$. Since $M_1 \subset M$ is precompact, there exists a finite set $E_2 \subset M_1$ with

$$M_1 = \bigcup_{y \in E_2} U_{1/4,M_1}(y).$$

Again, there exist a $y_2 \in E_2$ and an infinite set $I_2 \subset \mathbb{N}$ such that

$$x_n \in M_2 := U_{1/4,M_1}(y_2) \quad (n \in I_2).$$

Here, diam $(M_2) \leq 1/2$. Inductively, one obtains in this way a sequence $(M_j)_{j \in \mathbb{N}}$ of sets in X with $M_j \subset M_{j-1}$ and diam$(M_j) \leq 1/j$, and a sequence $(I_j)_{j \in \mathbb{N}}$ of infinite subsets of $\mathbb{N}$ such that $x_n \in M_j$ for all $n \in I_j$. If one chooses $n_0 := 1$ and $n_j \in I_j$ with $n_j > n_{j-1}$, then $(x_n)_{n \in I}$ with $I := \{n_j : j \in \mathbb{N}_0\}$ is a Cauchy subsequence of $(x_n)_{n \in \mathbb{N}}$.

$\Leftarrow$: Let $\varepsilon > 0$ be given. Assume that there does not exist a finite set $E \subset X$ with $M \subset \bigcup_{x \in E} U_\varepsilon(x)$. We define inductively a sequence $(x_n)_{n \in \mathbb{N}}$ in M such that $d(x_j, x_k) \geq \varepsilon$ for all j, k with $j \neq k$. For this, we choose $x_1 \in M$ arbitrarily and assume that we have already defined $x_1, \ldots, x_n \in M$ with $d(x_j, x_k) \geq \varepsilon$ for j, $k = 1, \ldots, n, j \neq k$. By assumption, there then exists an $x \in M \setminus \bigcup_{j=1}^{n} U_\varepsilon(x_j)$. With $x_{n+1} := x$, $d(x_{n+1}, x_j) \geq \varepsilon$ for $j = 1, \ldots, n$. Thus, (x_n) is as desired.

The sequence (x_n) constructed in this way has no Cauchy subsequence. Thus, there is a contradiction to the assumption.   □

Remark 6.1.5 In the case of a complete metric space (X, d), according to Theorem 6.1.4, a subset of X is precompact if and only if it is relatively compact. Since every compact metric space is complete, it also follows that a metric space is compact if and only if it is complete and precompact.

Definition 6.1.6 Let (X, d) be a metric space.

1. A family $(U_\iota)_{\iota \in I}$ of open sets in X is called an **open cover** of X, if

$$X = \bigcup_{\iota \in I} U_\iota .$$

(X, d) is called **covering compact**, if every open cover $(U_\iota)_{\iota \in I}$ of X contains a finite subcover, that is, if $(U_\iota)_{\iota \in I}$ is an open cover of X, then there exists a finite set $E \subset I$ with

$$X = \bigcup_{\iota \in E} U_\iota .$$

2. A subset M of X is called **covering compact**, if (M, d_M) is cover compact (or $M = \emptyset$).

We show that sequentially compactness and covering compactness are the same[1]:

Theorem 6.1.7 *Let (X, d) be a metric space. Then the following are equivalent:*

a) *X is covering compact.*
b) *All discrete subsets of X are finite.*
c) *X is (sequentially) compact.*

Proof. c) $\Rightarrow$ a): According to Remark 6.1.3, there exists a countable dense subset A of X. Then also

$$\mathscr{B} := \{U_{1/k}(x) : x \in A, k \in \mathbb{N}\}$$

is countable according to Theorem 1.4.22.

Let now $(U_\iota)_{\iota \in I}$ be an open cover of X and

$$\mathscr{B}_0 := \{B \in \mathscr{B} : B \subset U_\iota \text{ for some } \iota \in I\} .$$

For each $B \in \mathscr{B}_0$ we choose a $\iota_B \in I$ with $B \subset U_{\iota_B}$ and set $J := \{\iota_B : B \in \mathscr{B}_0\}$. Then $(U_\iota)_{\iota \in J}$ is countable (since $\mathscr{B}_0$ is countable). We show that $(U_\iota)_{\iota \in J}$ is an open cover of X.

To do this, let $x \in X$ be arbitrary. We choose an $\iota \in I$ with $x \in U_\iota$. Since U_ι is open, there exists an $\varepsilon > 0$ with $U_\varepsilon(x) \subset U_\iota$. Now we choose an $k \in \mathbb{N}$ with $1/k < \varepsilon/2$ and an $a \in A$ with $d(x, a) < 1/k$. Then

$$x \in U_{1/k}(a) \subset U_\varepsilon(x) \subset U_\iota ,$$

so $B := U_{1/k}(a) \in \mathscr{B}_0$. For ι_B one has $x \in B \subset U_{\iota_B}$. Consequently, $(U_\iota)_{\iota \in J}$ is a cover of X. (So far, we have shown: Every open cover has a *countable* subcover.)

Let $\{V_n : n \in \mathbb{N}\}$ be an enumeration of $\{U_\iota : \iota \in J\}$. We set

$$W_m := \bigcup_{j=1}^{m} V_j .$$

Then $W_m \subset W_{m+1}$ for $m \in \mathbb{N}$ and $\bigcup_{m \in \mathbb{N}} W_m = X$. It is sufficient to show: $W_m = X$ for a $m \in \mathbb{N}$.

Assume this is not the case. Then for all $n \in \mathbb{N}$ there exists an $x_n \in X \setminus W_n$. By assumption, then (x_n) has a convergent subsequence $(x_n)_{n \in I}$ with limit $x \in X$. We choose $m \in \mathbb{N}$ with $x \in W_m$. Then on the one hand $x_n \notin W_m$ for all $n \geq m$, but on the other hand also $x_n \in W_m$ for all sufficiently large $n \in I$, since W_m is open. Contradiction!

---

[1] In more general topological spaces, this is not always the case.

a) $\Rightarrow$ b): Assume, b) does not hold. Then there exists an infinite discrete sub-set D of X. Thus, for every $x \in D$, there exists an open neighborhood V_x of x with $D \cap V_x = \{x\}$. Since D is closed, $(U_x)_{x \in D}$ with $U_x := V_x \cup D^c$ is an open cover of X. If E is a finite subset of D, then

$$(D \setminus E) \cap \bigcup_{x \in E} U_x = \emptyset.$$

Therefore, $(U_x)_{x \in D}$ contains no finite subcover. Contradiction!

b) $\Rightarrow$ c): Let $(x_n)_{n \in \mathbb{N}}$ be a sequence in X. If $A := \{x_n : n \in \mathbb{N}\}$ is finite, then there exists a constant (thus convergent) subsequence $(x_n)_{n \in J}$ of $(x_n)_{n \in \mathbb{N}}$. If A is infinite, then A has, by assumption, an accumulation point $a \in X$. Then there exists a subse-quence $(x_n)_{n \in J}$ of $(x_n)_{n \in \mathbb{N}}$ with $x_n \to a$ $(n \to \infty, n \in J)$.

Definition 6.1.8 Let (X, d_X), (Y, d) be metric spaces and $\mathscr{F} \subset C(X, Y)$. Then $\mathscr{F}$ is called **equicontinuous** at the point $a \in X$, if for all $\varepsilon > 0$ a $\delta = \delta_\varepsilon > 0$ exists such that $d(f(x), f(a)) < \varepsilon$ for all $x \in X$ with $d_X(x, a) < \delta$ and all $f \in \mathscr{F}$. The family $\mathscr{F}$ is called **equicontinuous**, if $\mathscr{F}$ is equicontinuous at all $a \in X$. Furthermore, for $M \subset X$ we write in the following

$$\mathscr{F}|_M := \{f|_M : f \in \mathscr{F}\}.$$

We have already seen in Remark 2.6.9 that even in the case of compact spaces (X, d_X) and (Y, d) the space $(C(X, Y), d_\infty)$ is complete but not always compact. We now show

Theorem 6.1.9 (Arzelà-Ascoli) *Let (X, d_X) and (Y, d) becompact metric spaces. If $\mathscr{F} \subset C(X, Y)$ is equicontinuous, then $\mathscr{F}$ is relatively compact.*

Proof. According to Remark 6.1.5 it is sufficient to show that $\mathscr{F}$ is precompact. For this, let $\varepsilon > 0$ be given. For all $x \in X$ there exists a $\delta_{x,\varepsilon} > 0$ with

$$f\left(U_{\delta_{x,\varepsilon}}(x)\right) \subset U_\varepsilon\left(f(x)\right) \qquad (f \in \mathscr{F}).$$

Since X is compact and $(U_{\delta_{x,\varepsilon}})_{x \in X}$ is an open cover of X, there exists a finite subset E of X with

$$X = \bigcup_{x \in E} U_{\delta_{x,\varepsilon}}(x).$$

Since (Y, d) is compact, $(C(E, Y) = \mathrm{Map}(E, Y), d_\infty)$ is compact according to Remark 2.6.9. Therefore, $\mathscr{F}|_E$ is precompact (since relatively compact), that is, for a finite set $\mathscr{E} \subset \mathscr{F}$ the inclusion

$$\mathscr{F}|_E \subset \bigcup_{g \in \mathscr{E}} U_{\varepsilon, C(E,Y)}(g|_E)$$

holds. Let $f \in \mathscr{F}$ be given. Then there exists a $g \in \mathscr{E}$ with $f|_E \in U_{\varepsilon, C(E,Y)}(g|_E)$. If $x \in X$ is arbitrary, then $x \in U_{\delta_{a,\varepsilon}}(a)$ for some $a \in E$. For this a,

$$d\big(f(x), f(a)\big) < \varepsilon \quad \text{and} \quad d\big(g(x), g(a)\big) < \varepsilon,$$

so

$$d\big(f(x), g(x)\big) \leq d\big(f(x), f(a)\big) + d\big(f(a), g(a)\big) + d\big(g(a), g(x)\big) < 3\varepsilon.$$

Consequently, $d_\infty(f, g) < 3\varepsilon$ and thus $\mathscr{F} \subset \bigcup_{g \in \mathscr{E}} U_{3\varepsilon, C(X,Y)}(g)$.

Remark 6.1.10 From Theorem 6.1.9 also the following variant of the Arzelà-Ascoli theorem for $\mathbb{C}$-valued functions follows: If (X, d) is a compact metric space and $\mathscr{F} \subset C(X)$ is bounded in $(C(X), \|\cdot\|_\infty)$ and equicontinuous, then $\mathscr{F}$ is relatively compact.

Because: If $\|f\|_\infty \leq R$ for all $f \in \mathscr{F}$, then one can also consider $\mathscr{F}$ as a family in $C(X, B_R(0))$. Here, $\overline{B_R(0)}$ is compact. Thus, the assertion follows from Ttheorem 6.1.9.

In the following, we again want to examine functions on open sets in $\mathbb{C}$. For this, we equip $C(\Omega)$ and thus also $H(\Omega)$ with a suitable metric. We initially consider more generally open sets in $\mathbb{K}$.

Remark and Definition 6.1.11 Let $X \subset \mathbb{K}$ be open. We set

$$K_m := K_m(X) := B_m(0) \cap \{x \in X : \operatorname{dist}(x, \partial X) \geq 1/m\}.$$

Then the following holds

- $K_m \subset X$ is compact ($m \in \mathbb{N}$).
- $K_m \uparrow X$, that is $\bigcup_{m \in \mathbb{N}} K_m = X$ and $K_m \subset K_{m+1}$ and more precisely even $K_m \subset (K_{m+1})^\circ$.
- For all $K \subset X$ compact, $K \subset K_m$ for sufficiently large m.

We call $(K_m) = (K_m(X))$ the **standard exhaustion** of X.

Furthermore, let (Y, d) be a metric space. For $f, g \in C(X, Y)$ we define

$$d_{\mathrm{loc}}(f, g) := d_{\mathrm{loc}, Y}(f, g) := \sup_{m \in \mathbb{N}} \min\{1/m, d_{\infty, K_m}(f, g)\} \quad (\leq 1)$$

with

$$d_{\infty, K}(f, g) = d_{\infty, K, Y}(f, g) = \max_{x \in K} d(f(x), g(x))$$

for compact, non-empty sets $K \subset X$ (see Theorem 2.6.5) and $d_{\infty, \emptyset}(f, g) := 0$. It can be easily seen that d_{loc} is a metric on $C(X, Y)$.

Theorem 6.1.12 *Let $X \subset \mathbb{K}$ be open and (Y, d) a complete metric space. Then the following holds*

1. *A sequence* $(f_n)_{n \in \mathbb{N}}$ *in* $C(X, Y)$ *is* d_{loc}-*convergent if and only if it converges uniformly on all compact subsets of* X, *i.e., locally uniformly on* X.
2. *The metric space* $(C(X, Y), d_{\mathrm{loc}})$ *is complete.*

Proof.

1. $\Rightarrow$: If $f_n \to f$ in $(C(X, Y), d_{\mathrm{loc}})$ and $K \subset X$ is compact, we choose an $m \in \mathbb{N}$ with $K \subset K_m$. Then it follows

$$\min\{1/m, d_{\infty, K_m}(f, f_n)\} \leq d_{\mathrm{loc}}(f_n, f) \to 0 \qquad (n \to \infty),$$

thus also

$$d_{\infty, K_m}(f, f_n) \to 0 \qquad (n \to \infty)$$

and therefore

$$d_{\infty, K}(f, f_n) \to 0 \qquad (n \to \infty).$$

$\Leftarrow$: Let $\varepsilon > 0$ be given. Then there exists an $m_\varepsilon \in \mathbb{N}$ with $1/m_\varepsilon < \varepsilon$. Thus,

$$\sup_{m \geq m_\varepsilon} \min\left\{\frac{1}{m}, d_{\infty, K_m}(f, f_n)\right\} \leq \frac{1}{m_\varepsilon} < \varepsilon \qquad (n \in \mathbb{N}).$$

Furthermore, there exists an $R_\varepsilon > 0$ with

$$d_{\infty, K_{m_\varepsilon}}(f, f_n) < \varepsilon \qquad (n > R_\varepsilon).$$

Therefore,

$$\max_{1 \leq m \leq m_\varepsilon} \min\{1/m, d_{\infty, K_m}(f, f_n)\} < \varepsilon$$

for $n > R_\varepsilon$, and thus also $d_{\mathrm{loc}}(f, f_n) < \varepsilon$ for $n > R_\varepsilon$.

2. The considerations from 1. show that $(f_n)_{n \in \mathbb{N}}$ is a d_{loc}-Cauchy sequence exactly when $(f_n|_K)_{n \in \mathbb{N}}$ is a uniform Cauchy sequence on all compact subsets K of X, that is, the Cauchy condition from Theorem 2.6.7 is fulfilled for $M = K$.
   So if $(f_n)_{n \in \mathbb{N}}$ is a d_{loc}-Cauchy sequence, then for all compact, non-empty $K \subset X$ according to Remark 2.6.9 there exists a continuous function $f_K \colon K \to Y$ with $f_n \to f_K (n \to \infty)$ uniformly on K. By $f(x) := f_K(x)$, if $x \in K$, a limit function $f \in C(X, Y)$ is thus defined.   □

Remark and Definition 6.1.13 Let $X \subset \mathbb{K}$ be open and (Y, d) a complete metric space.

1. A family $\mathscr{F} \subset C(X, Y)$ is called **normal**, if $\mathscr{F}$ is relatively compact in the metric space $(C(X, Y), d_{\mathrm{loc}})$.
2. If $Y = \mathbb{C}$, we call a family $\mathscr{F} \subset C(X)$ **bounded around** $a \in X$, if a compact neighborhood U of a exists such that $\mathscr{F}|_U$ is bounded in $(C(U), \|\cdot\|_{\infty, U})$, i.e.,

$$\sup_{f \in \mathscr{F}} \|f\|_{\infty, U} < \infty.$$

holds. Furthermore, we briefly say, $\mathscr{F}$ is **locally bounded**, if $\mathscr{F}$ is bounded at all points $a \in X$. It is easily seen (Exercise 6.1.21.3): If $\mathscr{F}$ is normal in $C(X)$, then $\mathscr{F}$ is also locally bounded.

Theorem 6.1.14 *Let $X \subset \mathbb{K}$ be open and (Y, d) a complete metric space. For $\mathscr{F} \subset C(X, Y)$ the following statements are equivalent:*

a) *$\mathscr{F}$ is normal.*
b) *For all compact $K \subset X$ is $\mathscr{F}|_K$ relatively compact in $(C(K, Y), d_\infty)$.*
c) *For all $x \in X$ there exists an open neighborhood U of x such that $\mathscr{F}|_U$ is normal in $C(U, Y)$.*

Proof. c) $\Rightarrow$ b): Let $K \subset X$ be compact. For all $x \in K$ there exists an open neighborhood U_x of x such that $\mathscr{F}|_{U_x}$ is normal. Let $\delta_x > 0$ be such that $B_{\delta_x}(x) \subset U_x$ holds. Then $\left(U_{\delta_x}(x)\right)_{x \in K}$ is an open cover of K. Thus, there exist $x_1, \ldots, x_N \in K$ such that with $L_m := B_{\delta_{x_m}}(x_m)$

$$K \subset \bigcup_{m=1}^{N} L_m$$

holds. Now let $(f_n)_{n \in \mathbb{N}}$ be a sequence in $\mathscr{F}$. We set $I_0 := \mathbb{N}$. By assumption (remark $L_1 \subset U_{x_1}$ is compact) there exists a subsequence $(f_n)_{n \in I_1}$ of $(f_n)_{n \in I_0}$, which converges uniformly on L_1. Again by assumption, there exists a subsequence $(f_n)_{n \in I_2}$ of $(f_n)_{n \in I_1}$, which converges uniformly on L_2 (and thus also on K_2, where $K_m := L_1 \cup \ldots \cup L_m$). Inductively, for each $m \in \{1, \ldots, N\}$ there exists a subsequence $(f_n)_{n \in I_m}$ of $(f_n)_{n \in I_{m-1}}$, which converges uniformly on K_m. For $m = N$, the uniform convergence on $K_N \supset K$ is obtained.

b) $\Rightarrow$ a): Let (K_m) be the standard exhaustion of X. As in the first step of the proof, one can see: If $(f_n)_{n \in \mathbb{N}}$ is a sequence in $\mathscr{F}$, then (with $I_0 := \mathbb{N}$) for each $m \in \mathbb{N}$ there exists a subsequence $(f_n)_{n \in I_m}$ of $(f_n)_{n \in I_{m-1}}$, which converges uniformly on K_m. If one defines $n_0 := 1$ and chooses $n_j > n_{j-1}$ with $n_j \in I_j$, then the sequence $(f_{n_j})_j$ converges uniformly on all K_m and thus also uniformly on all compact sets $K \subset X$.

a) $\Rightarrow$ c): Clear. $\square$

Remark and Definition 6.1.15 If $X \subset \mathbb{K}$ is open and (Y, d) is a complete metric space, we say a family $\mathscr{F}$ in $C(X, Y)$ is **normal around** $a \in X$, if an open neighborhood U of a in X exists such that $\mathscr{F}|_U$ is normal. Again, we say that $\mathscr{F}$ is **locally normal**, if $\mathscr{F}$ is normal around every point $a \in X$. According to Theorem 6.1.14, $\mathscr{F}$ is normal if and only if $\mathscr{F}$ is locally normal. Normality is a local property!

Remark 6.1.16 Let $X \subset \mathbb{K}$ be open. From Theorem 6.1.14, the Arzelà-Ascoli theorem and Remark 6.1.10, it follows that each of the following conditions is sufficient for the normality of a family $\mathscr{F} \subset C(X, Y)$:

1. (Y, d) is compact and $\mathscr{F}$ is equicontinuous.
2. $Y = \mathbb{C}$ and $\mathscr{F} \subset C(X)$ is locally bounded and equicontinuous.

For continuously differentiable functions, the local boundedness of the derivatives is sufficient for uniform continuity. More precisely, it holds that

Theorem 6.1.17 *Let $X \subset \mathbb{K}$ be open and $\mathscr{F} \subset C(X)$ a family of continuously differentiable functions on X. If $\{f' : f \in \mathscr{F}\}$ is locally bounded in $C(X)$, then $\mathscr{F}$ is equicontinuous.*

Proof. Let $a \in X$ and $B_R(a) \subset X$ with

$$c := \sup_{f \in \mathscr{F}} \max_{B_R(a)} |f'| < \infty.$$

Then, according to the mean value inequality,

$$\left| f(x) - f(a) \right| \leq c \cdot |x - a| \qquad (x \in U_R(a)).$$

Therefore, $\mathscr{F}$ is equicontinuous at a. $\qquad\qquad\qquad\qquad\qquad\qquad\qquad\quad\square$

Now we again consider holomorphic functions.

Remark 6.1.18 Let $\Omega \subset \mathbb{C}$ be open. From Theorem 4.1.21 it follows that $H(\Omega)$ is closed in $(C(\Omega), d_{\mathrm{loc}})$. Therefore, $(H(\Omega), d_{\mathrm{loc}})$ is complete as a metric space.

From the Cauchy inequality, it now directly follows

Theorem 6.1.19 (Montel) *If $\Omega \subset \mathbb{C}$ is open, then every locally bounded family $\mathscr{F}$ in $H(\Omega)$ is normal.*

Proof. According to Remark 6.1.16.2 and Theorem 6.1.17, it is sufficient to show that the family $\{f' : f \in \mathscr{F}\}$ is locally bounded. To this end, let $a \in \Omega$ and $B_R(a) \subset \Omega$ be such that

$$c := \sup_{f \in \mathscr{F}} \max_{B_R(a)} |f| < \infty.$$

If $0 < r < R$, then for $f \in \mathscr{F}$ from the Cauchy inequality (4.1.2), applied to $g(w) := f(a + Rw)$,

$$\max_{B_r(a)} |f'| \leq \frac{1}{(1 - r/R)^2} \frac{c}{R} = \frac{R}{(R - r)^2} \cdot c.$$

Thus, $\{f' : f \in \mathscr{F}\}$ is also locally bounded. $\qquad\qquad\qquad\qquad\qquad\qquad\quad\square$

Remark 6.1.20 According to Montel's theorem and Remark 6.1.13, in the metric space $(H(\Omega), d_{\mathrm{loc}})$, local boundedness is equivalent to relative compactness.

Therefore, in $H(\Omega)$, the statement of the Heine-Borel theorem holds: $\mathscr{F}$ is compact if and only if $\mathscr{F}$ is (locally) bounded and closed.

Exercises 6.1.21

1. Let (X, d) be a metric space and $M \subset X$. Show that M is precompact if and only if for every $\varepsilon > 0$ there exists a finite set $E \subset X$ with

$$M \subset \bigcup_{x \in E} U_\varepsilon(x).$$

2. Show that finite unions of normal families are normal.
3. Let $X \subset \mathbb{K}$ be open. Show that every normal family in $C(X)$ is locally bounded.
4. For $n \in \mathbb{N}$ let

$$f_n(x) := \cos(nx) \qquad (x \in \mathbb{R}).$$

Show that the family $\{f_n : n \in \mathbb{N}\} \subset C^\omega(\mathbb{R})$ is locally bounded in $C(\mathbb{R})$, but is not normal.
Hint: Use Exercise 2.6.15.4
5. Let $X \subset \mathbb{K}$ be open, (Y, d) a metric space and $\mathscr{F}$ a normal family in $C(X, Y)$. Furthermore, let (f_n) be a sequence in $\mathscr{F}$. Show: It holds $f_n \to f$ in $C(X, Y)$ if and only if all *convergent* subsequences of (f_n) in $C(X, Y)$ converge to f.

6.2   Conformal Mappings and the Riemann Mapping Theorem

In the previous sections, we have at various points applied the affine-linear mapping $w \mapsto a + \rho w$ to transfer results from the unit disk $\mathbb{D}$ to arbitrary disks $U_\rho(a)$. In this section, we want to show that with the help of bijective holomorphic mappings, one can even deform arbitrary simply connected (strict) subdomains of $\mathbb{C}$ conformally to the unit disk.

Remark and Definition 6.2.1 Let G and D be domains in $\mathbb{C}$.

1. A bijective holomorphic function $\varphi : G \to D$ is called a **conformal** mapping from G to D. According to Theorem 5.3.5, φ' is free of zeros, and according to Remark 5.3.6, $\varphi^{-1} : D \to G$ is also a conformal mapping.[2] If $D = G$, we call φ a (conformal) **automorphism** .

---

[2] In fact, the conformity of a mapping is initially defined locally as angle preservation at a point. It can be shown that holomorphic functions are angle-preserving at a point z exactly when the derivative at z does not vanish, i.e., z is not critical. Therefore, bijective holomorphic functions are always angle-preserving at all points.

2. The domains G, D are called **conformally equivalent**, if a conformal mapping $\varphi : G \to D$ exists. According to 1. and the chain rule, this actually defines an equivalence relation on the set of all domains in $\mathbb{C}$.

Example 6.2.2

1. (**Rotations**) For $\theta \in \mathbb{R}$ let $\varphi(z) = \varphi_\theta(z) = e^{i\theta}z$ $(z \in \mathbb{D})$. Then $\varphi : \mathbb{D} \to \mathbb{D}$ is a conformal mapping with $\varphi(0) = 0$.
2. (Exponential function) Let $G = \{z \in \mathbb{C} : |\text{Im}(z)| < \pi\}$ and $D = \mathbb{C}^- = \mathbb{C} \setminus (-\infty, 0]$. Then $\varphi := \exp|_G : G \to D$ is a conformal mapping with $\varphi^{-1} = \log$ (see Example 5.1.18).
3. (Joukowski mapping) Let $G := \mathbb{C} \setminus \overline{\mathbb{D}}$ and $D := \mathbb{C} \setminus [-1, 1]$. Then the restriction $\varphi = j|_G$ of the Joukowski mapping on G, thus

$$\varphi(z) = \frac{1}{2} \left(z + \frac{1}{z} \right) \qquad (z \in G),$$

defines a conformal mapping $\varphi : G \to D$ (Exercise 6.2.12.1).

Remark and Definition 6.2.3 Let $a, b, c, d \in \mathbb{C}$ with $ad - bc \neq 0$. Then the mapping $\varphi : \mathbb{C}_\infty \to \mathbb{C}_\infty$ with

$$\varphi(z) := \begin{cases} \dfrac{az + b}{cz + d}, & \text{if } z \in \mathbb{C} \\ a/c, & \text{if } z = \infty \end{cases}$$

is called a **Möbius transformation**. Here we use the arithmetic of the Riemann sphere (see Remark 4.3.8).

Every Möbius transformation $\varphi : (\mathbb{C}_\infty, \chi) \to (\mathbb{C}_\infty, \chi)$ is a homeomorphism and it holds (as can be easily calculated)

$$\varphi^{-1}(w) = \begin{cases} \dfrac{dw - b}{a - cw}, & \text{if } w \in \mathbb{C} \\ -d/c, & \text{if } w = \infty \end{cases}.$$

Furthermore, $\varphi|_\mathbb{C}$ is meromorphic with a pole of order 1 at $-d/c$.[3]

Special classes of Möbius transformations result in important conformal mappings of open disk or half-planes onto the unit disk.

---

[3] With Möbius transformations, one has finally arrived in spherical geometry. For a deeper understanding, visualizations are extremely helpful. Excellent corresponding monographs exist, such as [8] and [12].

Theorem 6.2.4

1. *For $\alpha \in \mathbb{D}$, by*

$$\varphi(z) := \varphi_\alpha(z) := \frac{z - \alpha}{1 - \bar{\alpha}z} \qquad (z \in \mathbb{D})$$

   a conformal mapping $\varphi_\alpha : \mathbb{D} \to \mathbb{D}$ is defined with $\varphi_\alpha(\alpha) = 0$ and $\varphi_\alpha^{-1} = \varphi_{-\alpha}$, thus

$$\varphi_\alpha^{-1}(w) = \frac{w + \alpha}{1 + \bar{\alpha}w} \qquad (w \in \mathbb{D}).$$

2. *For $\beta \in \mathbb{H}$, by*

$$\varphi(z) := \varphi_\beta(z) := \frac{z - \beta}{z - \bar{\beta}} \qquad (z \in \mathbb{H})$$

   a conformal mapping $\varphi_\beta : \mathbb{H} \to \mathbb{D}$ is defined with $\varphi_\beta(\beta) = 0$ and

$$\varphi_\beta^{-1}(w) = \frac{\beta - \bar{\beta}w}{1 - w} \qquad (w \in \mathbb{D}).$$

Proof. We consider φ again as a Möbius transformation, defined on $\mathbb{C}_\infty$.

1. It holds for $|z| = 1$

$$|z - \alpha| = |z| \cdot |1 - \alpha\bar{z}| = |1 - \bar{\alpha}z|,$$

   so $|\varphi(z)| = 1$, that is $\varphi(\mathbb{S}) \subset \mathbb{S}$. Further, according to Remark 6.2.3

$$\varphi^{-1}(w) = \frac{w + \alpha}{1 + \bar{\alpha}w} \qquad (w \in \mathbb{D}),$$

   is therefore of the same form. This also applies to $\varphi^{-1}(\mathbb{S}) \subset \mathbb{S}$ and consequently $\varphi(\mathbb{S}) = \mathbb{S}$. This in turn implies $\varphi(\mathbb{C}_\infty \setminus \mathbb{S}) = \mathbb{C}_\infty \setminus \mathbb{S}$. From $\varphi(\alpha) = 0 \in \mathbb{D}$ and the domain preservation of φ (Remark 5.3.7) this results in $\varphi(\mathbb{D}) \subset \mathbb{D}$ and $\varphi(\mathbb{C}_\infty \setminus \overline{\mathbb{D}}) \subset \mathbb{C}_\infty \setminus \overline{\mathbb{D}}$ and thus also $\varphi(\mathbb{D}) = \mathbb{D}$.
2. For $x \in \mathbb{R}$ is $|x - \beta| = |x - \bar{\beta}|$, that is $\varphi(\mathbb{R}) \subset \mathbb{S}$. Also, $\varphi(\infty) = 1$, thus $\varphi(\mathbb{R}_\infty) \subset \mathbb{S}$, where $\mathbb{R}_\infty := \mathbb{R} \cup \{\infty\}$. Since $\varphi(\mathbb{R}_\infty)$ is connected and $\lim_{x \to \pm\infty} \varphi(x) = 1$, it follows $\varphi(\mathbb{R}_\infty) = \mathbb{S}$. From $\varphi(\beta) = 0 \in \mathbb{D}$ it follows as in 1. that also $\varphi(\mathbb{H}) = \mathbb{D}$. □

Example 6.2.5

1. (**Cayley Transformation**) By

$$\varphi(z) = \frac{z - i}{z + i} \qquad (z \in \mathbb{H})$$

   a conformal mapping from $\mathbb{H}$ to $\mathbb{D}$ is defined. It holds that $\varphi(i) = 0$ and

$$\varphi^{-1}(w) = i \cdot \frac{1+w}{1-w} \qquad (w \in \mathbb{D}).$$

2. Let $\varphi : \mathbb{D} \to \mathbb{C}$ be defined by

$$\varphi(z) := \frac{z}{1-z} = \sum_{\nu=1}^{\infty} z^{\nu} \quad (z \in \mathbb{D}).$$

According to 1., $z \mapsto (1+z)/(1-z)$ is a conformal mapping from $\mathbb{D}$ to the right half-plane $-i\mathbb{H}$. Because $2\varphi(z) = (1+z)/(1-z) - 1$, φ is therefore a conformal mapping from $\mathbb{D}$ to the half-plane $\{w \in \mathbb{C} : \mathrm{Re}(w) > -1/2\}$.

3. (**Koebe Mapping**) Let $\varphi : \mathbb{D} \to \mathbb{C}$ be defined by

$$\varphi(z) := \frac{z}{(1-z)^2} = \sum_{\nu=1}^{\infty} \nu z^{\nu} \quad (z \in \mathbb{D}).$$

Then φ is a conformal mapping from $\mathbb{D}$ to the slit plane $\mathbb{C} \setminus (-\infty, -1/4]$.

Because: It holds that

$$4\varphi(z) = \frac{4z}{(1-z)^2} = \left(\frac{1+z}{1-z}\right)^2 - 1 \quad (z \in \mathbb{D}).$$

Therefore, 4φ is the composition of the functions $z \mapsto (1+z)/(1-z)$ and $w \mapsto w^2 - 1$. Since $w \mapsto w^2$ maps the right half-plane $-i\mathbb{H}$ conformally onto $\mathbb{C} \setminus (-\infty, 0]$, the assertion follows from 2.

Figure 6.2 gives an impression of the behavior of the Koebe function. Each color corresponds to an image set $\varphi(V_{r,R}(0))$ for certain values $0 < r < R < 1$.

If $f : \mathbb{C} \to \mathbb{D}$ is holomorphic, then f is already constant according to Liouville's theorem. Therefore, in particular, there is no conformal mapping from $\mathbb{C}$ to $\mathbb{D}$, that is, $\mathbb{C}$ and $\mathbb{D}$ are not conformally equivalent. A major goal of this section is to prove the Riemann mapping theorem, which states that every simply connected region $G \neq \mathbb{C}$ is conformally equivalent to the unit disk $\mathbb{D}$. The theorem shows that the unit disk can be seen not only as an example, but—in this sense—as a model of any simply connected region.

We first prove an auxiliary result, which is also of interest in its own right.

Theorem 6.2.6 (Schwarz's Lemma) *Let $f \in H(\mathbb{D})$ with $f(\mathbb{D}) \subset \mathbb{D}$ and $f(0) = 0$. Then $M(r, f) \leq r$ for $0 < r < 1$ and $|f'(0)| \leq 1$. Moreover, equality in one of the two inequalities implies that f is a rotation, that is, there exists a real number θ such that $f(z) = e^{i\theta}z$ for $z \in \mathbb{D}$.*

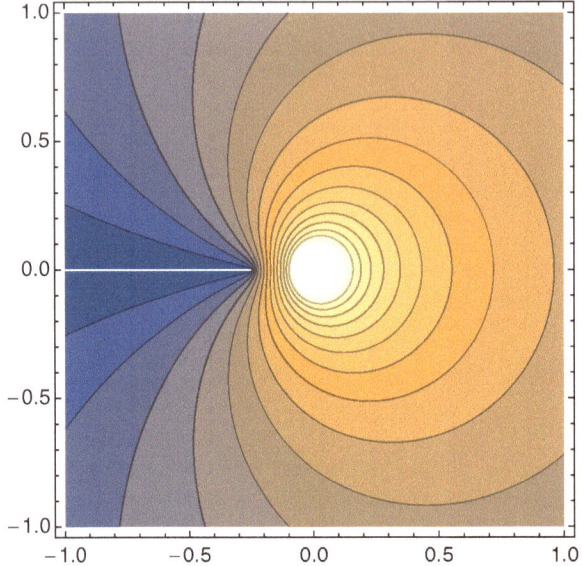

Fig. 6.2 Images of annular sectors under the Koebe function

Proof. From $f(0) = 0$ it follows that $g : \mathbb{D} \to \mathbb{C}$ with

$$g(z) := \begin{cases} f(z)/z, & \text{if } z \neq 0 \\ f'(0), & \text{if } z = 0 \end{cases}$$

is holomorphic in $\mathbb{D}$. Let $0 \leq r < 1$ be given. Since $r \mapsto M(r, g)$ is monotonically increasing according to the maximum principle (see Remark 4.1.17), for $r < s < 1$

$$M(r, g) \leq M(s, g) = M(s, f)/s \leq 1/s \to 1 \qquad (s \to 1^-).$$

Therefore, $M(r, g) \leq 1$ and thus $M(r, f) = rM(r, g) \leq r$ and $|f'(0)| \leq 1$.

If $M(r, f) = r$ for an r with $0 < r < 1$ or $|f'(0)| = 1$, then $|g|$ has a maximum in $\mathbb{D}$. According to the negative form of the maximum principle (Theorem 4.1.14), $g(z) = c$, where $|c| = 1$. Thus, f is a rotation.   □

Using Montel's theorem in the form of Remark 6.1.20, we show

Theorem 6.2.7 (Riemann Mapping Theorem) *Let $G \subset \mathbb{C}$, $G \neq \mathbb{C}$ be a simply connected domain and let $z_0 \in G$. Then there exists a conformal mapping $\varphi : G \to \mathbb{D}$ with $\varphi(z_0) = 0$. In particular, G and $\mathbb{D}$ are conformally equivalent.*

Proof.

1. We set

$$\mathscr{F} := \{\psi \in H(G) : \psi(G) \subset \mathbb{D}, \ \psi \text{ injective}, \ \psi(z_0) = 0\}.$$

Then $\mathscr{F}$ is not empty.

Because: Let $\zeta \in \mathbb{C} \setminus G$. Then, according to Remark 5.1.17 there exists an $f \in H(G)$ with $f^2(z) = z - \zeta$ for all $z \in G$. If $f(z_1) = \pm f(z_2)$, then

$$z_1 - \zeta = f^2(z_1) = f^2(z_2) = z_2 - \zeta,$$

is also $z_1 = z_2$. Thus, f is injective and from $w \in f(G) \setminus \{0\}$ it follows that $-w \notin f(G)$.

Since $f(G)$ is a domain, there exist a $c \in f(G)$ and a $r > 0$ with $B_r(c) \subset f(G) \setminus \{0\}$. But then $B_r(-c) \cap f(G) = \emptyset$. If $\psi := r/(f + c)$, then ψ is injective and $\psi(G) \subset \mathbb{D}$ (Note: $|f + c| > r$). The functions φ_α from Theorem 6.2.4.1 map $\mathbb{D}$ conformally onto $\mathbb{D}$ with $\varphi_\alpha(\alpha) = 0$. For $\alpha := \psi(z_0)$ it follows that $\varphi_\alpha \circ \psi \in \mathscr{F}$.

2. We show: If $\psi \in \mathscr{F}$ with $\psi(G) \neq \mathbb{D}$, then there exists a $\psi_1 \in \mathscr{F}$ with

$$\left|\psi_1'(z_0)\right| > \left|\psi'(z_0)\right|.$$

Because: Let $\alpha \in \mathbb{D} \setminus \psi(G)$. Then $\varphi_\alpha \circ \psi \in H(G)$ is injective and $(\varphi_\alpha \circ \psi)(G) \subset \mathbb{D} \setminus \{0\}$. Again, there exists a $g \in H(G)$ with $g^2 = \varphi_\alpha \circ \psi$. As in 1., one can see that g is injective. If $\psi_1 := \varphi_\beta \circ g$, where $\beta = g(z_0)$, then it follows $\psi_1 \in \mathscr{F}$ and

$$\psi = \varphi_{-\alpha} \circ g^2 = \varphi_{-\alpha} \circ (\varphi_{-\beta})^2 \circ \psi_1$$

(note that $\varphi_{-\alpha} = \varphi_\alpha^{-1}$ and correspondingly for β). If $h := \varphi_{-\alpha} \circ (\varphi_{-\beta})^2$, then $h(\mathbb{D}) \subset \mathbb{D}$ with

$$h(0) = \varphi_{-\alpha}(g^2(z_0)) = \psi(z_0) = 0$$

and h is *not injective*. According to the Schwarz Lemma, $|h'(0)| < 1$. Therefore, with the chain rule (note that: $\psi_1(z_0) = 0$)

$$|\psi'(z_0)| = |h'(0)| \cdot |\psi_1'(z_0)| < |\psi_1'(z_0)|.$$

3. Now let's make it pop: By definition, $\mathscr{F}_0 := \mathscr{F} \cup \{0\}$ is locally bounded in $H(G)$. Moreover, $\mathscr{F}_0$ is a closed subset of $H(G)$ according to Hurwitz's theorem, thus compact according to Remark 6.1.20. We consider the mapping $\ell : \mathscr{F}_0 \to \mathbb{R}$, defined by

$$\ell(\psi) = |\psi'(z_0)| \qquad (\psi \in \mathscr{F}_0).$$

According to Theorem 4.1.21, ℓ is continuous. Thus, ℓ becomes maximal, meaning there exists an $\varphi \in \mathscr{F}_0$ with

$$|\varphi'(z_0)| \geq |\psi'(z_0)| \qquad (\psi \in \mathscr{F}_0).$$

Here, $\varphi \neq 0$, since $\mathscr{F}$ is non-empty according to 1. From 2. $\varphi(G) = \mathbb{D}$ follows. Thus, φ is a conformal mapping from G to $\mathbb{D}$ with $\varphi(z_0) = 0$, as desired. $\square$

One can ask—in case of existence—"how many" conformal mappings φ between G and $\mathbb{D}$ with $\varphi(z_0) = 0$ exist.

Theorem 6.2.8 *Let $G \subset \mathbb{C}$ be a domain and $z_0 \in G$. If $\varphi : G \to \mathbb{D}$ is a conformal mapping with $\varphi(z_0) = 0$, then*

1. *For all $\theta \in \mathbb{R}$ $\psi = e^{i\theta}\varphi$ is also a conformal mapping from G to $\mathbb{D}$ with $\psi(z_0) = 0$.*
2. *If ψ is a conformal mapping from G to $\mathbb{D}$ with $\psi(z_0) = 0$, then there exists a $\theta \in \mathbb{R}$ with $\psi = e^{i\theta}\varphi$.*

Proof.

1. It holds that $\psi(z_0) = 0$. Since compositions of conformal mappings are again conformal, ψ is a conformal mapping of G onto $\mathbb{D}$ according to Example 6.2.2.1.
2. By assumption, $\sigma := \psi \circ \varphi^{-1}$ is a conformal mapping from $\mathbb{D}$ onto $\mathbb{D}$ with $\sigma(0) = 0$. Therefore, by applying the Schwarz Lemma to the functions σ^{-1} and σ,

$$|z| = |\sigma^{-1}(\sigma(z))| \le |\sigma(z)| \le |z| \qquad (z \in \mathbb{D})$$

holds and thus $|\sigma(z)| = |z|$. Again, by the Schwarz Lemma, σ is a rotation. Therefore, there exists a $\theta \in \mathbb{R}$ with $\psi = e^{i\theta}\varphi$.   □

Remark 6.2.9 Let $G \subset \mathbb{C}$ be a domain and $z_0 \in G$. According to Theorem 6.2.8.2, there exists at most one conformal mapping φ from G onto $\mathbb{D}$ with $\varphi(z_0) = 0$ and $\varphi'(z_0) > 0$. If $G \ne \mathbb{C}$ is simply connected, then according to the Riemann mapping theorem and Theorem 6.2.8.1, there exists *exactly one* conformal mapping $\varphi : G \to \mathbb{D}$ with $\varphi(z_0) = 0$ and $\varphi'(z_0) > 0$.

Remark and Definition 6.2.10 We set for regions $G \subset \mathbb{C}$

$$\mathrm{Aut}(G) := \{\varphi : G \to G \text{ conformal mapping}\}.$$

Then $(\mathrm{Aut}(G), \circ)$ is a group, the **automorphism group** of G.

Example 6.2.11

1. Let $\alpha \in \mathbb{D}$. According to Theorem 6.2.4.1 and Theorem 6.2.8, $\varphi : \mathbb{D} \to \mathbb{D}$ is conformal with $\varphi(\alpha) = 0$ exactly when φ has the form

$$\varphi(z) = \varphi_{\alpha,\theta}(z) := e^{i\theta}\frac{z - \alpha}{1 - \bar{\alpha}z} \qquad (z \in \mathbb{D})$$

for a $\theta \in \mathbb{R}$. Thus,

$$\mathrm{Aut}(\mathbb{D}) = \{\varphi_{\alpha,\theta} : \alpha \in \mathbb{D}, \theta \in \mathbb{R}\}.$$

2. For $G = \mathbb{C}$, $\mathrm{Aut}(\mathbb{C})$ is the set of affine-linear mappings $\varphi = \varphi_{a,c}$ of the form

$$\varphi_{a,c}(z) := a + cz \qquad (z \in \mathbb{C})$$

with $a \in \mathbb{C}$ and $c \in \mathbb{C}^*$ (Exercise 6.2.12.3).

Exercises 6.2.12

1. a) Show that the restriction $j|_G$ of the Joukowski mapping on $G := \mathbb{C} \setminus \overline{\mathbb{D}}$ maps the complement G of the closed unit disk conformally onto the complement $\mathbb{C} \setminus [-1, 1]$ of the interval $[-1, 1]$, and show that $j(K_R(0))$ for $R > 1$ is an ellipse.
   b) What is the image under $\cos$ of the line $\{\zeta : \operatorname{Im}(\zeta) = \beta\}$ for $\beta > 0$?
2. Let $\Omega \subset \mathbb{C}$ be open and $f : \Omega \to \mathbb{C}$. Then f is called **angle-preserving** at the point $a \in \Omega$, if an $\alpha \in \mathbb{R}$ exists such that

$$\lim_{r \to 0^+} \frac{\tau_a f(re^{i\theta})}{|\tau_a f(re^{i\theta})|} = e^{i(\theta + \alpha)}$$

holds for all $\theta \in \mathbb{R}$. Show: If $f \in H(\Omega)$, then f is angle-preserving at a if and only if $f'(a) = (\tau_a f)'(0) \neq 0$ holds.
3. Let f be entire. Show:
   a) If f is transcendental, then $f(\mathbb{D}) \cap f(\mathbb{C} \setminus \mathbb{D}) \neq \emptyset$ is true.
   b) If f is injective, then $a \in \mathbb{C}$ and $c \in \mathbb{C}^*$ exist such that $f(z) = a + cz$ for all $z \in \mathbb{C}$.

6.3 Theorems of Montel and Picard

Let $\Omega \subset \mathbb{C}$ be open. In the following, we will consider families $\mathscr{F} \subset M(\Omega)$ as families in $C(\Omega, \mathbb{C}_\infty)$, where the Riemann sphere $\mathbb{C}_\infty$ is equipped with the chordal metric χ. To distinguish normality of a family in $(C(\Omega, \mathbb{C}_\infty), d_{\mathrm{loc}, \mathbb{C}_\infty})$ from normality in $(C(\Omega), d_{\mathrm{loc}, \mathbb{C}})$, we call the family in the first case also **spherically normal**. Furthermore, in the case of convergence of sequences (a_n) in $\mathbb{C}_\infty$ we usually speak for clarity of **spherical convergence**, so for example in the case of sequences of functions in $(C(\Omega, \mathbb{C}_\infty), d_{\mathrm{loc}, \mathbb{C}_\infty})$ of spherically locally uniform convergence. Finally, we call a function $f \in M(\Omega)$ **extended holomorphic**, if f has no poles, so if for all $z \in \Omega$ either $f(z) \in \mathbb{C}$ or f is constant with value ∞ around z, and thus we set

$$H_\infty(\Omega) := \{f \in M(\Omega) : f \text{ extended holomorphic}\}.$$

Theorem 6.3.1 *Let $\Omega \subset \mathbb{C}$ be open. Then the following holds*

1. *If a sequence (f_n) in $C(\Omega, \mathbb{C}_\infty)$ converges spherically locally uniformly to a function $f \in C(\Omega)$, then for every compact subset K of Ω there exists a $n_K \in \mathbb{N}$ such that $(f_n|_K)_{n \geq n_K}$ in $(C(K), \|\cdot\|_\infty)$ converges to $f|_K$.*
2. *$M(\Omega)$ and $H_\infty(\Omega)$ are closed in $(C(\Omega, \mathbb{C}_\infty), d_{\mathrm{loc}, \mathbb{C}_\infty})$, thus complete as metric spaces.*

Proof.

1. If $K \subset \Omega$ is compact, then $f(K) \subset \mathbb{C}$ is compact, and therefore

$$\delta := \mathrm{dist}(f(K), \infty)/2 > 0.$$

Furthermore, $f_n|_K \to f|_K$ in $(C(K, \mathbb{C}_\infty), d_{\infty, K, \mathbb{C}_\infty})$ according to Remark 4.1.19. Thus, there exists an $n_K \in \mathbb{N}$ such that

$$\max_{z \in K} \chi \left(f_n(z), f(z) \right) < \delta \qquad (n \geq n_K)$$

and thus $\mathrm{dist}(f_n(K), \infty) \geq \delta$ for $n \geq n_K$. From the definition of the chordal metric, there exists a constant $c = c_\delta > 0$ with

$$|u - v| \leq c \cdot \chi(u, v) \qquad (u, v \in \mathbb{C} \setminus U_{\delta, \chi}(\infty)).$$

Therefore, $f_n|_K \to f|_K$ also holds in $(C(K), \|\cdot\|_\infty)$.

2. Let (f_n) be a sequence in $M(\Omega)$ with $f_n \to f$ in $C(\Omega, \mathbb{C}_\infty)$.

If $a \in \Omega$ with $f(a) \neq \infty$, then there exists an open neighborhood V of a with $f|_V \in C(V)$. If U is an open neighborhood of a with $\overline{U} \subset V$, then $f_n|_U \in H(U)$ for sufficiently large n and $f_n \to f$ (locally) uniformly on U according to 1. Therefore, $f|_U$ is holomorphic.

If $f(a) = \infty$, then $(1/f)(a) = 0$. From $f_n \to f$ in $C(\Omega, \mathbb{C}_\infty)$ it also follows that $1/f_n \to 1/f$ in $C(\Omega, \mathbb{C}_\infty)$, since $w \mapsto 1/w$ is an isometry on $\mathbb{C}_\infty$. As above, we see that $(1/f)|_U \in H(U)$ and thus $f|_U \in M(U)$ for a neighborhood U of a.

Let now $f_n \in H_\infty(\Omega)$ with $f_n \to f$ in $C(\Omega, \mathbb{C}_\infty)$. We need to show: If $a \in \Omega$ with $f(a) = \infty$, then f is constant (with value ∞) around a.

As before, there exists an open neighborhood U of a with $(1/f_n)|_U \in H(U)$ for sufficiently large n and $1/f_n \to 1/f$ locally uniformly on U. Without loss of generality, let U be a domain. Then either $1/f_n$ is constant with value 0 on U for infinitely many n or $1/f_n$ has no zeros in U for sufficiently large n. In both cases, $1/f$ is constant with value 0 on U, in the second case according to Hurwitz's theorem. □

The following theorem shows that in normal families of meromorphic functions, convergence in a certain way is inherited from small sets to the entire set:

Theorem 6.3.2 (Vitali) *Let $G \subset \mathbb{C}$ be a domain and $A \subset G$ with an accumulation point in G. If $\mathscr{F} \subset M(G)$ is a spherically normal family and (f_n) is a sequence in $\mathscr{F}$ such that $(f_n|_A)$ converges pointwise on A, then there exists an $f \in M(G)$ with $f_n \to f$ spherically locally uniformly.*

Proof. According to Exercise 6.1.21.5, it is sufficient to show that all spherically locally uniformly convergent subsequences of (f_n) have the same limit function. So if $f_n \to f$ $(n \to \infty, n \in I)$ and $f_n \to g$ $(n \to \infty, n \in J)$, then $f|_A = g|_A$ by assumption. Then, however, $f = g$ follows from the identity theorem (which also applies to meromorphic functions on domains). □

The distortion theorem shows that the distortion of holomorphic functions on lines can be estimated by the magnitude of the derivative. We now want to derive a corresponding result for meromorphic functions, where the magnitude of the classical derivative is replaced by a suitable spherical derivative.

Remark and Definition 6.3.3 Let $\Omega \subset \mathbb{C}$ be open and $f \in M(\Omega)$. If $z \in \Omega$ with $f(z) \neq \infty$, we set

$$f^{\#}(z) := |f'(z)|/(1 + |f(z)|^2)$$

and if f is constant $= \infty$ around z, we set $f^{\#}(z) = 0$. If a is a pole of order p, the Laurent expansion with respect to a implies the existence of a number $c \in \mathbb{C}^*$ with

$$f(a + h)h^p \to c \qquad (h \to 0)$$

and thus

$$(1 + |f^2(a + h)|)|h|^{2p} \to |c|^2 \qquad (h \to 0).$$

Furthermore,

$$f'(a + h)h^{p+1} \to -cp \qquad (h \to 0),$$

so overall

$$\frac{|f'(a + h)|}{1 + |f^2(a + h)|} \to \begin{cases} 1/|c|, & \text{if } p = 1 \\ 0, & \text{if } p > 1 \end{cases} \qquad (h \to 0).$$

Thus, the function $f^{\#}$ which is continuous on $U \setminus \{a\}$ for a neighborhood U of a has a limit at a in $[0, \infty)$. If we write $f^{\#}(a)$ for the limit, then $f^{\#}(z)$ is defined for all $z \in \Omega$. We call $f^{\#}(z)$ the **spherical derivative** of f at z. By definition, $f^{\#} \in C(\Omega)$ and $f^{\#} \geq 0$. In addition, the mapping

$$M(\Omega) \ni f \mapsto f^{\#} \in C(\Omega)$$

is continuous (Exercise 6.3.17.2).

Remark 6.3.4 The classical derivative f' does not change if a translation is applied to f, that is, if $w \in \mathbb{C}$ and if $\tau_w : \mathbb{C} \to \mathbb{C}$ is of the form $\tau_w(u) = u + w$, then $(\tau_w \circ f)' = f'$. The same applies in the case of $|f|$ for rotations $u \mapsto e^{i\theta}u$. A corresponding property can be derived for the spherical derivative and suitable rigid transformations of the sphere. For this, we consider for $w \in \mathbb{C}$ the Möbius transformation φ_w with

$$\varphi_w(u) := \frac{u - w}{1 + u\overline{w}} \qquad (u \in \mathbb{C}).$$

Then

$$(\varphi_w)'(u) = \frac{1 + |w|^2}{(1 + u\overline{w})^2} \qquad (u \in \mathbb{C} \setminus \{-1/\overline{w}\}),$$

so

$$(\varphi_w)^\#(u) = \frac{1 + |w|^2}{|1 + u\bar{w}|^2 + |u - w|^2} = \frac{1}{1 + |u|^2} = (\varphi_0)^\#(u) \qquad (u \in \mathbb{C}).$$

For $\varphi_\infty(u) := 1/u$ also $(\varphi_\infty)^\#(u) = 1/(1 + |u|^2)$ $(u \in \mathbb{C})$.

If $\Omega \subset \mathbb{C}$ is open, then by the chain rule for the spherical derivative (Exercise 6.3.17.1) for $w \in \mathbb{C}_\infty$ and $f \in M(\Omega)$

$$(\varphi_w \circ f)^\# = f^\#,$$

in particular $(1/f)^\# = f^\#$.

Remark 6.3.5 For $t \in \mathbb{R}$ (Exercise 3.1.34.6)

$$\sin(\arctan(t)) = \frac{t}{\sqrt{1 + t^2}}.$$

With the representation of the chordal metric from Remark 4.3.10 and with φ_w from Remark 6.3.4 results with a similar calculation as in Remark 6.3.4 for $u \in \mathbb{C}$ and $w \in \mathbb{C}_\infty$

$$\chi(u, w) = \sin\left(\arctan|\varphi_w(u)|\right),$$

where we set $\arctan|\infty| := \pi/2$. So with

$$\sigma(u, w) := \arctan|\varphi_w(u)|$$

in particular

$$\chi(u, w) \le \sigma(u, w) \le \frac{\pi}{2}\chi(u, w).$$

In fact, σ also defines a metric on $\mathbb{C}_\infty$, called **spherical metric**, which we will neither use nor prove. As with the chordal metric, $z \mapsto 1/z$ is an isometry with respect to σ. The spherical distance $\sigma(u, w)$ geometrically corresponds to the length of the smaller arc of a great circle on which the two pre-image points $\varphi(u)$ and $\varphi(w)$ under the stereographic projection lie.[4] Figure 6.3 gives an impression. Here, the corresponding great circle runs through the north and south pole.

We now show that the distortion of f as a mapping from $\Omega \subset \mathbb{C}$ into the Riemann sphere $\mathbb{C}_\infty$ can be estimated by the spherical derivative $f^\#$.

Theorem 6.3.6 (Spherical Distortion Theorem) *Let $\Omega \subset \mathbb{C}$ be open and $f \in M(\Omega)$. If $a \in \Omega$ and $h \in \mathbb{C}^*$ with $[a, a + h] \subset \Omega$, then*

$$\chi(f(a + h), f(a)) \le \max_{[a,a+h]} f^\# \cdot |h|.$$

---

[4] If the two points are not antipodal, there is exactly one such great circle; antipodal points always have the spherical distance $\pi/2$.

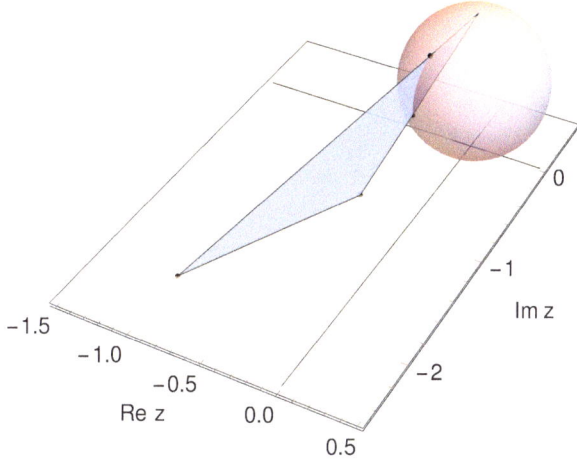

Fig. 6.3 spherical and Euclidean distance of the points $z = -1 - 2i$ and $w = (2/5)z$

Proof. If f is constant at a point $z \in [a, a + h]$, then f is constant on $[a, a + h]$ and thus the claim is clear. We can therefore assume that f is not constant at any point.

1. First, let f be such that $f(u) \neq f(a)$ and $f(u) \neq -1/\overline{f(a)}$ holds for all $u \in (a, a + h]$. Then $\psi : [0, 1] \to \mathbb{C}$ with

$$\psi(t) := (\varphi_{f(a)} \circ f)(a + th),$$

where φ_w as in Remark 6.3.4, is differentiable on $[0, 1]$ and has no zeros on $(0, 1)$. Thus, $|\psi| = \sqrt{\psi \cdot \overline{\psi}}$ is also differentiable on $(0, 1)$, and by the reverse triangle inequality,

$$\pm |\psi|' \leq |\psi'|.$$

Therefore, $g := \arctan \circ |\psi|$ is continuous on $[0, 1]$, differentiable on $(0, 1)$, and by the mean value theorem, there exists a $\tau \in (0, 1)$ with

$$\sigma(f(a + h), f(a)) = g(1) = g(1) - g(0) = g'(\tau) = \frac{|\psi|'(\tau)}{1 + |\psi|^2(\tau)} \leq \psi^\#(\tau).$$

Furthermore, with $\xi := a + \tau h$ according to Remark 6.3.4 and the chain rule,

$$\psi^\#(\tau) = f^\#(\xi) \cdot |h|$$

and thus

$$\chi(f(a + h), f(a)) \leq \sigma(f(a + h), f(a)) \leq f^\#(\xi) \cdot |h|.$$

2. Now, if f is arbitrary, then due to the compactness of $[a, a + h]$, there exist points $0 = s_0 < s_1 < \ldots < s_n = 1$ such that with $a_j := a + s_j h$ for $j = 0, \ldots, n$, the conditions from 1. are fulfilled on $[a_{j-1}, a_j]$. By applying 1. to the intervals $[a_{j-1}, a_j]$, using the triangle inequality for χ, the assertion is obtained (note that $\sum_{j=1}^{n} |a_j - a_{j-1}| = |h|$ holds). $\square$

From the spherical distortion theorem, we obtain a characterization of spherical normality for families of meromorphic functions:

Theorem 6.3.7 (Marty) *Let $\Omega \subset \mathbb{C}$ open and let $\mathscr{F} \subset M(\Omega)$. Then the following are equivalent:*

a) *$\mathscr{F}$ is spherically normal.*
b) *$\{ f^{\#} : f \in \mathscr{F} \}$ is normal in $C(\Omega)$.*
c) *$\{ f^{\#} : f \in \mathscr{F} \}$ is locally bounded in $C(\Omega)$.*

Proof c) $\Rightarrow$ a): Let $a \in \Omega$ and $R > 0$ with $B_R(a) \subset \Omega$ such that

$$c := \sup_{f \in \mathscr{F}} \max_{B_R(a)} f^{\#} < \infty.$$

If $z \in U_R(a)$, then for $f \in \mathscr{F}$ with the spherical distortion theorem

$$\chi(f(z), f(a)) \le \max_{[a,z]} f^{\#} \cdot |z - a| \le c|z - a|.$$

Therefore, $\mathscr{F}$ is uniformly continuous at a. Since $(\mathbb{C}_\infty, \chi)$ is compact, $\mathscr{F}$ is spherically normal according to Remark 6.1.16.

a) $\Rightarrow$ b): If (f_n) is any sequence in $M(\Omega)$, then from $f_n \to f$ in $C(\Omega, \mathbb{C}_\infty)$ also $f_n^{\#} \to f^{\#}$ in $C(\Omega)$ (Exercise 6.3.17.2). So if $\mathscr{F}$ is spherically normal, then $\{ f^{\#} : f \in \mathscr{F} \}$ is also normal in $C(\Omega)$.

b) $\Rightarrow$ c): follows from Remark 6.1.13. ☐

Example 6.3.8 We consider $\mathscr{F} = \{ f_n : n \in \mathbb{N} \} \subset H(\mathbb{C})$ with $f_n(z) = z^n$. Then

$$f_n^{\#}(z) = \frac{n|z|^{n-1}}{1 + |z|^{2n}} \le \begin{cases} n|z|^{n-1} & (z \in \mathbb{D}) \\ n/|z|^{n+1} & (z \in \mathbb{C} \setminus \overline{\mathbb{D}}) \end{cases}.$$

From this it follows that $\{ f_n^{\#} : n \in \mathbb{N} \}$ is locally bounded in $C(\mathbb{D})$ and in $C(\mathbb{C} \setminus \overline{\mathbb{D}})$. Thus, $\mathscr{F}|_{\mathbb{D}}$ and $\mathscr{F}|_{\mathbb{C} \setminus \overline{\mathbb{D}}}$ are spherically normal. Here

$$f_n(z) = z^n \to \begin{cases} 0 & \text{locally uniformly in } \mathbb{D} \\ \infty & \text{spherically locally uniformly in } \mathbb{C} \setminus \overline{\mathbb{D}} \end{cases}.$$

Note that $\mathscr{F}|_{\mathbb{D}}$ is also normal in $C(\mathbb{D})$, but $\mathscr{F}|_{\mathbb{C} \setminus \overline{\mathbb{D}}}$ not in $C(\mathbb{C} \setminus \overline{\mathbb{D}})$.

From $f_n^{\#}(z) = n/2$ for $z \in \mathbb{S}$ it follows that $\mathscr{F}$ is not spherically normal at all points $z \in \mathbb{S}$. It can be shown that

$$e^\zeta = \lim_{n \to \infty} \left(1 + \frac{\zeta}{n} \right)^n = \lim_{n \to \infty} (f_n \circ \varphi_n)(\zeta)$$

with $\varphi_n(\zeta) = 1 + \zeta/n$, where the convergence is locally uniform on $\mathbb{C}$ (Exercise 6.3.17.4). Thus, it can be seen that at the point $z = 1$ (where $\mathscr{F}$ is not spherically normal) a suitable scaling in the argument of f_n leads to a sequence that converges even on the entire plane $\mathbb{C}$.

A corresponding property applies in general for non-normal families, as we will now show. For this, we write

$$\text{Aut}_+(\mathbb{C}) := \{\varphi_{a,\rho} : a \in \mathbb{C}, \rho > 0\} \subset \text{Aut}(\mathbb{C})$$

(see Example 6.2.11).

Theorem 6.3.9 (Zalcman's Lemma) *Let $\Omega \subset \mathbb{C}$ be open, $a \in \Omega$ and $\mathscr{F} \subset M(\Omega)$ not spherically normal around a. Then there exist a sequence (f_n) in $\mathscr{F}$ and a sequence (φ_n) in $\text{Aut}_+(\mathbb{C})$ having the following properties:*

1. *$\varphi_n = \varphi_{a_n,\rho_n}$ with $\rho_n \to 0$ and $a_n \to a$ $(n \to \infty)$.*
2. *$f_n \circ \varphi_n \in M(\mathbb{D})$ with $(f_n \circ \varphi_n)^\#(0) = 1$ for all $n \in \mathbb{N}$.*
3. *For all $R > 0$ there exists an $n_R \in \mathbb{N}$ with $\varphi_n(U_R(0)) \subset \Omega$ for $n \geq n_R$ and*

$$\sup_{U_R(0)} (f_n \circ \varphi_n)^\# \to 1 \quad (n \to \infty, n \geq n_R).$$

Proof Let (r_n) be a sequence of positive numbers with $B_{r_n}(a) \subset \Omega$ and $r_n \to 0$. From Marty's theorem, the existence of a sequence (f_n) in $\mathscr{F}$ with

$$r_n \cdot \max_{B_{r_n/2}(a)} f_n^\# \to \infty \quad (n \to \infty)$$

is inductively derived. Then

$$R_n := \max_{z \in B_{r_n}(a)} f_n^\#(z)(r_n - |z - a|) \geq \frac{r_n}{2} \cdot \max_{B_{r_n/2}(a)} f_n^\# \to \infty \quad (n \to \infty).$$

By passing to a suitable subsequence, we can assume that (R_n) is increasing with $R_n \geq 1$ for all n. We choose $a_n \in U_{r_n}(a)$ with

$$f_n^\#(a_n)\left(r_n - |a_n - a|\right) = R_n$$

and set $\rho_n := 1/f_n^\#(a_n)$ for $n \in \mathbb{N}$. Then $a_n \to a$ $(n \to \infty)$ and

$$\rho_n R_n = \frac{R_n}{f_n^\#(a_n)} = r_n - |a_n - a|,$$

so in particular $\rho_n \to 0$. Moreover,

$$a_n + \rho_n U_{R_n}(0) = U_{\rho_n R_n}(a_n) \subset U_{r_n}(a).$$

With $\varphi_n := \varphi_{a_n,\rho_n}, f_n \circ \varphi_n$ is meromorphic in $\mathbb{D}$, and according to the spherical chain rule,

$$(f_n \circ \varphi_n)^\# = \rho_n f_n^\# \circ \varphi_n$$

holds in $\mathbb{D}$, so in particular $(f_n \circ \varphi_n)^\#(0) = \rho_n f_n^\#(a_n) = 1$ for all n.

Now let $R > 0$ be given, n_R chosen such that $R_{n_R} > R$ and $n \geq n_R$. If $\zeta \in U_R(0)$, then $w_n := a_n + \rho_n \zeta \in U_{r_n}(a)$ and thus

$$f_n^\#(w_n)(r_n - |a_n - a| - \rho_n|\zeta|) \leq f_n^\#(w_n)(r_n - |w_n - a|) \leq R_n,$$

therefore

$$(f_n \circ \varphi_n)^\#(\zeta) = \rho_n f_n^\#(w_n) \leq \frac{\rho_n R_n}{r_n - |a_n - a| - \rho_n|\zeta|} = \frac{\rho_n R_n}{\rho_n R_n - \rho_n|\zeta|} = \frac{1}{1 - |\zeta|/R_n}$$

and consequently $\sup_{U_R(0)} (f_n \circ \varphi_n)^\# \leq (1 - R/R_n)^{-1} \to 1 \, (n \to \infty)$.   □

Under the conditions of the Zalcman lemma, according to Marty's theorem, for every $R > 0$ the family $\{f_n \circ \varphi_n : n \geq n_R\}$ in $M(U_R(0))$ is spherically normal. Suitable (affine-linear) scaling in the argument thus leads to normality. This results in

Theorem 6.3.10 (Zalcman's Rescaling Theorem)   *Let $\Omega \subset \mathbb{C}$ be open, $a \in \Omega$ and $\mathcal{F} \subset M(\Omega)$ not spherically normal around a. Then there exist a sequence $(\varphi_n = \varphi_{a_n, \rho_n})$ in $\mathrm{Aut}_+(\mathbb{C})$ with $a_n \to a$ and $\rho_n \to 0$, a function $g \in M(\mathbb{C})$ with*

$$\max_{\mathbb{C}} g^\# = g^\#(0) = 1$$

and a sequence (f_n) in $\mathcal{F}$ such that for all $R > 0$ the sequence $(f_n \circ \varphi_n)_{n \geq n_R}$ for a suitable $n_R \in \mathbb{N}$ converges spherically locally uniformly on $U_R(0)$ to g. If $\mathcal{F} \subset H(\Omega)$, then g is an entire function.

Proof   Let (φ_n) and (f_n) be sequences as in the Zalcman lemma. Since, according to Marty's theorem, the family $\{f_n \circ \varphi_n : n \in \mathbb{N}\}$ is spherically normal in $M(\mathbb{D})$, we can assume, after passing to a suitable subsequence, that the sequence $(f_n \circ \varphi_n)$ converges in $M(\mathbb{D})$. We denote the limit function by $g_{\mathbb{D}}$. As already mentioned above, according to Marty's theorem, for every $R > 0$, the family $\{f_n \circ \varphi_n : n \geq n_R\}$ is spherically normal in $M(U_R(0))$. Therefore, according to Vitali's theorem, there exists a meromorphic extension g of $\mathbb{C}$ in $g_{\mathbb{D}}$ such that for all $R > 0$,

$$f_n \circ \varphi_n \to g \qquad (n \to \infty, \, n \geq n_R)$$

holds in $M(U_R(0))$. Due to the continuity of $h \mapsto h^\#$, it follows that $(f_n \circ \varphi_n)^\# \to g^\#$ in $C(U_R(0))$. Again, with the Zalcman lemma, we obtain

$$\max_{\mathbb{C}} g^\# = g^\#(0) = 1.$$

In particular, g is not constant. If $\mathcal{F}$ is a family in $H(\Omega)$, then the limit function is $g \in H_\infty(\mathbb{C})$ and not constant, thus a (non-constant) entire function.   □

With this, we prove the following central theorem.

Theorem 6.3.11 (Spherical Normality Theorem; Montel's Great Theorem) *Let* $\Omega \subset \mathbb{C}$ *be open and* $\mathscr{F} \subset M(\Omega)$. *If* $\mathscr{F}$ *omits three values, that is*

$$\#\left(\mathbb{C}_\infty \setminus \bigcup_{f \in \mathscr{F}} f(\Omega)\right) \geq 3,$$

then $\mathscr{F}$ *is spherically normal.*

Proof According to Remark 6.1.15, it suffices to show: $\mathscr{F}$ is spherically normal at all points $a \in \Omega$. We can therefore (after affine-linear scaling of the variables) assume without loss of generality $a = 0$ and $\Omega = \mathbb{D}$.

1. First, let the family $\mathscr{F}$ be such that

$$0, 1, \infty \notin \bigcup_{f \in \mathscr{F}} f(\mathbb{D})$$

holds. Then $\mathscr{F} \subset H(\mathbb{D})$ is true and from $0 \notin f(\mathbb{D})$ for $f \in \mathscr{F}$ the existence of m-th roots of f follows, that is, for all $m \in \mathbb{N}$ there exist $h = h_m \in H(\mathbb{D})$ with $h^m = f$ (see Remark 5.1.17). We set for $k \in \mathbb{N}$

$$\mathscr{F}_k := \{h \in H(\mathbb{D}) : h^{2^k} = f \text{ for some } f \in \mathscr{F}\}.$$

Assume, $\mathscr{F}$ is not spherically normal at the point 0. Then there exists a sequence $(f_n)_n$ in $\mathscr{F}$, which does not have a spherically locally uniformly convergent subsequence. If $k \in \mathbb{N}$ and $h_{n,k} \in \mathscr{F}_k$ with $(h_{n,k})^{2^k} = f_n$, then $(h_{n,k})_n$ also does not have a spherically locally uniformly convergent subsequence. Therefore, for all $k \in \mathbb{N}$ also $\mathscr{F}_k$ is not spherically normal at 0.

Let g_k be a limit function to $\mathscr{F}_k$ as in the rescaling theorem (there called g). Then g_k is entire and not constant. From Hurwitz's theorem, it follows that $g_k(\mathbb{C})$ contains no 2^k-th roots of unity (each function $h \in \mathscr{F}_k$ is such that $h(\mathbb{D})$ contains no such roots). Furthermore, $g_k^\# \leq 1$ on $\mathbb{C}$. Therefore, by Marty's theorem, $\{g_k : k \in \mathbb{N}\}$ is a spherically normal family. If $g \in H_\infty(\mathbb{C})$ is a locally uniform limit of a subsequence of $(g_k)_{k \in \mathbb{N}}$, then g is not constant, since $g^\#(0) = 1$ holds. Again, by Hurwitz's theorem, $g(\mathbb{C})$ contains no 2^k-th root of unity, now for all $k \in \mathbb{N}$, that is $g(\mathbb{C}) \cap W = \emptyset$, where

$$W := \{w \in \mathbb{S} : w^{2^k} = 1 \text{ for some } k \in \mathbb{N}\}.$$

Since W is dense in $\mathbb{S}$ (see Example 6.1.2 and Fig. 6.1) and since $g(\mathbb{C})$ is a domain, it follows that $g(\mathbb{C}) \subset \mathbb{D}$ or $g(\mathbb{C}) \subset \mathbb{C} \setminus \overline{\mathbb{D}}$. In the first case, g is constant according to Liouville's theorem. Contradiction. In the second case, applying Liouville's theorem to $1/g$ results in the same contradiction.

2. Now let $w_1, w_2, w_3 \in \mathbb{C}_\infty \setminus \bigcup_{f \in \mathscr{F}} f(\mathbb{D})$ be arbitrary. Then there exists a Möbius transformation $\varphi : \mathbb{C}_\infty \to \mathbb{C}_\infty$, which maps the three values w_1, w_2, w_3 to 0, 1, ∞ ($\varphi(z) = \frac{w_2-w_3}{w_2-w_1} \cdot \frac{z-w_1}{z-w_3}$ is suitable in the case of $w_1, w_2, w_3 \in \mathbb{C}, \varphi(z) = \frac{z-w_1}{w_2-w_1}$ for $w_3 = \infty$). According to 1., the family $\{\varphi \circ f : f \in \mathscr{F}\}$ is spherically normal. Since $\varphi^{-1} : \mathbb{C}_\infty \to \mathbb{C}_\infty$ is continuous and $f = \varphi^{-1} \circ \varphi \circ f$ for $f \in \mathscr{F}$ holds, $\mathscr{F}$ is also spherically normal. $\qquad\square$

Remark and Definition 6.3.12 Let $\Omega \subset \mathbb{C}$ be open, $a \in \Omega$ and f holomorphic in $\Omega\backslash\{a\}$. If f has an essential singularity at a, then it follows from the Casorati-Weierstraß theorem that $f(U_\delta^*(a))$ is dense in $\mathbb{C}_\infty$ for all $\delta > 0$ (with $U_\delta^*(a) \subset \Omega$). A point $w \in \mathbb{C}_\infty$ with

$$w \notin f\left(U_\delta^*(a)\right)$$

for some $\delta > 0$ is called a **Picard exceptional value** or simply **exceptional value** of f at the point a. We also use this term in the case $f \in M(\Omega \backslash\{a\})$ and write $E_\infty(f, a)$ for the set of exceptional values of f at a. For functions $f \in H(\Omega \backslash\{a\})$, the value ∞ is obviously always an exceptional value. In this case, we write $E(f, a) := E_\infty(f, a) \cap \mathbb{C}$.

Example 6.3.13 Let $f(z) = e^{1/z}$ for $z \in \mathbb{C}^*$. Since 0 is not taken as a value, it follows that $E(f, 0) = \{0\}$ with Example 4.3.13. Also, f with $f(z) = e^{1/z}(z- 1)$ for $z \in \mathbb{C}^*$ has the Picard exceptional value 0 at the point 0, even though here $0 = f(1)$ holds, so 0 is taken as a value.

Theorem 6.3.14 (Great Picard Theorem) *Let $\Omega \subset \mathbb{C}$ be open and $a \in \Omega$. If $f \in M(\Omega \backslash\{a\})$ with $\#(E_\infty(f, a)) \geq 3$, then f is meromorphically extendable to Ω. If $f \in H(\Omega \backslash\{a\})$ with $\#(E(f, a)) \geq 2$, then f has a removable singularity or a pole at a.*

Proof. Without loss of generality, we can assume that $a= 0$, $\mathbb{D} \subset \Omega$ and $0, 1, \infty \notin f(\mathbb{D}^*)$ with $\mathbb{D}^* := \mathbb{D} \setminus \{0\}$ holds. (Otherwise, one can again precede with an affine-linear mapping and follow with a Möbius transformation, cf. the proof of Theorem 6.3.11.)

Then, in particular, f and $1/f$ are in $H(\mathbb{D}^*)$. We define $g_n \in H(\mathbb{D}^*)$ by

$$g_n(z) := f(z/n) \qquad (n \in \mathbb{N}, z \in \mathbb{D}^*).$$

Then, $0, 1, \infty \notin g_n(\mathbb{D}^*)$ for all n. From the great theorem of Montel, it follows that $\{g_n : n \in \mathbb{N}\}$ is spherically normal. Therefore, there exists a subsequence $(g_n)_{n \in I}$ of (g_n) and a $g \in H_\infty(\mathbb{D}^*)$ with $g_n \to g$ $(n \to \infty, n \in I)$ locally uniformly on $\mathbb{D}^*$.

If g is not constant with value ∞, then $g \in H(\mathbb{D}^*)$. In this case, there exists an $M > 0$ such that

$$\left|f(z)\right| \leq M \quad \text{and} \quad \left|g(z)\right| \leq M - 1 \qquad \left(|z| = \frac{1}{2}\right).$$

Then $|g_n(z)| \leq M$ for $|z| = 1/2$ and $n \in I$ sufficiently large, thus

$$\left|f(z)\right| \leq M \qquad \left(|z| = \frac{1}{2n}\right)$$

for $n \in I$ sufficiently large. According to the maximum principle, even

$$\left|f(z)\right| \leq M \qquad \left(\frac{1}{2n} \leq |z| \leq \frac{1}{2}\right)$$

for $n \in I$ sufficiently large and therefore also $|f| \leq M$ on $U^*_{1/2}(0)$. Consequently, f has a removable singularity at 0 according to the Riemann removable singularity theorem.

If g is constant $= \infty$, one argues correspondingly with $1/f$ instead of f. Then $1/f$ has a removable singularity at 0 which is a zero and consequently f has a pole at 0. $\square$

Remark and Definition 6.3.15 Let $f \in M(\mathbb{C})$. A point $w \in \mathbb{C}_\infty$ is called a **Picard exceptional value** or shortly **exceptional value** of f, if w is an exceptional value of the function $g \in M(\mathbb{C}^*)$ with $g(z) := f(1/z)$ for $z \in \mathbb{C}^*$ at the point 0. This is exactly the case when $f^{-1}(\{w\})$ is finite (according to the identity theorem, there are no accumulation points of w-points in $\mathbb{C}$).

If f is entire and transcendental, then $g \in H(\mathbb{C}^*)$ has an essential singularity at the point 0. If we set $E(f) := E(g, 0) \subset \mathbb{C}$, then $E(f)$ is at most a single-point set according to the Great Picard Theorem. In particular, $\mathbb{C} \setminus f(\mathbb{C})$ is either empty or single-pointed. According to the Fundamental Theorem of Algebra, this also applies to non-constant polynomials and thus to any non-constant entire function f. This statement is also referred to as the **Little Picard Theorem**.

Example 6.3.16

1. For $f = \exp$ is $0 \notin \exp(\mathbb{C})$, thus 0 is a Picard exceptional value and for $f(z) = ze^z$, $0 \in E(f)$, thus also a Picard exceptional value, although $0 \in f(\mathbb{C})$.
2. According to Exercise 2.3.32.17, $\cos(\mathbb{C}) = \mathbb{C}$. Since $\cos$ is 2π-periodic, $\cos$ has no Picard exceptional values in $\mathbb{C}$.

Exercises 6.3.17

1. (**Spherical Chain Rule**) Let $\Omega \subset \mathbb{C}$ open, $f \in H(\Omega)$ and $g \in M(U)$ for an open set $U \supset f(\Omega)$. Show:

$$(g \circ f)^\# = (g^\# \circ f) \cdot |f'|.$$

2. Let $\Omega \subset \mathbb{C}$ be open. Show that the mapping $M(\Omega) \ni f \mapsto f^\# \in C(\Omega)$ is continuous.
3. Let $\mathscr{F} := \{f_n : n \in \mathbb{N}\} \subset M(\mathbb{C})$ with $f_n(z) := \cos(nz)$ for $z \in \mathbb{C}$.
   a) Calculate $f_n^\#$ and show that $f_n^\#(0) = 0$ for all $n \in \mathbb{N}$.
   b) Show that $\mathscr{F}$ is not spherically normal around 0.
4. Let $f_n(z) = (1 + z/n)^n$ for $z \in \mathbb{C}$ and $n \in \mathbb{N}$. Show:

$$f_n \to \exp \quad (n \to \infty) \quad \text{locally uniformly on } \mathbb{C}.$$

Hint: Use Exercise 3.1.34.10 and the Vitali's theorem.
5. Find a family $\mathscr{F} \subset M(\mathbb{C})$ such that $\mathscr{F}$ is not normal and omits two values, so $\#(\mathbb{C}_\infty \setminus \bigcup_{f \in \mathscr{F}} f(\mathbb{C})) \geq 2$.
6. a) Let (φ_n) be a sequence in $\text{Aut}_+(\mathbb{C})$ such that for all $R > 0$ there exists a n_R with $\varphi_n(U_R(0)) \subset \mathbb{D}$ for all $n \geq n_R$. Show: If $\varphi_n(\zeta) = a_n + \rho_n \zeta$, then (a_n) is bounded and $\rho_n \to 0$ $(n \to \infty)$.

b) Let $\Omega \subset \mathbb{C}$ be open with $\overline{\mathbb{D}} \subset \Omega$ and $\mathscr{F} \subset M(\Omega)$ spherically normal. Show: If $U \subset \mathbb{C}$ is a domain, (φ_n) as in a) and (f_n) in $\mathscr{F}$ with $f_n \circ \varphi_n \in M(U)$ and $f_n \circ \varphi_n \to g$ spherically locally uniformly on U, then g is constant.

6.4   Complex Dynamics

If X is a non-empty set, then $(\mathrm{Abb}(X, X), \circ)$ is a monoid. For $f \in \mathrm{Abb}(X, X)$ we consider the powers $f^{\circ 0} := \mathrm{id}_X$ and

$$f^{\circ n} := f^{\circ(n-1)} \circ f \qquad (n \in \mathbb{N})$$

with respect to the composition $\circ$ on $\mathrm{Abb}(X, X)$. In this context, we speak of a (discrete) dynamical system (X, f) (or briefly f) and consider the behavior of the sequence $(f^{\circ n})$ for large n.

Remark and Definition 6.4.1 Let X be a set and $f : X \to X$. Then a set $A \subset X$ is called

1. **forward-invariant** or shortly **invariant** (under f), if $f(A) \subset A$,
2. **backward-invariant** (under f), if $f^{-1}(A) \subset A$,
3. **completely invariant** (under f), if $f(A) \subset A$ and $f^{-1}(A) \subset A$.

It can be easily deduced (Exercise 6.4.25.2): A is completely invariant if and only if A and $X \setminus A$ are invariant. Furthermore, for $x \in X$

$$O^+(x) := O^+(f, x) := \{f^{\circ n}(x) : n \in \mathbb{N}\}$$

is called **forward orbit** or shortly **orbit** of x. In addition, we set

$$O^-(x) := O^-(f, x) := \bigcup_{n \in \mathbb{N}} (f^{\circ n})^{-1}(\{x\})$$

and $O(x) := O(f, x) := O^+(x) \cup O^-(x) \cup \{x\}$ and for $A \subset X$

$$O^+(A) := O^+(f, A) := \bigcup_{n \in \mathbb{N}} f^{\circ n}(A).$$

Thus, A is invariant if and only if $O^+(A) \subset A$ holds. If this is the case, the system can be restricted to the subset A, that is (A, f_A) with $f_A : A \to A$, defined by $f_A(x) := f(x)$ for $x \in A$, is also a dynamical system.

For the local dynamical behavior of smooth functions, fixed points are particularly important.

Definition 6.4.2 Let $X \subset \mathbb{K}$ be open and $f : X \to \mathbb{C}$ continuously differentiable. Furthermore, let $x^* \in X$ be a **fixed point** of f, so $f(x^*) = x^*$. With $\lambda := f'(x^*)$ x^* is called

1. **super attracting**, if $\lambda = 0$,
2. **attracting**, if $0 < |\lambda| < 1$,
3. **neutral**, if $|\lambda| = 1$,
4. **repelling**, if $|\lambda| > 1$.

Theorem 6.4.3 *Let $X \subset \mathbb{K}$ be open and $f : X \to \mathbb{K}$ continuously differentiable. If x^* is an attractive or super attracting fixed point, then for every α with $|\lambda| < \alpha < 1$ there exists an $r > 0$ such that $f(U) \subset U$ for $U := U_r(x^*)$ with*

$$|f^{\circ n}(x) - x^*| \le \alpha^n |x - x^*| \qquad (x \in U, n \in \mathbb{N}).$$

In particular, $f^{pn} \to x^* \ (n \to \infty)$ converges uniformly on U.

Proof. Let $r > 0$ be such that $B_r(x^*) \subset X$ and $\max\limits_{B_r(x^*)} |f'| \le \alpha$. Then, according to the distortion theorem

$$\left| f(x) - f(y) \right| \le \alpha |x - y| \qquad (x, y \in U)$$

and specifically for $y = x^*$

$$\left| f(x) - x^* \right| = \left| f(x) - f(x^*) \right| \le \alpha |x - x^*|,$$

thus $f(U) \subset U$. Inductively, it also follows that

$$|f^{\circ n}(x) - x^*| \le \alpha^n |x - x^*| \qquad (x \in U, n \in \mathbb{N}).$$

$\square$

Example 6.4.4 If $f(z) = z^2$ for $z \in \mathbb{C}$, then $z^* = 0$ is a super attracting fixed point. For $0 < \alpha < 1$, $U := U_\alpha(0)$ is invariant under f, and it holds that $f^{\circ n}(z) = z^{2^n} \to 0$ with uniform convergence on U. It can be seen that the sequence $(f^{pn}(z))_n$ converges to the fixed point 0 at a "super geometric" rate.

We further investigate the global behavior of (f^{pn}) for dynamical systems $(\mathbb{C}, f)$ in the case of entire functions f and especially in the case of polynomials.[5]

Remark and Definition 6.4.5 We write $\mathscr{E}$ for the set of non-affine-linear entire functions, so

$$\mathscr{E} := H(\mathbb{C}) \setminus \{f : f(z) = a + bz : a, b \in \mathbb{C}\}.$$

Then for $f \in \mathscr{E}$

$$F := F(f) := \{z \in \mathbb{C} : \{f^{\circ n} : n \in \mathbb{N}\} \text{ spherically normal at } z\}$$

---

[5] Most of the literature treats the theory for polynomials within the framework of iterating rational functions on the sphere. Standard works in which the corresponding theory is elaborated are, for example, [2, 4, 7] and [11].

is called the **Fatou set** of f. Also, $J := J(f) := \mathbb{C} \setminus F$ is called the **Julia set** of f.
Note: From the definition it follows that F is open and thus J is closed in $\mathbb{C}$. Also,
$\{f^{\circ n}|_F : n \in \mathbb{N}\}$ is spherically normal according to Remark 6.1.15. Finally, we set
$I := I(f) := \{z \in \mathbb{C} : f^{\circ n}(z) \to \infty \ (n \to \infty)\}$ and call $I(f)$ the $(\infty-)$**attraction set**
of f.

Example 6.4.6 Let again $f(z) = z^2$, so $f^{\circ n}(z) = z^{2^n}$. From Example 6.3.8 it fol-
lows that $\{f^{\circ n} : n \in \mathbb{N}\}$ is spherically normal around z if and only if $|z| \neq 1$. So here
$J = \mathbb{S}$ and $F = \mathbb{C} \setminus \mathbb{S}$. Also, $I(f) = \mathbb{C} \setminus \overline{\mathbb{D}}$.

Remark 6.4.7 If z^* is an attracting or super-attracting fixed point of $f \in \mathscr{E}$, then
$z^* \in F(f)$ follows from Theorem 6.4.3. If, on the other hand, z^* is repelling, then
always $z^* \in J(f)$.

Because: By definition,

$$(f^{\circ n})^{\#}(z^*) = |(f^{\circ n})'(z^*)|/(1 + |z^*|^2)$$

and with the chain rule $(f^{\circ n})'(z^*) = (f'(z^*))^n = \lambda^n$ for $n \in \mathbb{N}$. Thus

$$(f^{\circ n})^{\#}(z^*) = |\lambda|^n/(1 + |z^*|^2) \to \infty \qquad (n \to \infty).$$

According to Marty's theorem, $\{f^{\circ n} : n \in \mathbb{N}\}$ is not spherically normal around z^*.

Example 6.4.8 For $f(z) = \lambda \sin z$, where $\lambda \in \mathbb{C}$, $z^* = 0$ is a fixed point with $f'(0)$
$= \lambda$. So $0 \in F(f)$ if $|\lambda| < 1$, and $0 \in J(f)$ if $|\lambda| > 1$.

Theorem 6.4.9 *For $f \in \mathscr{E}$ $F(f)$, $J(f)$ and $I(f)$ are* completely invariant.

Proof.

1. We show $f^{-1}(F) \subset F$. For this, let an arbitrary compact set $K \subset f^{-1}(F)$ be given.
   Then $f(K) \subset f(f^{-1}(F)) \subset F$ is compact. With Marty's Theorem and the spherical
   chain rule, for all $n \in \mathbb{N}$ on K

$$(f^{\circ n})^{\#} = ((f^{\circ(n-1)})^{\#} \circ f)|f'| \leq \sup_{n \in \mathbb{N}} \max_{f(K)} (f^{\circ(n-1)})^{\#} \cdot \max_K |f'| < \infty.$$

Again, according to Marty's Theorem, $\{f^{\circ n} : n \in \mathbb{N}\}$ is spherically normal on
$f^{-1}(F)$.

2. We show: $f(F) \subset F$. For this, let $c \in f(F)$ be given. Then there exist $a \in F$ and
   $\delta > 0$ with $f(a) = c$ and $K := B_\delta(a) \subset F$. Since f is an open mapping, $f(K)$ is
   a neighborhood of c. Let $(n_j)_{j \in \mathbb{N}}$ be a sequence in $\mathbb{N}$. Then the sequence
   $(f^{\circ(n_j+1)})_{j \in \mathbb{N}}$ has a subsequence $(f^{\circ(n_j+1)})_{j \in L}$ that converges spherically uniformly
   on K. For each $w \in f(K)$ there exists a $z \in K$ with $f(z) = w$. Therefore,

$$\max_{w \in f(K)} \chi\left(f^{\circ n}(w), f^{\circ m}(w)\right) \leq \max_{z \in K} \chi\left(f^{\circ(n+1)}(z), f^{\circ(m+1)}(z)\right) \quad (n, m \in \mathbb{N}).$$

Therefore, $(f^{\circ n_j})_{j \in L}$ is a spherically uniform Cauchy sequence on $f(K)$ and thus, according to the Cauchy criterion for uniform convergence, also spherically uniformly convergent on $f(K)$. Therefore, $c \in F$.

3. By 1. and 2., $F(f)$ is completely invariant and thus according to Remark 6.4.1 also $J(f) = \mathbb{C} \setminus F(f)$. The invariance of $I(f)$ easily follows from the definition of $I(f)$. $\qquad \square$

Remark 6.4.10 Let (X, d) be a metric space and $f \colon X \to X$ continuous. If $U \subset X$ is open and completely invariant under f, then ∂U is invariant.

Because: On the one hand, $f(\partial U) \subset f(X \setminus U) \subset X \setminus U$, and on the other hand, due to the continuity of f

$$f(\partial U) \subset f(\overline{U}) \subset \overline{f(U)} \subset \overline{U}.$$

Therefore, $f(\partial U) \subset \partial U$.

If $f \in \mathscr{E}$ has a fixed point z^*, then certainly $z^* \notin I(f)$, so in any case $I(f) \neq \mathbb{C}$. This applies in particular to polynomials f in $\mathscr{E}$, because according to the fundamental theorem of algebra, $f(z) - z$ has a root, so f has a fixed point. With this we show

Theorem 6.4.11 If $f \in \mathscr{E}$ is a polynomial, then $J(f)$ is non-empty and compact.

Proof. Let $f(z) = \sum\limits_{\nu=0}^{d} c_\nu z^\nu$ with $d \geq 2$ and $c_d \neq 0$. Then

$$f(z)/z^d \to c_d \qquad (|z| \to \infty)$$

and thus there exists an $R > 0$ such that $|f(z)| \geq 2|z|$ for $z \in V := V_{R,\infty}(0)$. Inductively, $|f^n(z)| \geq 2^n|z|$ for $n \in \mathbb{N}$ and $z \in V$. Thus, $V \subset I(f)$, so

$$\emptyset \neq I(f) \neq \mathbb{C}$$

and therefore also $\partial I(f) \neq \emptyset$. Moreover, by definition of $I(f)$

$$I(f) = \bigcup_{k \in \mathbb{N}} (f^{\circ k})^{-1}(V).$$

Consequently, $I(f)$ is open. Moreover, $f^n \to \infty$ uniformly on $(f^{\circ k})^{-1}(V)$ for all $k \in \mathbb{N}$. By definition of $F(f)$, $I(f) \subset F(f)$, so $J(f) \subset \mathbb{C} \setminus V$ is bounded and as a closed set according to the Heine-Borel theorem consequently compact. According to Theorem 6.4.9 and Remark 6.4.10, $\partial I(f)$ is also invariant, so $f^n(\partial I(f)) \subset \partial I(f)$ for all $n \in \mathbb{N}$. Thus, $(f^n(z))_n$ is bounded in $\mathbb{C}$ for all $z \in \partial I(f)$ (since $\partial I(f)$ is bounded).

If $a \in \partial I(f)$, then $\{f^{\circ n}|_U : n \in \mathbb{N}\}$ is not spherically normal for all open neighborhoods U of a, since $(f^n(a))_n$ is bounded and $f^n(z) \to \infty$ on $U \cap I(f)$. Thus, $a \in J(f)$. $\square$

While every polynomial $f \in \mathscr{E}$ has a fixed point, this is not the case for arbitrary transcendental functions in $\mathscr{E}$. For example, $z \mapsto e^z + z$ has no fixed point. We will now apply the theorems of Montel and Picard to make further statements also for general entire functions.

Definition 6.4.12 If $X \neq \emptyset$ is a set and $f : X \to X$, then $x_0 \in X$ is called a **periodic point** of f, if $f^{\circ p}(x_0) = x_0$ for some $p \in \mathbb{N}$, that is, if x_0 is a fixed point of an iterate $f^{\circ p}$ of f. Every p with this property is called a **period** of x_0, and the minimum of all periods is called the **minimal period** of x_0. If p is the minimal period, then the periodic point x_0 is called **(super-)attracting** or **neutral** or **repelling**, if x_0 as a fixed point of $f^{\circ p}$ has the corresponding property.

Theorem 6.4.13 *Every $f \in \mathcal{E}$ has a periodic point of period 2.*

Proof. Without restriction, let f (and thus also f') be transcendental. Suppose f has no periodic point of period 2, i.e., no fixed point and no periodic point of minimal period 2. Then

$$ g := \frac{f^{\circ 2} - \mathrm{id}}{f - \mathrm{id}} $$

is an entire function with $0, 1 \notin g(\mathbb{C})$. According to the (little) Picard theorem, g is constant $= c \in \mathbb{C} \setminus \{0, 1\}$. From $f^{\circ 2} - \mathrm{id} = c(f - \mathrm{id})$ it follows $(f' \circ f)f' - 1 = c(f' - 1)$, thus

$$ ((f' \circ f) - c)f' = 1 - c. $$

Since $1 - c \neq 0$, it follows $0 \notin f'(\mathbb{C})$ and $c \notin f'(f(\mathbb{C}))$. According to the little Picard theorem, $\mathbb{C} \setminus f(\mathbb{C})$, thus also $(f')^{-1}(\{c\})$ is at most a single-point set, and thus 0 and c are Picard exceptional values of f', contradicting Remark 6.3.15 □

Remark 6.4.14 Let $f \in \mathcal{E}$. Since $F(f) = F(f^{\circ p})$ holds (see Exercise 6.4.25.3), (super-)attracting periodic points of f lie in $F(f)$ and repelling ones in $J(f)$. According to Theorem 6.4.13, f always has periodic points. If a periodic point is repelling, then in particular $J(f)$ is not empty. In fact, one can show that the Julia set for every f is not empty. A proof (to be found for example in [1, 3], Section 3, and [10], Section 1) would, however, somewhat exceed our scope.

In the following, we will deal with the—complicated—structure of Julia sets and the no less complicated dynamics of a function on its Julia set.

Theorem 6.4.15 *Let $f \in \mathcal{E}$. Then the following holds*

1. *If $U \subset \mathbb{C}$ is open with $U \cap J(f) \neq \emptyset$, then $\mathbb{C} \setminus O^+(U)$ is at most single-pointed.*
2. *Either $J(f) = \mathbb{C}$ or $J(f)$ has no interior points.*

Proof. The first statement follows directly from Montel's Great Theorem. We show: If $J(f)$ has an interior point a, then $F(f) = \emptyset$.

   To this end, let $U := U_\delta(a) \subset J(f)$. Since $J(f)$ is invariant, it follows that $O^+(U) \subset J(f)$, thus $F(f) \subset \mathbb{C} \setminus O^+(U)$. According to 1., this means that $F(f)$ is at most single-pointed and as an open set then empty.   □

Remark 6.4.16 For polynomials f, $J(f)$ has no interior points according to Theorem 6.4.11 and Theorem 6.4.15. On the other hand, there are transcendental functions whose Julia set is $\mathbb{C}$. For example, it can be shown that $J(\exp) = \mathbb{C}$ holds. A proof can be found, for example, in [5], Section 3.9.

Remark 6.4.17 Let $f \in \mathcal{E}$. If U is an open set with $U \cap J(f) \neq \emptyset$ and if $w \in \mathbb{C} \setminus O^+(U)$, then also $O^-(w) \cap O^+(U) = \emptyset$.

Because: Assume there exists a $z \in \mathbb{C}$ with $f^n(z) = w$ and $z \in f^k(U)$ for certain $n, k \in \mathbb{N}$. Then

$$w = f^{\circ n}(z) \in f^{\circ n}(f^{\circ k}(U)) = f^{\circ(n+k)}(U) \subset O^+(U).$$

Contradiction!

According to Theorem 6.4.15.1, $\mathbb{C} \setminus O^+(U)$ is at most single-pointed, so

$$O^-(w) \subset \{w\}.$$

A point $w \in \mathbb{C}$ with $O^-(w) \subset \{w\}$ we call a **Montel exceptional value** of f. We write $M(f)$ for the set of Montel exceptional values. Thus: If U is open with $U \cap J(f) \neq \emptyset$, then

$$O^+(U) \supset \mathbb{C} \setminus M(f).$$

If w is a Montel exceptional value, then either $O^-(w) = \emptyset$, that is $w \notin f(\mathbb{C})$ (like for example $w = 0$ for $f(z) = e^z$) or $O^-(w) = \{w\}$ and thus w is a fixed point with $O(w) = \{w\}$ (like for example $w = 0$ for $f(z) = ze^z$ or also for $f(z) = z^2$).

For transcendental f, $M(f) \subset E(f)$, so every Montel's exceptional value is also a Picard's, of which there is at most one in $\mathbb{C}$. If $p \in \mathcal{E}$ is a polynomial, then $p(\mathbb{C}) = \mathbb{C}$ according to the fundamental theorem of algebra. If $w \in \mathbb{C}$ with $O(w) = \{w\}$, then the equation $p(z) = w$ only has the solution w. Thus,

$$p(z) - w = c(z - w)^d$$

with $d = \deg(p) \geq 2$ and a constant $c \in \mathbb{C}^*$. These are therefore the only polynomials where a Montel's exceptional value exists. In this case, w is also a super-attracting fixed point and thus $w \in F(p)$. According to these considerations, $M(f)$ is always *at most single-pointed* and $M(f) \subset F(f)$, if f is a polynomial.

Theorem 6.4.18 Let $f \in \mathcal{E}$ and $U \subset \mathbb{C}$ be open with $U \cap J(f) \neq \emptyset$. Then the following holds

1. $J(f) \setminus M(f) \subset O^+(U \cap J(f)) \subset J(f)$.
2. If f is a polynomial, then $\bigcup_{n=1}^{N} f^{\circ n}(U \cap J(f)) = J(f)$ is true for N sufficiently large.

Proof. The first statement results from Remark 6.4.17 and the fact that

$$O^+(J(f) \cap U) = J(f) \cap O^+(U)$$

(note that $J(f)$ is completely invariant). If f is a polynomial, then $M(f) \subset F(f)$ and thus

$$\bigcup_{n\in\mathbb{N}} f^{\circ n}(U \cap J(f)) = O^+(U \cap J(f)) = J(f).$$

Since $f^{pn}(U)$ is open for all $n \in \mathbb{N}$ in $\mathbb{C}$, $f^{pn}(U \cap J(f)) = J(f) \cap f^{pn}(U)$ is open in $J(f)$ for all n. From the covering compactness of $J(f)$ follows

$$\bigcup_{n=1}^{N} f^{\circ n}(U \cap J(f)) = J(f)$$

for all sufficiently large N.   □

Remark 6.4.19 Let $f : X \to X$ and $a \in X$. For $A \subset X$, $a \in O^+(A)$ if and only if $A \cap O^-(a)$ is not empty. Thus, From Theorem 6.4.18 it follows: If $f \in \mathscr{E}$ and if $z \in J(f) \setminus M(f)$, then $O^-(z)$ is dense in $J(f)$. If f is a polynomial, this applies to all $z \in J(f)$. The density of $O^-(z)$ in $J(f)$ can be used to generate pictures of $J(f)$. One starts with any $z \in J(f) \setminus M(f)$ and successively calculates the corresponding pre-image sets. The union of these sets densely fills $J(f)$.[6]

We will now show that for many $z \in J(f)$ the forward orbit $O^+(z)$ is dense in $J(f)$. We assume that the possibly existing Montel exceptional value is not an isolated point of $J(f)$, which is of course particularly fulfilled if no Montel exception value exists or if the exceptional value is in $F(f)$.

In preparation, we show using Baire's theorem

Theorem 6.4.20 (Universality Criterion) *Let (X, d_X) be a complete metric space and (Y, d_Y) a separable metric space. Further, let (T_n) be a sequence of continuous mappings $T_n : X \to Y$. If $D := \{x \in X : \{T_n(x) : n \in \mathbb{N}\}$ dense in $Y\}$, then D is a G_δ-set and the following statements are equivalent:*

a) *For all open, non-empty sets $U \subset X$ and $V \subset Y$ there exists a $n \in \mathbb{N}$ with*

$$T_n(U) \cap V \neq \emptyset \qquad (\Leftrightarrow T_n^{-1}(V) \cap U \neq \emptyset).$$

b) *D is dense in X.*[7]

Proof. Let $\{y_j : j \in \mathbb{N}\}$ be dense in Y and

$$\mathscr{U} := \{U_{1/k}(y_j) : j \in \mathbb{N}, k \in \mathbb{N}\}.$$

Then every open, non-empty subset of Y contains a set from $\mathscr{U}$. Let $(W_m)_{m\in\mathbb{N}}$ be an enumeration of $\mathscr{U}$. Then $x \in D$ if and only if for every $m \in \mathbb{N}$ there exists a $n \in \mathbb{N}$ with $T_n(x) \in W_m$, that is $x \in T_n^{-1}(W_m)$. Thus,

---

[6] Excellent graphics can be found in the technical literature and on the internet.

[7] Elements of D are referred to as universal elements with respect to (T_n). Hence the name Universality Criterion (see [6]).

$$D = \bigcap_{m \in \mathbb{N}} O_m,$$

where $O_m := \bigcup_{n \in \mathbb{N}} T_n^{-1}(W_m)$ is open for all $m \in \mathbb{N}$. In particular, D is a G_δ-set.

Furthermore, a) holds if and only if O_m is dense in X for all $m \in \mathbb{N}$. Thus, the implication b) $\Rightarrow$ a) is clear and the implication a) $\Rightarrow$ b) is a consequence of Baire's theorem. $\qquad\square$

Theorem 6.4.21 *Let* $f \in \mathcal{E}$. *If* $J(f)$ *is not empty and* $J(f) \setminus M(f)$ *is dense in* $J(f)$, *then the set of points* $z \in J(f)$ *whose orbit* $O^+(z)$ *is dense in* $J(f)$ *is a dense* G_δ-*set in* $J(f)$. *In particular, this applies to all polynomials in* $\mathcal{E}$.

Proof.

1. Since $J(f)$ is closed in $\mathbb{C}$, $J(f)$ (with the absolute value metric) is a complete metric space. Furthermore, $J(f)$ is separable according to Remark 6.4.19. Let T_n: $J(f) \to J(f)$ be defined by $T_n(z) := f^{\circ n}(z)$ for $z \in J(f)$. According to Theorem 6.4.18,

$$\bigcup_{n \in \mathbb{N}} T_n(U \cap J(f)) = O^+(U \cap J(f)) \supset J(f) \setminus M(f)$$

   holds for all open sets $U \subset \mathbb{C}$ with $U \cap J(f) \neq \emptyset$. Since $J(f) \setminus M(f)$ is dense in $J(f)$, condition a) from the universality criterion is fulfilled (Remark: according to Exercise 3.2.31.11, A is open in $J(f)$ if and only if $A = U \cap J(f)$ for an open set $U \subset \mathbb{C}$). The claim follows from the universality criterion.
2. Let f now be a polynomial. According to Theorem 6.4.11, $J(f)$ is then nonempty and according to Remark 6.4.17, $M(f) \subset F(f)$, so $J(f) \setminus M(f) = J(f)$. $\qquad\square$

Example 6.4.22 We consider once again $f(z) = z^2$, so $f^{\circ n}(z) = z^{2^n}$. Here there is a dense set of periodic points in $J(f) = \mathbb{S}$, because it holds for $z \in \mathbb{S}$ and $n \in \mathbb{N}$

$$f^{\circ n}(z) = z^{2^n} = z$$

exactly when

$$z^{2^n - 1} = 1$$

so exactly when z is a $(2^n - 1)$-th root of unity. The set

$$\{z \in \mathbb{S} : z^{2^n - 1} = 1 \text{ for some } n \in \mathbb{N}\}$$

of periodic points of f is dense in $\mathbb{S}$. Forward orbits of periodic points are—as finite sets—not dense in $\mathbb{S}$. However, according to Theorem 6.4.21 there also exists a dense G_δ-set of points $z \in \mathbb{S}$ such that the forward orbit $\{z^{2^n} : n \in \mathbb{N}\}$ is dense in $\mathbb{S}$. It thus turns out that smallest changes in the initial values z can cause completely different behavior of the iterated sequence (z^{2^n}). The dynamics on the Julia set prove to be extremely complicated.

We have already seen that Julia sets $J(f) \neq \mathbb{C}$ have no inner points in $\mathbb{C}$. It can be shown that Julia sets, on the other hand, are always "rich". We will only prove the theorem for polynomials. However, the statement actually applies to all $f \in \mathscr{E}$ (see again for example [1], Section 3 in [3] or Section 1 in [10]). This also implies that the extra conditions in Theorem 6.4.21 are always met. We consider $J(f)$ again as a metric space with the absolute value metric.

Theorem 6.4.23 *If $f \in \mathscr{E}$ is a polynomial, then $J(f)$ is perfect and locally uncountable.*

Proof. According to Theorem 6.4.11, $J(f)$ is non-empty and compact, so $J(f)$ as a metric space is in particular complete. According to Remark 2.4.32, it is therefore sufficient to show that $J(f)$ is perfect.

We first show: For every $z \in J(f)$ there exists a $\zeta \in J(f)$ with $z \in O^+(\zeta)$ and $\zeta \notin O^+(z)$, so $\zeta \in O^-(z) \setminus O^+(z)$.

Because: If z is not periodic, then one can arbitrarily choose $\zeta \in O^-(z)$ ($O^-(z)$ is not empty according to the Fundamental Theorem of Algebra). So let z be periodic and $p \in \mathbb{N}$ be the minimal period of z. According to Exercise 6.4.25.3, $z \in J(f^p)$ and according to Remark 6.4.17 not a Montel exceptional value of the polynomial f^p. Thus, there exist a $\zeta \in J(f) \setminus \{z\}$ and an $m \in \mathbb{N}$ with $f^{\circ(mp)}(\zeta) = z$. Here, $\zeta \notin O^+(z)$, because otherwise $\zeta = f^k(z)$ for a $1 \leq k < p$ (because $f^{\circ p}(z) = z$) and thus

$$f^{\circ k}(z) = f^{\circ k}\left(f^{\circ(mp)}(z)\right) = f^{\circ(mp)}\left(f^{\circ k}(z)\right) = f^{\circ(mp)}(\zeta) = z$$

in contradiction to the minimality of p.

Let now $V \subset J(f)$ be a neighborhood of z. Then there exists an open neighborhood U of z in $\mathbb{C}$ with $U \cap J(f) \subset V$. If ζ is as above, then according to Theorem 6.4.18 there exists an $n \in \mathbb{N}$ with $\zeta \in f^n(U \cap J(f))$. If $\eta \in U \cap J(f)$ with $f^n(\eta) = \zeta$, then $\eta \neq z$ since $\zeta \notin O^+(z)$. Thus, z is an accumulation point of $J(f)$. $\square$

As a further application of the rescaling theorem, we finally prove that there always exist many periodic points in the Julia set. According to the above remarks, the additional condition of perfectness proves to be redundant.

Theorem 6.4.24 *Let it be $f \in \mathscr{E}$. If $J(f)$ is perfect, then $J(f)$ is the closure of the set of repelling periodic points.*

Proof. We consider the set D of all points in $J(f)$, whose forward orbit is dense in $J(f)$. Since, by assumption, in particular $J(f) \setminus M(f)$ is dense in $J(f)$, D is dense in $J(f)$ according to Theorem 6.4.21. Since repelling periodic points always lie in $J(f)$, it suffices to show: If $a \in D$ and U is an open neighborhood of a, then there exists a repelling periodic point in U.

So let $a \in D$ and U be an open neighborhood of a. According to the rescaling theorem, there exist sequences (a_k) with $a_k \to a$, (ρ_k) with $\rho_k \to 0$, (n_k) and a non-constant entire function g such that for $\varphi_k = \varphi_{a_k, \rho_k}$

$$f^{\circ n_k} \circ \varphi_k \to g \quad (k \to \infty)$$

holds uniformly on every compact set $K \subset \mathbb{C}$ (for k sufficiently large depending on K).

Since $J(f)$ is perfect and g omits at most one value according to Picard's little theorem, there exists an $m \in \mathbb{N}$ with $f^{\circ m}(a) \in U \cap g(\mathbb{C})$. Let $w \in \mathbb{C}$ with $g(w) = f^{\circ m}(a)$. Then there exists an open neighborhood V of w with $g(V) \subset U$ and $g'(\zeta) \neq 0$ for $\zeta \in V \setminus \{w\}$. Since $g(V \setminus \{w\})$ is open, there also exist a $\ell \in \mathbb{N}$ and a ζ_0 in $V \setminus \{w\}$ with $g(\zeta_0) = f^{\circ l}(a)$. Since ζ_0 is a zero of $g - f^{\circ l}(a)$ and

$$f^{\circ n_k} \circ \varphi_k - f^{\circ \ell} \circ \varphi_k \to g - f^{\circ \ell}(a) \quad (k \to \infty)$$

uniformly on every compact subset of $\mathbb{C}$, according to Hurwitz's theorem, there exist points ζ_k with $\zeta_k \to \zeta_0$ and

$$f^{\circ n_k}(\varphi_k(\zeta_k)) = f^{\circ \ell}(\varphi_k(\zeta_k))$$

for sufficiently large k. Thus,

$$\eta_k := f^{\circ \ell}(\varphi_k(\zeta_k))$$

is a fixed point of $f^{\circ(n_k - \ell)}$ and therefore a periodic point of f for sufficiently large k. From $\varphi_k(\zeta_k) \to a$ it follows that $\eta_k \to f^{\circ l}(a) = g(\zeta_0) \in U$ for $k \to \infty$, so $\eta_k \in U$ for sufficiently large k.

Finally, for sufficiently large k,

$$(f^{\circ n_k} \circ \varphi_k)'(\zeta_k) = (f^{\circ(n_k - \ell)} \circ f^{\circ \ell} \circ \varphi_k)'(\zeta_k) = (f^{\circ(n_k - \ell)})'(\eta_k) \cdot (f^{\circ \ell})'(\varphi_k(\zeta_k)) \cdot \rho_k.$$

From

$$(f^{\circ \ell})'(\varphi_k(\zeta_k)) \cdot \rho_k \to (f^{\circ \ell})'(a) \cdot 0 = 0 \quad (k \to \infty)$$

and

$$(f^{\circ n_k} \circ \varphi_k)'(\zeta_k) \to g'(\zeta_0) \neq 0 \quad (k \to \infty)$$

it follows that $(f^{\circ(n_k - \ell)})'(\eta_k) \to \infty$ for $k \to \infty$ and thus the η_k are repelling for sufficiently large k. $\square$

Exercises 6.4.25

1. (Newton Iteration) Let $\Omega \subset \mathbb{C}$ be open, $g \in H(\Omega)$ and a a simple zero of g. Show that an open neighborhood U of a exists such that

$$f(z) = z - g(z)/g'(z)$$

   is defined for all $z \in U$ and that a is a super attracting fixed point of f.
2. Let X be a set and $f : X \to X$. Furthermore, let $A \subset X$. Show that the following statements are equivalent:
   a) A is completely invariant,
   b) $f^{-1}(A) = A$,
   c) A and $X \setminus A$ are invariant.

3. Let $f \in \mathcal{E}$ and $m \in \mathbb{N}$. Show: $F(f^m) = F(f)$ and $J(f^m) = J(f)$.

   Hint: $F(f)$ is invariant and $\{f^{\circ n} : n \in \mathbb{N}\} = \bigcup_{j=0}^{m-1} \{f^{\circ(km+j)} : k \in \mathbb{N}\}$ holds.

4. Let $f \in \mathcal{E}$ with non-empty Julia set and $M(f) \subset F(f)$. Prove that $J(f)$ is perfect and locally uncountable.

   Hint: Consider that the proof of Theorem 6.4.23 can essentially be transferred.

5. Let there be $U, V \subset \mathbb{C}$ areas and $f : U \to U$ as well as $g : V \to V$ holomorphic. Then f and g are called conjugated, if a conformal mapping $\varphi : U \to V$ exists with

$$g \circ \varphi = \varphi \circ f.$$

   a) Show that $f : \mathbb{C} \setminus \overline{\mathbb{D}} \to \mathbb{C} \setminus \overline{\mathbb{D}}$ and $g : \mathbb{C} \setminus [-2, 2] \to \mathbb{C} \setminus [-2, 2]$ with

$$f(z) = z^2 \qquad (z \in \mathbb{C} \setminus \overline{\mathbb{D}})$$

   and

$$g(\zeta) = \zeta^2 - 2 \qquad \left(\zeta \in \mathbb{C} \setminus [-2, 2]\right)$$

   are conjugated.

   Hint: Consider $\varphi := 2j|_{\mathbb{C} \setminus \overline{\mathbb{D}}}$, where j denotes the Joukowski mapping.

   b) Determine $F(g)$ and $J(g)$.

6.5   Concepts V: Contractions and Fixed Points

Particularly efficient methods for approximately solving equations often rely on iterative processes. In this case, the given problem is typically reformulated into a fixed point problem for a self-mapping $\varphi : X \to X$. We have learned a first example with the *Heron method* for the approximate calculation of roots. The corresponding mapping φ, given by

$$\varphi(x) = \frac{1}{2}\left(x + \frac{c}{x}\right),$$

proves to be a suitable self-mapping, for example on the interval $(0, \infty)$. We secured the convergence of the method with suitable monotonicity considerations, which require an order structure and are therefore essentially limited to the real axis. More widely applicable is the concept of contractions, i.e., self-mappings on metric spaces (X, d), where the distances of the function values are smaller than those of the arguments, in the sense that a constant $\alpha < 1$ exists with

$$d(\varphi(x), \varphi(x')) \leq \alpha \cdot d(x, x') \quad (x, x' \in X).$$

If the metric space is complete, contractions always have exactly one fixed point and with any starting point $x \in X$ the sequence of iterates $x_n = \varphi^{\circ n}(x)$ converges with geometric speed α^n towards the fixed point. This is the statement of the important *Banach's Fixed Point Theorem* (see for example [9]), which opens up a multitude of application fields due to its quite general prerequisites, such as in connection with the local invertibility of continuously differentiable functions of several variables or with the existence and uniqueness of solutions to ordinary differential equations with initial conditions.

The problem, however, often lies in the prerequisite of the global contraction property, a condition that in many interesting cases is not fulfilled or at least difficult to prove in this form. Often, by applying the *Distortion Theorem*, a local contraction behavior can be guaranteed. If a function is continuously differentiable with a derivative less than 1 in magnitude at a fixed point, the mapping is contracting in a suitable neighborhood of the fixed point, with the contraction constant α essentially given by the magnitude of the derivative. We have learned a corresponding local result in Theorem 6.4.3. In Exercise 6.4.25.1 we also outlined the *Newton method* for the approximate calculation of roots (in the one-dimensional situation). The self-mapping underlying the Newton method is locally given by

$$\varphi(x) := x - f(x)/f'(x)$$

in the case of simple roots of f and allows an extension to functions of several variables with a suitable definition of the derivative.

The particular appeal of the Newton method lies in the fact that the fixed point is super attracting in the sense that the derivative of φ at the root vanishes. This results locally in an extremely fast, so-called quadratic convergence. We have seen the effect in the Heron method, which turns out to be a special case of the Newton method for determining the positive root of $f(x) = x^2 - c$. A major problem is— besides the possibly complex calculation of derivatives—the fact that only a very good *local* behavior can be guaranteed. So the tricky question arises as to how close the starting point of the iteration must be to the root in a concrete situation to guarantee convergence of the method. An impression of how complicated the global behavior can be even in the simplest cases is given by the results on the dynamics of entire functions in the last section.

Similar challenges are faced in many applications of analytical methods: Analysis provides efficient and universally applicable tools for local investigations, but the localization typically requires specially tailored arguments.

References

1. Bargmann, D.: Simple proofs of some fundamental properties of the Julia set. Ergodic Theory Dyn. Syst. **19**, 553–558 (1999)
2. Beardon, A.F.: Iteration of Rational Functions. Complex Analytic Dynamical Systems. Graduate Texts in Mathematics, Vol. 132. Springer, New York (1991)
3. Bergweiler, W.: Iteration of meromorphic functions. Bull. Am. Math. Soc. (N. S.) **29**, 151–188 (1993)
4. Carleson, L., Gamelin, T.W.: Complex Dynamics. Springer, New York (1993)
5. Devaney, R.L.: An Introduction to Dynamical Systems, 2nd ed. Addison-Wesley, Redwood City (1989)
6. Grosse-Erdmann, K.-G., Peris-Manguillot, A.: Linear Chaos. Springer, London (2011)

7. Milnor, J.: Dynamics in One Complex Variable, Vol. 160, 3rd ed. Annals of Mathematics Studies, Princeton University Press, Princeton (2006)
8. Needham, T.: Visual Complex Analysis. Oxford University, New York (1997)
9. Pöschel, J.: Etwas mehr Analysis. Springer Spektrum, Wiesbaden (2014)
10. Schleicher, D.: Dynamics of entire functions. In: Holomorphic Dynamical Systems. Lecture Notes in Mathematics, 1998, Vol. 295–339. Springer, Berlin (2010)
11. Steinmetz, N.: Rational Iteration. Complex Analytic Dynamical Systems. Walter de Gruyter, Berlin (1993)
12. Wegert, E.: Visual Complex Functions—An Introduction with Phase Portraits, Birkhäuser, Basel (2012)

In the last chapter, the Cauchy theorem is taken up within the framework of Runge's theory to prove various results about the uniform and local uniform approximation of holomorphic functions by rational functions or polynomials. Runge's theorem for polynomial approximation is applied, among other things, in the last section. There, an impression is given in two situations of how complicated the conditions typically become when it comes to the boundary behavior of holomorphic functions. Essential tools for the proofs of the corresponding results are Baire's theorem in the form of the universality criterion and results on simultaneous approximation. The comparatively special results presented here have more the character of an appendix.

7.1 Runge's Theory

If $\Omega \subset \mathbb{C}$ is open and $f \in H(\Omega)$, then for $a \in \Omega$ and $R := \mathrm{dist}(a, \partial\Omega)$ the polynomial sequence (p_n) with

$$p_n(z) := \sum_{v=0}^{n} \frac{f^{(v)}(a)}{v!} (z-a)^v = T_n(f, a)(z-a)$$

converges uniformly to f on every compact subset of $U_R(a)$.

If $V_{r,R}(a) \subset \Omega$ for an $a \in \mathbb{C}$ and $0 \le r < R \le \infty$, then f has a Laurent expansion in $V_{r,R}(a)$. If

$$q_n(z) := \sum_{v=-n}^{n} c_{v,r,R}(f, a)(z-a)^v$$

J. Müller, *Concepts of Function Theory*, Mathematics Study Resources 12,
https://doi.org/10.1007/978-3-662-69115-1_7

is the n-th partial sum of the Laurent expansion of f with respect to a and the pair of radii (r, R), then $q_n \to f$ uniformly on all compact sets $K \subset V_{r,R}(a)$. In particular, f is uniformly approximable on such K by rational functions with poles exclusively in a and ∞, where a rational function q has a pole at ∞ if $|q(z)| \to \infty$ for $|z| \to \infty$ holds.

We want to investigate to what extent f can be approximated (uniformly) on more general compact sets by polynomials or rational functions. If Ω is open with holes, then in general not every function in $H(\Omega)$ can be approximated uniformly by polynomials on all compact subsets:

Example 7.1.1. Let $\Omega = \mathbb{C}^*$ and $f(z) = 1/z$ for $z \in \Omega$. If $r > 0$, then there is no sequence (p_n) of polynomials with

$$p_n \to f \text{ uniformly on } K_r(0).$$

Because assume there is otherwise. Then

$$0 = \int_{k_r(0)} p_n(\zeta)d\zeta \to \int_{k_r(0)} \frac{d\zeta}{\zeta} = 2\pi i \qquad (n \to \infty).$$

Contradiction!

Remark and Definition 7.1.2. Let $K \subset \mathbb{C}$ be compact. For $M \subset C(K)$, we denote in the following with $\overline{\operatorname{span}}(M)$ the closure of the linear span of M in $(C(K), || \cdot ||_{\infty,K})$. For a function $f \in C(K)$ it thus holds that $f \in \overline{\operatorname{span}}(M)$ if and only if for every $\varepsilon > 0$ a finite set $E \subset M$ and scalars λ_g $(g \in E)$ exist such that

$$||f - \sum_{g \in E} \lambda_g g||_{\infty,K} < \varepsilon,$$

in other words, f can be approximated arbitrarily well in the supremum norm by linear combinations of functions from M. Fejér's theorem shows that for every continuous function $f : \mathbb{S} \to \mathbb{C}$ the arithmetic means $f * F_n$ of the Fourier partial sums converge uniformly on $\mathbb{S}$ to f, that is

$$||f - f * F_n||_{\infty,\mathbb{S}} \to 0 \qquad (n \to \infty).$$

For $M := \{e_k : k \in \mathbb{Z}\}$ it holds that $f * F_n \in \operatorname{span}(M)$ and thus in particular

$$C(\mathbb{S}) = \overline{\operatorname{span}}(M).$$

Remark and Definition 7.1.3. Let $K \subset \mathbb{C}$ be compact. Then we denote by

$$P(K) := \overline{\operatorname{span}}\{z \mapsto z^n : n \in \mathbb{N}_0\}$$

the closure of the polynomials in $C(K)$, that is, the set of all functions in $C(K)$ that can be approximated arbitrarily closely by polynomials in the supremum norm on K. If $g_a : \mathbb{C} \setminus \{a\} \to \mathbb{C}$ is defined for $a \in \mathbb{C}_\infty$ by

$$g_a(z) := \begin{cases} (a - z)^{-1}, & \text{if } a \in \mathbb{C} \\ z, & \text{if } a = \infty \end{cases},$$

we set for $A \subset \mathbb{C}_\infty \setminus K$

$$R_A(K) := \overline{\text{span}}\{(g_a|_K)^n : a \in A, \, n \in \mathbb{N}_0\}.$$

With these notations, $P(K) = R_{\{\infty\}}(K)$. Using partial fraction decomposition, one can show (Exercise 7.1.15.4) that $\text{span}\{(g_a|_K)^n : a \in A, \, n \in \mathbb{N}_0\}$ corresponds to the set of rational functions with poles only in A. Thus, $R_A(K)$ is the set of all functions in $C(K)$ that can be approximated arbitrarily closely in the supremum norm on K by rational functions with poles exclusively in A.

Example 7.1.4. If $K := B_R(a)$ for an $a \in \mathbb{C}$, then for all open sets $\Omega \supset B_R(a)$ and $f \in H(\Omega)$ from the local uniform convergence of the Taylor series

$$f|_K \in P(K).$$

For $a \in \mathbb{C}, 0 < r \le R < \infty$ and the annulus

$$K := \{z : r \le |z - a| \le R\} = B_R(a) \setminus U_r(a)$$

follows for all functions f holomorphic on an open superset of K from the Laurent expansion

$$f|_K \in R_{\{a,\infty\}}(K).$$

Specifically, if $a = 0$ and $r = R = 1$, i.e., $K = \mathbb{S}$, then Fejér's theorem shows that actually $C(\mathbb{S}) = R_{\{0,\infty\}}(\mathbb{S})$ holds. For $g_0 \in H(\mathbb{C}^*)$ from Example 7.1.1 also follows $g_0|_\mathbb{S} \notin P(\mathbb{S})$.

We first prove, in preparation for what follows, an approximation theorem for Cauchy integrals.

Theorem 7.1.5. *Let $K \subset \mathbb{C}$ be compact and $\tau := s_u^v$ an oriented line segment in $\mathbb{C} \setminus K$. Then for all continuous functions $f : [u, v] \to \mathbb{C}$*

$$C_\tau f \in \overline{\text{span}}\{g_a|_K : a \in [u, v]\}.$$

Proof. Let $\varepsilon > 0$ be given. Since

$$[u, v] \times K \ni (\zeta, z) \mapsto \frac{f(\zeta)}{\zeta - z} \in \mathbb{C}$$

is (uniformly) continuous on the compact set $[u, v] \times K \subset \mathbb{C}^2$, there exists a $\delta > 0$ such that

$$\left| \frac{f(\zeta)}{\zeta - z} - \frac{f(\zeta')}{\zeta' - z} \right| < \varepsilon \qquad (\zeta, \zeta' \in [u, v], |\zeta - \zeta'| < \delta, \, z \in K).$$

Furthermore, there exist $0 = t_0 < t_1 < \ldots < t_n = 1$ such that for $\tau_j := s_u^v|_{[t_{j-1}, t_j]}$

$$|\zeta - \zeta'| < \delta \qquad (\zeta, \zeta' \in \tau_j^*, \, j = 1, \ldots, n)$$

holds. If we choose $a_j \in \tau_j^* \subset [u, v]$ and set

$$c_j := f(a_j) \int_{\tau_j} d\zeta \qquad (j = 1, \ldots, n),$$

then for $z \in K$

$$\left| 2\pi i \, C_\tau f(z) - \sum_{j=1}^n c_j \frac{1}{a_j - z} \right| = \left| \sum_{j=1}^n \left(\int_{\tau_j} \frac{f(\zeta)}{\zeta - z} \, d\zeta - \int_{\tau_j} \frac{f(a_j)}{a_j - z} \, d\zeta \right) \right|$$

$$\leq \varepsilon \cdot \sum_{j=1}^n L(\tau_j) = \varepsilon \cdot L(\tau) = \varepsilon |v - u|.$$

Therefore,

$$\max_K \left| C_\tau f - \frac{1}{2\pi i} \sum_{j=1}^n c_j g_{a_j} \right| \leq \frac{|v - u|}{2\pi} \varepsilon. \qquad \qquad \square$$

The following theorem is of central importance for the further considerations in this section.

Theorem 7.1.6. *Let* $\Omega \subset \mathbb{C}$ *be open and* $K \subset \Omega$ *compact. Then there exists a null-homologous cycle* $\gamma = (\gamma_l)_{l \in P}$ *in* Ω *consisting of oriented (axis-parallel) segments* γ_l *such that*

$$\mathrm{ind}_\gamma(z) = 1 \qquad (z \in K).$$

Proof. Let

$$\delta := \mathrm{dist}(K, \partial\Omega) \, (> 0).$$

We consider the grid $d\mathbb{Z} + id\mathbb{Z}$, where the mesh size d is chosen so small that $d\sqrt{2} < \delta$. Let $Q_1, \ldots, Q_N$ be the (finitely many) compact squares with corners in $d\mathbb{Z} + id\mathbb{Z}$, which hit K (i.e., have non-empty intersection with K). For $L := \bigcup_{n=1}^N Q_n$ then

$$K \subset L^\circ \subset L \subset \Omega.$$

Because: From the definition of the Q_n it follows $K \subset L$ and $K \cap \partial L = \emptyset$. If $n \in \{1, \ldots, N\}$ and $z_n \in Q_n \cap K$, then $U_\delta(z_n) \subset \Omega$. Since $\mathrm{diam}(Q_n) = d\sqrt{2} < \delta$, it follows $Q_n \subset \Omega$.

If Q is any compact square with corners a, b, c, d (positively oriented), then $\partial Q = \gamma_Q^*$, where

$$\gamma_Q := \left(s_a^b, s_b^c, s_c^d, s_d^a \right)$$

is a closed path. We now consider those (in this way oriented) lines, whose trace belongs to the edge ∂L of L, and denote these with γ_ι ($\iota \in I$). The construction of the γ_ι shows that for a suitable decomposition $(I_\kappa)_{\kappa \in M}$ of I the chains $(\gamma_\iota)_{\iota \in I_\kappa}$ are closed paths (Task 7.1.15.3; it is important: each corner is equally often the starting as well as the endpoint). Thus, $\gamma := (\gamma_\iota)_{\iota \in I}$ is a cycle and it holds

$$\gamma^* = \bigcup_{\iota \in I} \gamma_\iota^* = \partial L \subset \Omega \setminus K.$$

For arbitrary compact squares Q it holds (cf. Task 7.1.15.2)

$$\mathrm{ind}_{\gamma_Q}(z) = \begin{cases} 1 & \text{if } z \in Q^\circ \\ 0, & \text{if } z \notin Q \end{cases}.$$

According to the above construction, this results in

$$\mathrm{ind}_\gamma(z) = \frac{1}{2\pi i} \sum_{\iota \in I} \int_{\gamma_\iota} \frac{d\zeta}{\zeta - z} = \sum_{n=1}^N \mathrm{ind}_{\gamma_{Q_n}}(z) = \begin{cases} 1 & \text{if } z \in \bigcup_{n=1}^N Q_n^\circ \\ 0, & \text{if } z \notin L \end{cases}$$

(remark: the "missing" lines on the left side are each traversed twice in opposite orientation and thus the corresponding integrals vanish).

For reasons of continuity, $\mathrm{ind}_\gamma(z) = 1$ also holds for any $z \in L^\circ$ and thus in particular for all $z \in K$. $\qquad \Box$

If A, B are sets, we say that A hits the set B if $A \cap B$ is not empty.

Theorem 7.1.7 (Runge's Theorem for Compact Sets). *Let $K \subset \mathbb{C}$ be compact and f holomorphic on an open set $\Omega \supset K$.*

1. *If $A \subset \mathbb{C}_\infty \setminus K$ meets every component of $\mathbb{C}_\infty \setminus K$, then $f|_K \in R_A(K)$.*
2. *If $\mathbb{C}_\infty \setminus K$ is connected, then $f|_K \in P(K)$.*

Proof. It suffices to prove the first statement. The second follows as a special case of the first with $A = \{\infty\}$.

Let $\gamma = (\gamma_\iota)_{\iota \in I}$ as in Theorem 7.1.6. Then, according to the Cauchy theorem

$$f|_K = (C_\gamma f)|_K = \sum_{\iota \in I} (C_{\gamma_\iota} f)|_K.$$

Since $\gamma^* \subset \mathbb{C} \setminus K$ holds, f is the sum of Cauchy integrals as in Theorem 7.1.5. Therefore, according to Theorem 7.1.5, it suffices to show that $g_a|_K \in R_A(K)$ for all $a \in \mathbb{C} \setminus K$. For this, let G be a component of $\mathbb{C} \setminus K$. We show: For all $a \in G$, $g_a|_K \in R_A(K)$.

If $\varphi : G \to C(K)$ is defined by

$$\varphi(a) := g_a|_K \qquad (a \in G),$$

then φ is continuous.

Because: Let $a \in G$ and $\delta := \text{dist}(a, K)$. If (a_n) is a sequence in G with $a_n \to a$, then $\text{dist}(a_n, K) \geq \delta/2$ for n sufficiently large. From

$$\frac{1}{a_n - z} - \frac{1}{a - z} = \frac{a - a_n}{(a_n - z)(a - z)}$$

it follows

$$\|g_{a_n} - g_a\|_{\infty, K} \leq \frac{2}{\delta^2} |a - a_n| \to 0 \qquad (n \to \infty).$$

We set $V := \varphi^{-1}(R_A(K))$. Then $V \subset G$ is closed, since $R_A(K) \subset C(K)$ is closed.

We show: V is non-empty and open. Since G is a domain, then already $V = G$ and thus $g_a|_K \in R_A(K)$ for all $a \in G$.

1. By definition, $A \cap G \subset V$. If G is bounded, then G is also a component of $\mathbb{C}_\infty \setminus K$. By assumption, there exists $b \in A \cap G(\subset V)$. If G is the unbounded component of $\mathbb{C} \setminus K$, then $G_\infty := G \cup \{\infty\}$ is the component of $\mathbb{C}_\infty \setminus K$ that contains ∞. If $b \in A \cap G_\infty$, then in the case $b \neq \infty$ again $b \in A \cap G \subset V$. If $b = \infty$, then for $|a| > \max_{z \in K} |z|$

$$g_a(z) = \frac{1}{a - z} = \frac{1}{a} \frac{1}{1 - z/a} = \sum_{v=0}^{\infty} \frac{1}{a^{v+1}} z^v$$

with uniform convergence on K. Thus, $g_a|_K \in P(K) = R_{\{\infty\}}(K) \subset R_A(K)$. Therefore, $V \neq \emptyset$.

2. Let $b \in V$, that is, $g_b \in R_A(K)$. If $\delta := \text{dist}(b, K)$, then for $a \in U_{\delta/2}(b)$

$$g_a(z) = \frac{1}{b - z} \cdot \frac{1}{1 - \frac{b-a}{b-z}} = \sum_{v=0}^{\infty} (b - a)^v \frac{1}{(b - z)^{v+1}} = \sum_{v=0}^{\infty} (b - a)^v g_b^{v+1}(z)$$

holds with uniform convergence on K (note: $\left|\frac{b-a}{b-z}\right| \leq 1/2$). Since $R_A(K)$ is an algebra (Exercise 7.1.15.4), with $g_b|_K$ also $(g_b|_K)^n$ is in $R_A(K)$ for all $n \in \mathbb{N}_0$. Therefore, $g_a|_K \in R_A(K)$ as well. □

Example 7.1.8. There exists a sequence (p_n) of polynomials with

$$p_n \to 0 \quad (n \to \infty) \quad \text{pointwise on } \mathbb{C}$$

and such that

$$\max_{B_\delta(a)} |p_n| \to \infty \qquad (n \to \infty)$$

for all $\delta > 0$ and all $a \in \mathbb{R}$. In particular, (p_n) is not uniformly convergent at any point $a \in \mathbb{R}$.

Because: For $n \in \mathbb{N}$ let

$$L_n := B_n(0) \cap \left(\{\mathrm{Im}\, z \leq 0\} \cup \{\mathrm{Im}\, z \geq 2/n\} \right)$$

and

$$M_n := B_n(0) \cap \{\mathrm{Im}\, z = 1/n\}$$

as well as $K_n := L_n \cup M_n$. Then $K_n \subset \mathbb{C}$ is compact and $\mathbb{C}_\infty \setminus K_n$ is connected. If $U_n \supset M_n$ is open with $L_n \cap U_n = \emptyset$ and $\Omega_n := U_n \cup (\mathbb{C} \setminus \overline{U_n})$, then

$$f_n := n \mathbb{1}_{U_n, \Omega_n} \in H(\Omega_n).$$

So, according to Theorem 7.1.7.2, there exists a polynomial p_n with

$$\max_{K_n} |f_n - p_n| < \frac{1}{n}.$$

If $z \in \mathbb{C}$, then $z \in L_n$ for n sufficiently large. From $f_n(z) = 0$ for n sufficiently large, it follows that $p_n(z) \to 0$ $(n \to \infty)$. If $\delta > 0$ and $a \in \mathbb{R}$, then $B_\delta(a) \cap M_n \neq \emptyset$ for n sufficiently large. If one chooses $z_n \in B_\delta(a) \cap M_n$, then

$$\max_{B_\delta(a)} |p_n| \geq |p_n(z_n)| \geq |f_n(z_n)| - |f_n(z_n) - p_n(z_n)| \geq n - \frac{1}{n} \to \infty$$

holds for $n \to \infty$.

Runge's theorem in the above form provides information about the uniform approximability on compact sets K for functions f which are holomorphic on an open neighborhood of K. This condition is in general not necessary, as shown by the Fejér's theorem in the case of $\mathbb{K} = \mathbb{S}$.[1]

A more natural question in the case of holomorphic functions is the one about locally uniform approximability, for example by polynomials or rational functions on open sets Ω.

Theorem 7.1.9. *Let $\Omega \subset \mathbb{C}$ be open and $(K_m) = (K_m(\Omega))$ the standard exhaustion of Ω. Then each component of $\mathbb{C}_\infty \setminus K_m$ contains a component of $\mathbb{C}_\infty \setminus \Omega$.*

Proof. Let G be a component of $\mathbb{C}_\infty \setminus K_m$.

1. We show that $G \setminus \Omega$ is not empty: If G is the component that contains ∞, then $\infty \in G \setminus \Omega$. If G is another component (if it exists), then $G \subset B_m(0)$. By definition of K_m, there exists a $\zeta \in \mathbb{C} \setminus \Omega$ for $z \in G$ with $|z - \zeta| < 1/m$. Then $z \in U_{1/m}(\zeta) \subset \mathbb{C} \setminus K_m$. Since $U_{1/m}(\zeta)$ is connected, it follows that $U_{1/m}(\zeta) \subset G$ and therefore in particular $\zeta \in G$.
2. If $\zeta \in G \setminus \Omega$, then from the definition of components

$$G_{\mathbb{C}_\infty \setminus \Omega}(\zeta) \subset G_{\mathbb{C}_\infty \setminus K_m}(\zeta) = G. \qquad \square$$

---

[1] The theory of uniform approximation on compact subsets of $\mathbb{C}$ by polynomials or rational functions is profound and multifaceted, as can be seen in [2].

Theorem 7.1.10 (Runge's Theorem for Open Sets). *Let $\Omega \subset \mathbb{C}$ be open.*

1. *If $A \subset \mathbb{C}_\infty \setminus \Omega$ meets every component of $\mathbb{C}_\infty \setminus \Omega$, then for every $f \in H(\Omega)$ there exists a sequence (r_n) of rational functions with poles only in A and*

$$r_n \to f \quad \text{locally uniformly on } \Omega.$$

2. *If $\mathbb{C}_\infty \setminus \Omega$ is connected, then for every $f \in H(\Omega)$ there exists a sequence (p_n) of polynomials with*

$$p_n \to f \quad \text{locally uniformly on } \Omega.$$

Proof. It is sufficient to show the first statement (then the second follows with $A := \{\infty\}$).

For this, let (K_m) be the standard exhaustion of Ω. If $m \in \mathbb{N}$ and G is a component of $\mathbb{C}_\infty \setminus K_m$, then according to Theorem 7.1.9 there exists a component L of $\mathbb{C}_\infty \setminus \Omega$ with $L \subset G$. Thus, $\emptyset \neq A \cap L \subset A \cap G$. According to Theorem 7.1.7 there exists a function $r_m \in \mathrm{span}\{g_a^\nu : a \in A, \nu \in \mathbb{N}_0\}$ (so r_m is rational with poles only in A) and

$$\|f - r_m\|_{\infty, K_m} < 1/m.$$

If $K \subset \Omega$ is compact, then $K \subset K_m$ for m large enough, so

$$\|f - r_m\|_{\infty, K} < 1/m$$

for m large enough. $\qquad\qquad\qquad\qquad\qquad\qquad\qquad\qquad\qquad\qquad\qquad\qquad\qquad\square$

In the following, we write $P(\Omega)$ for the closure of the polynomials in $(H(\Omega), d_{\mathrm{loc}})$. From Example 7.1.1 it follows in particular that $P(\mathbb{C}^*) \neq H(\mathbb{C}^*)$ is. More precisely, one can show

Theorem 7.1.11. *For an open set $\Omega \subset \mathbb{C}$ the following statements are equivalent:*

a) $\mathbb{C}_\infty \setminus \Omega$ *is connected.*
b) $P(\Omega) = H(\Omega)$.
c) *If $f \in H(\Omega)$ and γ is a cycle in Ω, then*

$$\int_\gamma f = 0.$$

d) *Every cycle in Ω is Ω-nullhomologous.*

Proof. a) $\Rightarrow$ b) is Theorem 7.1.10.2.

b) $\Rightarrow$ c): For polynomials p, according to Theorem 5.1.6 always $\int_\gamma p = 0$ (p has an antiderivative on $\mathbb{C}$). If p_n are polynomials with $p_n \to f$ locally uniform on Ω, then by interchanging integral and limit also $\int_\gamma f = 0$.

c) $\Rightarrow$ d) follows from the Cauchy theorem.

d) $\Rightarrow$ a): Assume, $M := \mathbb{C}_\infty \setminus \Omega$ is not connected. Then there exist two disjoint, closed sets $K, L \subset M$ with $M = L \cup K$. Since M is closed in $\mathbb{C}_\infty$, K, L are also closed in $\mathbb{C}_\infty$. Without loss of generality, let $\infty \in L$. If $U := \mathbb{C} \setminus L$, then U is open in $\mathbb{C}$ and $K \subset U$ is compact. According to Theorem 7.1.6 there exists a cycle γ in $U \setminus K = \Omega$ with $\operatorname{ind}_\gamma(K) = \{1\}$. Thus, γ is not Ω-nullhomologous, contradicting the assumption. $\qquad\square$

Remark 7.1.12. If $\Omega \subset \mathbb{C}$ has no holes, then according to Remark 5.1.14 every cycle in Ω is also Ω-nullhomologous. Therefore, all statements from Theorem 7.1.11 apply. In particular, $\mathbb{C}_\infty \setminus \Omega$ is connected. It can be shown that conversely, open sets $\Omega \subset \mathbb{C}$, for which $\mathbb{C}_\infty \setminus \Omega$ is connected, have no holes. This sounds very plausible, but it is not so easy to prove.[2]

Definition 7.1.13. Let $\Omega \subset \mathbb{C}$ be open and $\varphi_w \in \operatorname{span}\{g_0^n : n \in \mathbb{N}\}$ for $w \in \Omega$, thus φ_w is a rational function with poles only in 0 or $\varphi_w = 0$. If $\{w \in \Omega : \varphi_w \neq 0\}$ is discrete in Ω, then $(\varphi_w)_{w \in \Omega}$ is called a **principal part distribution** in Ω.

The question arises whether for every principal part distribution in Ω there exists a meromorphic function in Ω such that for all $w \in \Omega$ the Laurent expansion around w has the principal part φ_w. If $A := \{w \in \Omega : \varphi_w \neq 0\}$ is finite, this is clear (the rational function $\sum_{w \in A} \varphi_w(z - w)$ is suitable). Using Runge's theorems, it can be shown that the question can be answered positively in general.

Theorem 7.1.14 (Mittag-Leffler). *Let $\Omega \subset \mathbb{C}$ be open. Then for every principal part distribution $(\varphi_w)_{w \in \Omega}$ there exists a function $f \in M(\Omega)$ with principal parts φ_w at w for all $w \in \Omega$.*

Proof. Let again $(K_m)_m$ be the standard exhaustion of Ω. For $m \in \mathbb{N}_0$ we set with $K_0 := \emptyset$ and $A := \{w \in \Omega : \varphi_w \neq 0\}$

$$A_m := A \cap (K_{m+1} \setminus K_m)$$

and (with $\sum_\emptyset := 0$)

$$q_m(z) := \sum_{w \in A_m} \varphi_w(z - w) \qquad (z \in \mathbb{C} \setminus A_m)$$

(note: A_m is finite, since A is discrete in Ω). Then $q_m \in H(\mathbb{C} \setminus A_m)$. Since each component of $\mathbb{C}_\infty \setminus K_m$ contains a point from $\mathbb{C}_\infty \setminus \Omega$,

$$q_m|_{K_m} \in R_{\mathbb{C}_\infty \setminus \Omega}(K_m)$$

according to Theorem 7.1.7, that is, there exists a rational function $r_m \in H(\Omega)$ (thus with poles only in $\mathbb{C}_\infty \setminus \Omega$) such that

---

[2] More details on this and on Runge's theory in general can be found, for example, in [5].

$$\|q_m|_{K_m} - r_m\|_{\infty, K_m} < \frac{1}{m^2 + 1}$$

(where $r_m = 0$, if $K_m = \emptyset$). According to the Weierstrass criterion, the function series $\sum_{v=m}^{\infty} (q_v - r_v)$ converges uniformly on K_m. Therefore, a function $f \in H(\Omega \setminus A)$ is defined by

$$f(z) := \sum_{v=0}^{\infty} \left(q_v(z) - r_v(z) \right) \qquad (z \in \Omega \setminus A)$$

(remark: the series converges locally uniformly on $\Omega \setminus A$). For all $m \in \mathbb{N}$, $f = g_m + \sum_{v=0}^{m-1} q_v$ with

$$g_m := \sum_{v=m}^{\infty} (q_v - r_v) - \sum_{v=0}^{m-1} r_v \in H(\Omega \setminus \bigcup_{v=m}^{\infty} A_v).$$

Since $K_m \subset \Omega \setminus \bigcup_{v=m}^{\infty} A_v$, f in K_m has exactly the prescribed poles with corresponding principal parts. Since m was arbitrary, the assertion follows.   □

Exercises 7.1.15.

1. Prove that $P(\overline{\mathbb{D}}) = A(\overline{\mathbb{D}})$ and $P(\mathbb{S}) = \{f|_\mathbb{S} : f \in A(\overline{\mathbb{D}})\}$ hold.
   Hint: Use Fejér's theorem.
2. Let $a = 1 - i$ and $b = 1 + i$ and

   $$\gamma = \left(s_a^b, \, s_b^{-a}, \, s_{-a}^{-b}, \, s_{-b}^a \right).$$

   Then γ^* is the boundary of the square Q with the corners $\pm a$ and $\pm b$. Show that

   $$\mathrm{ind}_\gamma(z) := \frac{1}{2\pi i} \int_\gamma \frac{d\zeta}{\zeta - z} = 1$$

   holds for all $z \in Q^\circ$.
3. Let $(\gamma_\iota)_{\iota \in I}$ be a family of oriented segments $\gamma_\iota = s_{u_\iota}^{v_\iota}$ such that

   $$\#\{\iota \in I : w = u_\iota\} = \#\{\iota \in I : w = v_\iota\}$$

   for each $w \in \mathbb{C}$. Convince yourself that then for a suitable partition $(I_\kappa)_{\kappa \in M}$ of I the chains $(\gamma_\iota)_{\iota \in I_\kappa}$ are closed paths.
   Hint: The proof essentially has a combinatorial character; see [6] Sect. 13.4.
4. Let $K \subset \mathbb{C}$ be compact and $A \subset \mathbb{C}_\infty \setminus K$. Show:
   a) If K is infinite and p/q is a rational function, then all poles of p/q are in A exactly when

   $$(p/q)|_K \in \mathrm{span}\{(g_a|_K)^n : a \in A, \, n \in \mathbb{N}_0\}.$$

   Hint: Partial fraction decomposition.
   b) For $f, g \in R_A(K), f \cdot g \in R_A(K)$ is also true (thus $R_A(K)$ is a $\mathbb{C}$-algebra).

5. Prove the following variant of Runge's theorem for approximation on compact sets $K \subset \mathbb{C}$: If $A \subset \mathbb{C} \setminus K$ has an accumulation point in each component of $\mathbb{C} \setminus K$, then

$$f|_K \in \overline{\operatorname{span}}\{g_a|_K : a \in A\}$$

holds for all holomorphic functions f on an open superset of K.

6. Let $G = \mathbb{D} \setminus B_{2/3}(1/3)$ and $K := \overline{G}$. Prove:
   a) $H(G) = P(G)$.
   b) There exist $f \in H(\mathbb{C}^*)$ with $f|_K \notin P(K)$.

7. Let $G \subset \mathbb{C}$ be a region. Show that the following statements are equivalent:
   a) $\mathbb{C}_\infty \setminus G$ is connected.
   b) $H(G) = P(G)$.
   c) For all $f \in H(G)$ and all closed paths γ in G, $\int_\gamma f = 0$.
   d) Every $f \in H(G)$ has antiderivative on G.
   e) If $g \in H(G)$ is zero-free, then there exists an $f \in H(G)$ with $e^f = g$.
   f) If $g \in H(G)$ is zero-free, then there exists an $f \in H(G)$ with $f^2 = g$.
   g) $G = \mathbb{C}$ or G is conformally equivalent to $\mathbb{D}$.
   h) Every closed path in G is G-nullhomologous.

7.2 Boundary Behavior

As an application of Runge's theorems for polynomial approximation, we deduce the existence of entire functions that show extremely complicated behavior near ∞.

Remark 7.2.1. We consider the space $(H(\Omega), d_{\text{loc}})$ from Remark 6.1.18 and set for $f \in H(\Omega)$, $\varepsilon > 0$ and $K \subset \Omega$ compact

$$V_\varepsilon(f, K) := \{g \in H(\Omega) : \|f - g\|_{\infty, K} < \varepsilon\} = f + V_\varepsilon(0, K).$$

If $f \in H(\Omega)$ and $U \subset H(\Omega)$, then it follows: U is a neighborhood of f exactly when an $\varepsilon > 0$ and a compact set $K \subset \Omega$ exists with $V_\varepsilon(f, K) \subset U$.

Because: If U is a neighborhood of f, then by assumption there exists an $\varepsilon > 0$ with $U_\varepsilon(f) \subset U$. We choose $m_\varepsilon \in \mathbb{N}$ with $1/m_\varepsilon < \varepsilon$. If $g \in V_\varepsilon(f, K_{m_\varepsilon})$, then $\|f - g\|_{\infty, K_{m_\varepsilon}} < \varepsilon$, and therefore also

$$\max_{1 \le m \le m_\varepsilon} \min\{1/m, \|f - g\|_{\infty, K_m}\} < \varepsilon$$

and thus $d_{\text{loc}}(f, g) < \varepsilon$. Consequently, $V_\varepsilon(f, K_{m_\varepsilon}) \subset U_\varepsilon(f)$.
If, conversely, $V_\varepsilon(f, K) \subset U$, we choose an $m \in \mathbb{N}$ with $K \subset K_m$. For $\varepsilon' := \min\{1/m, \varepsilon\}$ then holds: If $g \in U_{\varepsilon'}(f)$, then $\min\{1/m, \|f - g\|_{\infty, K_m}\} < \varepsilon'$, and therefore also $\|f - g\|_{\infty, K_m} < \varepsilon'$, and thus

$$\|f - g\|_{\infty, K} \le \|f - g\|_{\infty, K_m} < \varepsilon' \le \varepsilon.$$

So $U_{\varepsilon'}(f) \subset V_\varepsilon(f, K)$ and thus U is a neighborhood of f.

If (a_n) is a sequence in $\mathbb{R}$ with $a_n \to -\infty$, then $e^{z+a_n} \to 0$ holds locally uni-formly on $\mathbb{C}$ for $n \to \infty$. If we define for $a \in \mathbb{C}$ the **translation operator** $T_a : H(\mathbb{C}) \to H(\mathbb{C})$ by

$$(T_a f)(z) := f(z + a) \qquad (z \in \mathbb{C}),$$

then

$$T_{a_n} \exp \to 0 \quad (n \to \infty)$$

holds in $H(\mathbb{C})$. We show that the behavior under translations is much more compli-cated for many entire functions.

Theorem 7.2.2 (Birkhoff).[3] *Let (a_n) be an unbounded sequence in $\mathbb{C}$. Then there exists a dense G_δ-set of functions $f \in H(\mathbb{C})$ such that the set of translations $T_{a_n} f$ is dense in $H(\mathbb{C})$.*

Proof. First, $(H(\mathbb{C}), d_{\text{loc}})$ is (complete and) separable, since the polynomials with coefficients in $\mathbb{Q} + i\mathbb{Q}$ form a dense subset in $H(\mathbb{C})$ (see Exercise 7.2.7.1). Furthermore, it is easy to see that $T_a : H(\mathbb{C}) \to H(\mathbb{C})$ is continuous. We prove a) from the universality criterion for the sequence (T_{a_n}). The equivalent condition b) and the fact that D is a G_δ-set then yield the assertion.

Let $\emptyset \neq U, V \subset H(\mathbb{C})$ be open. Then there exist $g, h \in H(\mathbb{C})$ and $\varepsilon > 0$, $K \subset \mathbb{C}$ compact (without restriction $K = B_R(0)$ for an $R > 0$) with

$$V_\varepsilon(g, K) \subset U, \qquad V_\varepsilon(h, K) \subset V.$$

So it is enough to show: There exists an n with

$$T_{a_n}(V_\varepsilon(g, K)) \cap V_\varepsilon(h, K) \neq \emptyset.$$

Let n be such that $R_n := |a_n|/2 > R$. Then $K \cap (K + a_n) = B_R(0) \cap B_R(a_n) = \emptyset$ and $K \cup (K + a_n) \subset \mathbb{C}$ is compact. In addition, $\mathbb{C}_\infty \setminus K$ is connected. If we define

$$\varphi(z) := \begin{cases} g(z), & z \in U_{R_n}(0) \\ h(z - a_n) = (T_{-a_n} h)(z), & z \in U_{R_n}(a_n) \end{cases},$$

then φ is holomorphic in $U_{R_n}(0) \cup U_{R_n}(a_n)$. According to Theorem 7.1.7.2, there exists a polynomial p with

$$\|\varphi - p\|_{\infty, K \cup (K + a_n)} < \varepsilon.$$

So on the one hand, $\|g - p\|_{\infty, K} < \varepsilon$ (so $p \in V_\varepsilon(g, K)$) and on the other hand,

---

[3]We are dealing here with a typical example of linear chaos. A systematic introduction to the exciting theory can be found in [3].

$$\|h - T_{a_n}p\|_{\infty,K} = \|T_{-a_n}h - p\|_{\infty,K+a_n} < \varepsilon,$$

that is, $T_{a_n}p \in V_\varepsilon(h,K)$. $\square$

The question arises for examples of entire functions with correspondingly complicated behavior near ∞. Although the universality criterion guarantees that the set of these functions is large in a certain sense in $H(\mathbb{C})$, all "concrete" examples show a much more regular behavior.[4]

We want to show that a comparable phenomenon of extremely complicated boundary behavior occurs in Taylor partial sums $s_n f$ for functions $f \in A(\overline{\mathbb{D}})$. As is well known, the partial sums

$$(s_n f)(z) := T_n(f,0)(z) = \sum_{\nu=0}^{n} c_\nu z^\nu$$

converge locally uniformly on $\mathbb{D}$ to f. In addition, $\sum_{\nu=0}^{\infty} |c_\nu|^2 < \infty$ follows from Parseval's equation, and in particular $(c_k) = (\widehat{f}(k))$ is a null sequence. If even $\sum_{\nu=0}^{\infty} |c_\nu| < \infty$ is true, then according to the Weierstrass criterion, there is uniform convergence of $(s_n f)$ on $\overline{\mathbb{D}}$. Furthermore, Abel's theorem shows that for functions in $A(\overline{\mathbb{D}})$, in the case of convergence of $(s_n f(\zeta))$ at a boundary point $\zeta \in \mathbb{S}$, convergence to $f(\zeta)$ always holds.

One might think that the sequence $(s_n f)$ converges to f for all $f \in A(\overline{\mathbb{D}})$ on $\overline{\mathbb{D}}$, perhaps even uniformly. We show that—quite the contrary—for every countable set $E \subset \mathbb{S}$ there exist functions in $A(\overline{\mathbb{D}})$ for which the sequence $(s_n f)$ diverges dramatically on E.

In the rest of this chapter, let $\|\cdot\|_\infty = \|\cdot\|_{\infty,\overline{\mathbb{D}}}$ always be the supremum norm with respect to the closed unit disk $\overline{\mathbb{D}}$. First, we prove

Theorem 7.2.3. *For the sequence (f_n) of polynomials of degree $2n$, given by*

$$f_n(z) := \sum_{\nu=0}^{n-1} \frac{z^\nu}{n-\nu} - \sum_{\nu=n+1}^{2n} \frac{z^\nu}{\nu-n} \qquad (z \in \mathbb{C}, \ n \in \mathbb{N}),$$

it holds

$$\sup_{n \in \mathbb{N}} \|f_n\|_\infty < \infty.$$

---

[4] In this sense, one could speak of the set of such complicated objects as a kind of "dark matter" in mathematics.

Proof. According to the maximum principle, $\max\limits_{\mathbb{S}} |f_n| = \|f_n\|_\infty$. For $1 \leq r \leq n$ and $z \in \mathbb{S}$

$$\sum_{v=n-r}^{n-1} \frac{z^v}{n-v} - \sum_{v=n+1}^{n+r} \frac{z^v}{v-n} = \frac{z^{n-r}}{r} + \ldots + \frac{z^{n-1}}{1} - \left(\frac{z^{n+1}}{1} + \ldots + \frac{z^{n+r}}{r} \right)$$

$$= \sum_{k=1}^{r} \frac{1}{k} \left(z^{n-k} - z^{n+k} \right) = z^n \sum_{k=1}^{r} \frac{1}{k} \left(z^{-k} - z^{k} \right)$$

and thus for $\theta \in [-\pi, \pi]$ and $z = e^{i\theta}$ from $|\sin(k\theta)| \leq k|\theta|$

$$\left| \sum_{v=n-r}^{n-1} \frac{z^v}{n-v} - \sum_{v=n+1}^{n+r} \frac{z^v}{v-n} \right| \leq \sum_{k=1}^{r} \frac{1}{k} |z^k - z^{-k}| = \sum_{k=1}^{r} \frac{2}{k} |\sin(k\theta)| \leq 2r|\theta|$$

follows. If $n \leq \lfloor \pi/|\theta| \rfloor$, then $|f_n(z)| \leq 2\pi$ with $r = n$.

For $r < n$ applies with Abel's partial summation

$$\sum_{v=0}^{n-r-1} \frac{z^v}{n-v} = \sum_{k=1}^{n-r} \frac{1}{n-k+1} z^{k-1}$$

$$= \frac{1}{r+1} \sum_{\ell=1}^{n-r} z^{\ell-1} - \sum_{k=1}^{n-r-1} \left(\sum_{\ell=1}^{k} z^{\ell-1} \right) \left(\frac{1}{n-k} - \frac{1}{n-k+1} \right).$$

Thus, with $|e^{i\theta} - 1| \geq 2\theta/\pi$ and the geometric sum formula for $\theta \neq 0$

$$\left| \sum_{v=0}^{n-r-1} \frac{z^v}{n-v} \right| \leq \frac{2}{r+1} \frac{1}{|e^{i\theta} - 1|} + \frac{2}{|e^{i\theta} - 1|} \sum_{j=r+1}^{n-1} \left(\frac{1}{j} - \frac{1}{j+1} \right) \leq \frac{2\pi}{|\theta|} \cdot \frac{1}{r+1}.$$

The same applies for $\displaystyle\sum_{v=n+r+1}^{2n} \frac{z^v}{v-n}$. This results in total

$$|f_n(z)| \leq 2r|\theta| + \frac{4\pi}{(r+1)|\theta|}.$$

If now $n > \lfloor \pi/|\theta| \rfloor$, then $|f_n(z)| \leq 2\pi + 4$ with $r = \lfloor \pi/|\theta| \rfloor$.   □

The polynomials f_n are called **Fejér polynomials**. Note that the n-th Taylor partial sums $s_n f_n$ at the point 1 coincide with the nth partial sums of the harmonic series. In particular, the sequence $(s_n f_n(1))$ is unbounded.

Theorem 7.2.4. *Let* $E \subset \mathbb{C} \setminus \mathbb{D}$ *be finite,* $c \in \mathbb{C}^E$ *and* $g \in A(\overline{\mathbb{D}})$. *Then for every* $\varepsilon > 0$ *there exists a polynomial* p *and an* $n \in \mathbb{N}$ *with*

$$\|g - p\|_\infty < \varepsilon \quad \text{and} \quad s_n p|_E = c.$$

Proof.

1. We first show: If $B \subset \mathbb{C} \setminus \{1\}$ is finite, then there exists a sequence $(p_{n,B})$ of polynomials with $\|p_{n,B}\|_\infty \to 0 \ (n \to \infty)$ and

$$s_n p_{n,B}|_B = 0, \qquad (s_n p_{n,B})(1) = 1 \qquad (n > \#(B)).$$

For this, let f_n be the n-th Fejér polynomial. We set $m := \#(B)$ and

$$f_{n,m}(z) := f_n(z) - \sum_{\nu=n-m}^{n-1} \frac{z^\nu}{n-\nu} = \sum_{\nu=0}^{n-m-1} \frac{z^\nu}{n-\nu} - \sum_{\nu=n+1}^{2n} \frac{z^\nu}{\nu-n} \qquad (z \in \mathbb{C}, \ n > m).$$

Since $(\|f_n\|_\infty)_n$ is bounded, the sequence $(\|f_{n,m}\|_\infty)_{n>m}$ is also bounded. Furthermore,

$$(s_n f_{n,m})(1) = \sum_{k=m+1}^n \frac{1}{k} \to \infty \qquad (n \to \infty).$$

We define $\varphi_B(z) := \prod_{w \in B} (z - w)$ for $z \in \mathbb{C}$ and

$$p_{n,B} := \frac{\varphi_B f_{n,m}}{\varphi_B(1)(s_n f_{n,m})(1)} \qquad (n > m).$$

Then $\|p_{n,B}\|_\infty \to 0 \ (n \to \infty)$. Since $f_{n,m}$ does not contain any of the powers $z^{n-m}, \ldots, z^n$, it follows that

$$s_n p_{n,B} = \frac{\varphi_B}{\varphi_B(1)(s_n f_{n,m})(1)} \cdot s_n f_{n,m}$$

and therefore $s_n p_{n,B}|_B = 0$ and $(s_n p_{n,B})(1) = 1$ for all $n > m$.

2. From Fejér's theorem, it follows that the set of polynomials is dense in $A(\overline{\mathbb{D}})$. Therefore, it is sufficient to prove the statement for polynomials instead of general functions g.

Let g be a polynomial. We set $d := C - g|_E$ and choose a $\delta > 0$ such that

$$\delta \cdot \sum_{w \in E} |d(w)| < \varepsilon.$$

by step 1 of the proof, applied to $B_w := w^{-1}(E \setminus \{w\})$ for $w \in E$, there exist an $n \geq \deg(g)$ and for $w \in E$ polynomials $p_w := p_{n,B_w}$ with $\|p_w\|_\infty < \delta$ and

$$s_n p_w|_{B_w} = 0, \qquad (s_n p_w)(1) = 1.$$

Since $|w| \geq 1$, for $q_w(z) := p_w(z/w)$ we also get $\|q_w\|_\infty < \delta$ and moreover

$$s_n q_w|_{E \setminus \{w\}} = 0, \qquad (s_n q_w)(w) = 1.$$

If we define $p := g + \sum_{w \in E} d(w) \cdot q_w$, then

$$\|p - g\|_\infty \leq \sum_{w \in E} |d(w)| \|q_w\|_\infty < \varepsilon,$$

and from $n \geq \deg(g)$ it follows that $s_n g = g$. Thus,

$$(s_n p)(w) = g(w) + \sum_{w' \in E} d(w') s_n q_{w'}(w) = g(w) + d(w) = c(w)$$

holds for $w \in E$. □

Theorem 7.2.5. *If $E \subset \mathbb{C} \setminus \mathbb{D}$ is countable, then there exists a dense G_δ-set of functions $f \in A(\overline{\mathbb{D}})$ with the property that for all $h : E \to \mathbb{C}$ a subsequence of $(s_n f)$ converges pointwise on E to h.*[5]

Proof.

1. Let us first assume that $E \subset \mathbb{C} \setminus \mathbb{D}$ is finite. For $n \in \mathbb{N}$ let $T_n : A(\overline{\mathbb{D}}) \to \mathbb{C}^E = C(E)$ be defined by

$$T_n f := s_n f|_E \quad (f \in A(\overline{\mathbb{D}})).$$

   Then for $R \geq 1$ with $E \subset B_R(0)$ and $f, g \in A(\overline{\mathbb{D}})$

$$||s_n f - s_n g||_{\infty, E} \leq \sum_{\nu=0}^{n} |\widehat{f}(\nu) - \widehat{g}(\nu)| R^\nu \leq (n+1) R^n \, ||f - g||_\infty.$$

   Thus, T_n is continuous. Moreover, $(A(\overline{\mathbb{D}}), || \cdot ||_\infty)$ is a Banach space and $(\mathbb{C}^E, || \cdot ||_{\infty, E})$ is separable.

   If $U \subset A(\overline{\mathbb{D}})$ and $V \subset C(E)$ are open and non-empty, then $\varepsilon > 0$ and $g \in A(\overline{\mathbb{D}})$ as well as $c \in \mathbb{C}^E$ exist with $U_\varepsilon(g) \subset U$ and $c \in V$. According to Theorem 7.2.4, a polynomial p and an $n \in \mathbb{N}$ exist with $p \in U_\varepsilon(g)$ and $T_n p = s_n p|_E = c$. In particular, condition a) from the universality criterion is fulfilled. According to the universality criterion, for a dense G_δ-set of functions $f \in A(\overline{\mathbb{D}})$, the set of partial sums $\{s_n f|_E : n \in \mathbb{N}\}$ is dense in $C(E)$.

2. Let $E =: \{\zeta_j : j \in \mathbb{N}\}$ be countable and $E_m := \{\zeta_1, ..., \zeta_m\}$ for $m \in \mathbb{N}$. According to 1., for each $m \in \mathbb{N}$, a dense G_δ-set $A_m \subset A(\overline{\mathbb{D}})$ exists such that for all $f \in A_m$ the set $\{s_n f|_{E_m} : n \in \mathbb{N}\}$ is dense in $C(E_m)$. Then also $\bigcap_{m \in \mathbb{N}} A_m$ is a dense G_δ-set in $A(\overline{\mathbb{D}})$ (see Exercise 2.4.33.7). Every function $f \in \bigcap_{m \in \mathbb{N}} A_m$ has the property that for all m the set of partial sums $\{s_n f|_{E_m} : n \in \mathbb{N}\}$ is dense in $C(E_m)$. For such f one has: If $h : E \to \mathbb{C}$ and if we set $n_0 := 1$, then for every $j \in \mathbb{N}$ there exists an $n_j \in \mathbb{N}$ with $n_j > n_{j-1}$ and

$$||s_{n_j} f - h||_{\infty, E_j} < 1/j.$$

Now if $\zeta \in E$ and $\varepsilon > 0$, then there exists an $m \in \mathbb{N}$ with $1/j < \varepsilon$ and $\zeta \in E_j$ for $j \geq m$ and thus $|s_{n_j}f(\zeta) - h(\zeta)| < \varepsilon$ for $j \geq m$. Therefore, $s_{n_j}f \to h$ $(j \to \infty)$ pointwise on E.

$\square$

Example 7.2.6. If E is the set of unit roots, then E is countable and thus the statement of the preceding sentence applies to E. It is easy to see that E is dense in $\mathbb{S}$. Therefore, functions exist $f \in A(\overline{\mathbb{D}})$ such that the sequence $(s_n f)$ diverges dramatically on a set dense in $\mathbb{S}$.[6] Similar to the case of Birkhoff's theorem, it can be proven that for many functions in the disc algebra, the partial sums $s_n f$ have an extremely complicated boundary behavior in the above sense, but all known examples behave much more moderately.

Exercises 7.2.7.

1. Show:
    a) If $\mathbb{Q} + i\mathbb{Q}$ is the set of Gaussian rational numbers, i.e., the set of complex numbers with rational real and imaginary parts, then the set of polynomials with coefficients in $\mathbb{Q} + i\mathbb{Q}$ is countable.
    b) If p is any polynomial, then for every compact set $K \subset \mathbb{C}$ and for every $\varepsilon > 0$, there exists a polynomial q with coefficients in $\mathbb{Q} + i\mathbb{Q}$ and

$$\max_{K} |p - q| < \varepsilon.$$

2. Let $f \in A(\overline{\mathbb{D}})$. Prove: For all $r \in [0, 1)$ and all $n \in \mathbb{N}$

$$\max_{B_r(0)} |s_n f| \leq \frac{1}{1 - r} \|f\|_\infty.$$

3. Show:
    a) If (c_n) is a sequence in $\mathbb{C}$ with $\sum_{\nu=0}^{\infty} |c_\nu|^2 < \infty$, then

$$\sum_{\nu=0}^{\infty} \frac{|c_\nu|}{\nu + 1} < \infty.$$

    Hint: Apply the Cauchy-Schwarz inequality to the partial sums.

    b) If $f \in H(\mathbb{D})$ with $f'(z) = \sum_{\nu=0}^{\infty} c_\nu z^\nu$ for $z \in \mathbb{D}$ and $\sum_{\nu=0}^{\infty} |c_\nu|^2 < \infty$, then the Taylor series $(s_n f)$ of f converges uniformly on $\overline{\mathbb{D}}$. This is particularly the case when f' has a continuous extension to $\overline{\mathbb{D}}$.

---

[6] The famous and profound theorem of Carleson implies that for all $f \in A(\overline{\mathbb{D}})$ the partial sum sequence $(s_n f)$ converges almost everywhere on $\mathbb{S}$ to f.

References

1. Erdös, P., Herzog, F., Piranian, G.: On Taylor series of functions regular in Gaier regions. Arch. Math. **5**, 39–52 (1954)
2. Gaier, D.: Vorlesungen über Approximation im Komplexen. Birkhäuser, Basel (1980)
3. Grosse-Erdmann, K.-G., Peris-Manguillot, A.: Linear Chaos. Springer, London (2011)
4. Herzog, G., Kunstmann, P.C.: Universally divergent Fourier series via Landau's extremal functions. Comment. Math. Univ. Carol. **56**, 159–168 (2015)
5. Remmert, R.: Funktionentheorie II. Springer, Berlin (1991)
6. Rudin, W.: Real and Complex Analysis. 3rd ed. McGraw-Hill, New York (1987)